POWER SEMICONDUCTOR CIRCUITS

POWER SEMICONDUCTOR CIRCUITS

S. B. Dewan

Associate Professor of Electrical Engineering
University of Toronto

A. Straughen

Associate Professor of Electrical Engineering
University of Toronto

A WILEY-INTERSCIENCE PUBLICATION

JOHN WILEY & SONS

New York • Chichester • Brisbane • Toronto

1257136

Library of Congress Cataloging in Publication Data:

Dewan, S B
    Power semiconductor circuits.

    "A Wiley-Interscience publication."
    Bibliography: p.
    Includes index.
    1. Electric current converters. 2. Thyristors.
3. Diodes, Semiconductor. 4. Electric switchgear.
I. Straughen, A., joint author. II. Title.
TK7872.C8D48   621.3815'32   75-8911
ISBN 0-471-21180-X

Printed in the United States of America

10 9 8 7 6 5 4

Bestir thyself therefore on this occasion; for, though we will always lend thee proper assistance in difficult places, as we do not, like some others, expect thee to use the arts of divination to discover our meaning, yet we shall not indulge thy laziness where nothing but thy own attention is required; for thou art highly mistaken if thou dost imagine that we intended, when we began this great work, to leave thy sagacity nothing to do; or that, without sometimes exercising this talent, thou wilt be able to travel through our pages with any pleasure or profit to thyself.

Henry Fielding
The History of Tom Jones

PREFACE

The design of a power semiconductor converter (rectifier, inverter, etc.) starts with the design of the power circuit of that converter, including the specification of the required thyristors and diodes as well as the necessary protective and commutating circuit elements. During the design of the power circuit, the required control signals are specified, and the design of the control circuits then follows.

This book is concerned with the first stage of the design procedure and is based on a great deal of experience in the consulting field. Thus the methods described in the following pages have been successfully applied to the design of industrial equipment. While the approach is practical, however, a sound basis of theoretical knowledge is a prerequisite to a full understanding of the subject matter. In particular, an acquaintance with transient circuit analysis, Fourier analysis, electric machine theory, and elementary control system theory are needed by the reader.

Discussion of semiconductor physics has been avoided as far as possible, since it is only exceptionally needed by the systems designer, and many books at all levels have been published on this subject. Simiiarly the details of thyristor construction, a knowledge of which may be helpful when a special converter is to be built, are dealt with in an abundance of manufacturers' literature.

The various topics are treated in the order in which each develops naturally out of those preceding it. After a brief introductory discussion in Chapter 1, intended to convey a general idea of the types of system to be considered as well as of the physical appearance of the devices employed, Chapter 2 applies linear circuit analysis to the discontinuously linear circuits that result from the introduction of diodes. In Chapter 3 the discussion is extended to circuits including thyristors. Here, the method of rating thyristors is discussed in some detail, and the problem of commutation is introduced. This leads logically to a simple scheme of classification of the various kinds of converter. The individual classes of converter are then treated chapter by chapter.

Our constant aim has been to explain principles both of analysis and design rather than to attempt the impossible task of discussing every type of converter circuit that has been built. The reader should then be able to apply these principles to circuits other than those actually dealt with in this book.

At the University of Toronto, graduate courses in Power Electronics may be grouped under the headings:

1. Power semiconductor physics
2. Power circuits of converters
3. Control circuits of converters

This book, in draft form, has been employed as the main text for a first-level graduate course in the second of these groups. It has also been employed in this and other institutions for undergraduate instruction. The following sections have been covered in a one-semester course for senior-year undergraduates in electrical engineering:

Chapter 1	All sections
Chapter 2	All sections
Chapter 3	All sections
Chapter 4	Sections 4.1 to 4.4 inclusive
Chapter 5	Sections 5.1 to 5.2.1 inclusive
Chapter 6	Sections 6.1 to 6.2.4 inclusive and Sections 6.3 to 6.3.1
Chapter 7	Sections 7.1 to 7.3

At least half of the problems were worked by the class during a tutorial period that alternated each week with laboratory work.

We are indebted to many undergraduate and graduate students for the correction of errors to which an undertaking of this nature is prone. We have also been very fortunate indeed in having a virtuoso of the typewriter to help us. Mrs. Sarah Cherian has not contented herself with merely typing a gruelingly difficult manuscript. She has also pointed out omitted or repeated equation numbers, erroneous cross-references, inconsistent notation, etc. The errors that remain are our very own.

S. B. DEWAN
A. STRAUGHEN

Toronto, Canada
March 1975

CONTENTS

LIST OF SYMBOLS

So far as possible, lower case letters have been employed to signify instantaneous values of the quantities concerned. Upper case letters may signify constant, direct, rms, or average values. In each case the symbols may be employed with subscripts referring to particular circuit elements.

a_n, b_n, c_n, d_n	Fourier coefficients
f	Frequency
i	Current
i_A	Anode current
i_D	Diode current
i_F	Forced component of current
i_G	Cathode-to-gate current
i_M	Magnetizing current
i_N	Natural component of current
i_O	Output current
i_Q	Thyristor current
j	Complex number operator, $\sqrt{-1}$
m	Ratio of V_C to $\sqrt{2V}$, $\sin\alpha$
n	Turns ratio of a transformer, $N_1|N_2$
n	Order of harmonic voltage or current
q	Charge
s	Complex root
s	Induction motor slip
t	Time
t_{off}	Turn-off time
t_q	Time available for turn-off
v	Voltage
v_{AK}	Anode-to-cathode voltage
v_D	Diode voltage
v_G	Cathode-to-gate voltage

v_O	Output voltage
C	Capacitance
D_R	Derating factor
I	Current
I_N	Normalized value of average current
I_{nR}	rms value of n'th harmonic current
I_O	Average output current
I_Q	Average thyristor current
I_{QR}	rms thyristor current
I_R	rms output current
I_{RI}	rms value of current harmonics
I_{RN}	Normalized value of rms current
K_i	Current ripple factor
K_v	Voltage ripple factor
L	Inductance
N	Number of turns
P	Power
R	Resistance
R_Q	Thyristor rating of a controller
S	Apparent power
T	Periodic time
T	Temperature
T	Torque
V	Voltage
V_C	Load-circuit counter emf
V_O	Average output voltage
V_R	rms output voltage
V_{RI}	Ripple voltage
W	Energy
X	Reactance at source frequency
Z	Impedance at source frequency
Z_n	Impedance at n'th harmonic frequency
α	Delay angle, angle of retard
β	Extinction angle
γ	Conduction angle
ε	Base of natural logarithms
ζ	Damping factor
η	$\sin^{-1} m$
θ	Thermal impedance
λ	Flux linkage
τ	Time constant
ϕ	Impedance angle at source frequency

ϕ_n	Impedance angle at n'th harmonic frequency
ψ	Phase angle
ω	Angular frequency
ω	Angular speed of rotation
ω_m	Motor speed
ω_o	Resonant frequency
ω_r	Ringing frequency
ω_{syn}	Synchronous speed of a motor
Φ	Flux per pole
Ω	Angular speed of rotation
Ω_m	Motor speed

POWER SEMICONDUCTOR CIRCUITS

ONE

INTRODUCTION

The thyristor or silicon controlled rectifier (SCR) was developed in the laboratories of the General Electric Company and became commercially available in 1960 to 1961 in ratings of approximately 200 A and 1000 V. The SCR was clearly a revolutionary device, and its appearance therefore led to predictions of an immediate revolution in industrial control equipment in particular. However, the revolution was delayed for about 5 years by unexpected difficulties in applying thyristors. Unexplained equipment failures took place, and the use of thyristors was consequently restricted to relatively simple and unsophisticated applications until it was realized that some of the characteristics of thyristors that must be understood to apply them successfully were either not known or not specified in the manufacturers' data sheets. As a consequence the equipment designer led a hard life until the range of data which he required was made generally available.

Once these initial difficulties were understood and surmounted, the revolution began in real earnest, and a large and rapidly developing branch of industry was soon established.

1.1 THYRISTOR APPLICATIONS

Devices and system components that are being superseded in many applications by thyristors or thyristor systems include:

1. Thyratrons.
2. Mercury-arc rectifiers.
3. Saturable-core reactors.
4. Relays and contactors.
5. Rheostats and motor starters.

1

    6.   Constant-voltage transformers.

    7.   Autotransformers.

    8.   Mechanical speed changers.

Examples of the successful application of thyristors include:

    1.   Electric drives for rapid-transit systems, in which wasteful control resistors are eliminated.

    2.   Rolling-mill drives, in which the motor-generator set of the conventional Ward-Leonard system is replaced by a controlled rectifier-inverter.

    3.   Frequency changers, superseding motor-generator sets.

    4.   Multiple-drive systems for textile and paper mills.

    5.   Machine tool controls.

    6.   Aircraft power supplies.

    7.   Uninterruptible power supplies (UPS) for computers, and in general a very wide range of variable-speed control systems.

Development work is also proceeding in the application of thyristors to electric road vehicles, high-speed ground transportation, d-c power transmission, and many other fields.

1.2 THYRISTOR SYSTEMS OR CONVERTERS

The main parts of a converter are shown in the block diagram of Fig. 1.1. They are as follows:

    1.   The power circuits, the output of which may be a variable direct or alternating voltage source or may be an alternating source of variable voltage and frequency.

    2.   The digital circuits, which in response to the signals from the controlling system switch the thyristors of the power circuits on and off at appropriate instants.

    3.   The controlled system, which may simply be a rotating machine and

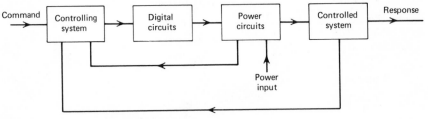

Fig. 1.1 Block diagram of a typical converter.

driven load with appropriate feedback output, or may be something considerably more complicated.

4. The controlling system, which in response to the command and feedback signals issues the appropriate control signals to the digital circuits.

In the chapters that follow attention is confined to the power circuits of the converter and the controlled system, including the specification of the signals required from the digital circuits. Occasionally it has been thought necessary to indicate in broad outline the arrangement of the control signal circuits. However, there are several alternative and rapidly developing technologies by means of which the control signals themselves may be generated, so that the subject of control circuit design is large enough to warrant a book equal in size to this one.

1.3 DIODES AND THYRISTORS

Before discussing circuits and methods of circuit analysis, it is desirable that the reader should have some idea of what the devices concerned look like, and to this end Figs. 1.2 to 1.5 are included.

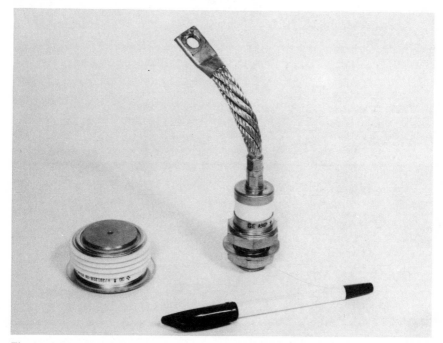

Fig. 1.2 Stud-type and disk-type diodes.

In Fig. 1.2 two configurations of diode are shown. The one on the right is called a "stud-type" or "stud-mounted" diode, the "stud" being the threaded part at the bottom that, in the example shown here, forms the cathode terminal of the diode. The braided cable at the top forms the anode terminal. Current may therefore pass from cable to stud, but not in the opposite direction. This terminal arrangement may also be reversed. The device on the left in Fig. 1.2 is called a "disk," "press-pak," or "hockey-puck" type of diode. The anode and cathode terminals are formed by the round flat faces on top and bottom of the disk, and contact with these is made by clamping appropriate matching surfaces against them.

In Fig. 1.3 three common configurations of thyristor are shown. In the stud-type thyristor on the left the necessary control or "gate" terminal is shown on top adjacent to the larger cathode terminal. The stud forms the anode terminal in stud-type thyristors. When a small current is passed from the gate terminal to the cathode, the thyristor will conduct, provided that the anode is at a higher potential than the cathode. Once conduction

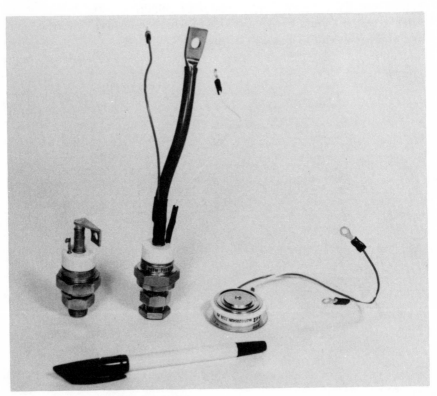

Fig. 1.3 Stud-type and disk-type thyristors.

has begun, the gate circuit loses control, and the device continues to conduct until the potential of the anode becomes equal to or less than that of the cathode for a brief interval, at the end of which the gate circuit is once again in control.

In the stud-type thyristor in the middle of Fig. 1.3, the gate connection may be seen entering the top of the case at the right of the cathode lead, and for convenience of wiring, a long gate lead is provided. The braided cathode lead is covered with a plastic insulating sleeve, and an additional thin cathode lead for use in the gate circuit is provided. This lead is needed because at full load current a small but significant voltage drop may take place in the braided cathode lead which could upset the operation of the gate circuit.

On the right of Fig. 1.3 is shown a disk-type thyristor. Here also the gate connection enters the case separately and is provided with a long (white) lead. An additional cathode connection made directly to the terminal of the thyristor is provided for use in the gate circuit. This again avoids the introduction into the gate circuit of any contact voltage drop between the cathode terminal of the thyristor and the external circuit connection.

The anode-to-cathode resistance of a diode or of a thyristor in the conducting state is very low; nevertheless, it is not zero, and passage of the large anode current produces heat sufficient to destroy the device very quickly if some means of removing the heat is not provided. In Fig. 1.4 is shown a type of heat sink suitable for stud-mounted diodes or thyristors. The devices are mounted in the sink by means of the threaded studs and nuts (in some cases press-fit studs are provided) and the sink then forms the anode terminal. Heat is conducted from the case of the device to the heat sink, from which it is radiated and conducted to atmosphere. The sinks are mounted with the fins vertical so that convection currents of air may flow past them, and the fins are painted black to improve radiation. Frequently two or more devices are mounted on a single heat sink that provides convenient electrical connection between them.

In Fig. 1.5 are shown a disk-type thyristor and a diode clamped between two heat sinks in such a way that they conduct in opposite directions, this being a common circuit arrangement. As may be seen from the positions of the thyristor leads, the lower terminal of the thyristor is the cathode. The lower terminal of the diode is therefore its anode. The clamps holding the two heat sinks together, and maintaining contact between them and the terminals of the devices, are necessarily insulated from one heat sink—in this illustration, the lower one. It may also be remarked that for the particular type of thyristor shown here, the pressure on the terminals must exceed a certain specified value before the connections are made and the thyristor can conduct.

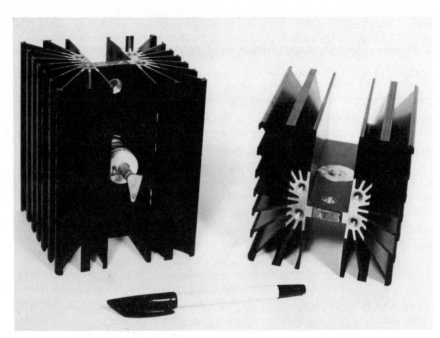

Fig. 1.4 Heat sinks for stud-type devices.

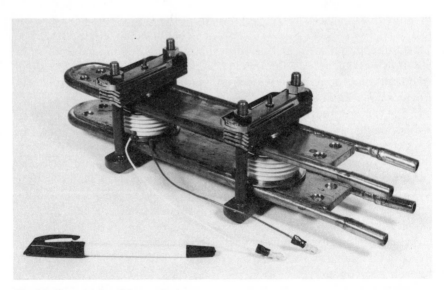

Fig. 1.5 Heat sink for disk-type devices.

6

The two heat sinks shown in Fig. 1.5 are water cooled by means of tubes round the edges of each sink. Sinks and tubes are made of heavy gauge copper or aluminum to provide a thermal capacity capable of smoothing out any rapid fluctuations of temperature that might otherwise take place at the cases of the thyristor and diode.

In Fig. 1.6 is shown the open cabinet of a variable-speed drive system designed to control a squirrel-cage induction motor. Six thyristors mounted in heat sinks may be seen in the cabinet. Systems such as this may be used with motors of up to 100 hp and provide a range of speed that depends on the type of load to be driven. They are described in detail in Chapter 4 under the heading "AC Voltage Controllers."

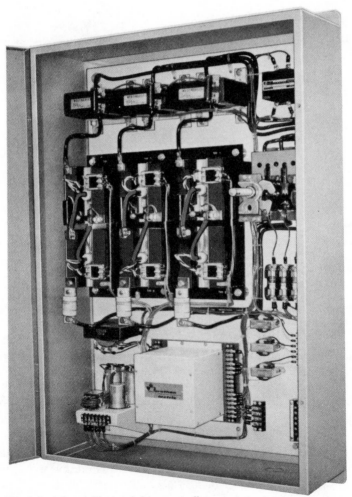

Fig. 1.6 AC variable-speed drive controller.

1.4 TYPES OF CONVERTER SYSTEMS

Any power semiconductor system employed for rectifying, inverting, or
otherwise modulating the power output of an ac or dc energy source is
called a "converter system" or "power conditioning system." These systems may be classified according to the function that they perform and the
method of commutation that they employ. The principal types are listed by
function in Table 1.1 in the order in which they are discussed in later
chapters. It should be noted that some converter systems are capable of
varying more than one property of their output; for example, a controlled
inverter may have an output of which both frequency and voltage are
variable.

Table 1.1 Types of Converter Systems

System	Conversion Function
1. AC voltage controllers	Fixed voltage ac to variable voltage ac (line commutation)
2. Rectifiers (uncontrolled)	Fixed voltage ac to fixed voltage dc (line commutation)
3. Rectifiers (controlled)	Fixed voltage ac to variable voltage dc (line commutation)
4. DC-to-dc converters (choppers)	Fixed voltage dc to variable voltage dc (load or forced commutation)
5. Inverters (uncontrolled)	Fixed voltage dc to fixed voltage ac (line, load, or forced commutation)
6. Inverters (controlled)	Fixed voltage dc to variable voltage ac (line, load, or forced commutation)
7. Cycloconverters	Fixed frequency ac to variable frequency and variable voltage ac (line or forced commutation)

1.5 DESIGN OF CONVERTER SYSTEMS

In the chapters that follow, various converters are analyzed in order that
their operation may be understood and that quantitative predictions of
their performance may be made as a necessary part of a design procedure.

The circuits that are employed for this purpose are idealized to the extent that components included in the practical circuit to protect the devices are not shown. Normally the parameters of these components are such that they make only a very slight difference to the overall performance of the converter. However, it must not be forgotten that these components must be fitted and that the determination of their necessary magnitudes must follow on after the design of the power circuits described in Chapters 4 to 8.

A further stage of idealization is also often employed in the form of the neglect of certain parameters of components such as transformers when these parameters are known from experience to be of little practical significance. Idealizations of this kind are mentioned in each analysis as they are made.

Nevertheless, "many a pickle makes a mickle," and the cumulative effect of these two types of idealization is appreciable and often shows up in some difference between predicted and measured performance. The system designer can only be absolutely confident in the acceptability of his design after he has seen a prototype built, wired, mounted in its enclosure, so that all "make-up" inductances and capacitances are present, and operated under the conditions that apply in practice.

1.6 HARMONIC DISTORTION AND INTERFERENCE

Power semiconductor converters are such that they tend to introduce current and voltage harmonics into the supply system and the controlled system. These can cause serious problems of interference with communications systems and distortion of supply voltage waveforms. It is therefore often necessary to introduce filters on the input side of the converter to reduce supply-system harmonics to acceptable amplitudes. The introduction of current and voltage harmonics into the circuit of the controlled system may also be undesirable in some applications, so that steps must be taken to filter these out also. The design of suitable filter circuits is a specialized topic for which there is an abundant literature. No attempt is made therefore to introduce it into later chapters.

Further precautions are sometimes necessary to avoid radio-frequency interference caused by electromagnetic radiation from the converter circuits themselves. These precautions may include enclosing the offending parts of the system by grounded shielding.

TWO

CIRCUITS WITH SWITCHES AND DIODES

Any circuit, no matter how complicated and nonlinear it may be, can be analyzed and its response to a given excitation predicted, but probably at considerable expense. It is part of the business of the designer of power semiconductor systems to reduce this work of analysis to the essential minimum necessary to ensure that the designed system will function as required.

In this chapter some circuits made up of ideal elements are discussed. These elements include ideal switches that present either infinite or zero resistance to current and are capable of instantaneous transition from one state to the other. They also include ideal diodes.

2.1 SWITCHES AND DIODES

The circuit symbol for a semiconductor diode is marked D in Fig. 2.1a. An ideal diode has zero resistance to positive anode current i_A, but infinite resistance to current in the reverse direction. Thus the diode conducts if the source voltage v is positive, and the anode-to-cathode voltage v_{AK} is then zero. The diode does not conduct if the source voltage is negative, when v_{AK} also is negative. In the diagram of i_A versus v_{AK} shown in Fig. 2.1b, the operating point of the diode may thus lie on the positive axis of i_A in the range $0 \leqslant i_A \leqslant \sqrt{2} \, V/R$ or on the negative axis of v_{AK} in the range $0 \geqslant v_{AK} \geqslant \sqrt{2} \, V$ for the circuit of Fig. 2.1a.

The purpose of analyzing such ideal circuits as are discussed in this and in the next chapter is to give the reader the ability to look at a circuit embodying power semiconductor devices and to envisage approximately by inspection how that circuit functions. This should also show what

10

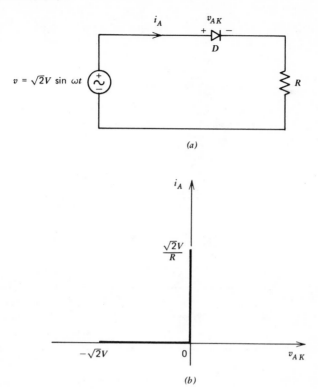

Fig. 2.1 Ideal diode.

measures must be taken to protect the practical devices from destruction. The operation of a switch in a network may:

1. Apply an energy source.
2. Remove an energy source.
3. Change the configuration of the network in other ways.

In the following sections of this chapter, simple switched circuits are first discussed.

2.2 SWITCHED DC SOURCE

The effect of applying a step function of voltage by means of a switch to circuits of different parameters is discussed in this section. The purpose of doing this is to arrive at conclusions that may be applied to similar circuits when they are switched by means of power semiconductor devices.

2.2.1 Resistive Load Circuit   In the circuit of Fig. 2.2a, when switch SW is closed at $t = 0$, the current rises instantaneously to the value

$$i = \frac{V}{R} \quad \text{A} \tag{2.1}$$

When SW is opened at $t = t_1$, the current falls instantaneously to zero as illustrated in Fig. 2.2b. The voltage across the open switch is $v_S = V$.

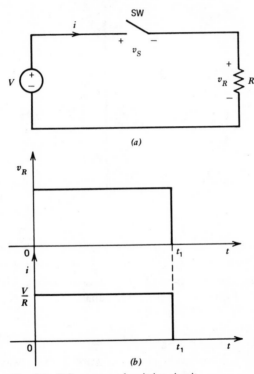

(a)

(b)

Fig. 2.2   DC source and resistive circuit.

2.2.2 RC Load Circuit   In the circuit of Fig. 2.3a, when SW is closed at $t = 0$, by Kirchhoff's voltage law

$$V = v_C + v_R = \frac{1}{C} \int_0^t i \, dt + Ri \quad \text{V} \tag{2.2}$$

Solution of equation 2.2 gives an expression for the time variation of

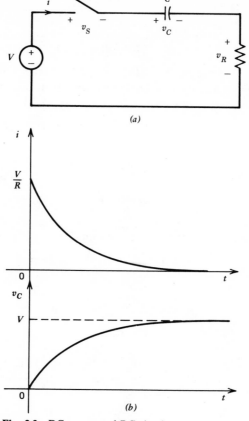

(a)

(b)

Fig. 2.3 DC source and RC circuit.

current i, and hence also for the voltages v_C and v_R. Differentiation of equation 2.2 yields

$$\frac{di}{dt} + \frac{1}{RC}i = 0 \quad \text{A/s} \qquad (2.3)$$

so that

$$i = A\epsilon^{-t/RC} \quad A \qquad (2.4)$$

where A is a constant of integration that must be determined from the initial conditions.

The capacitor is initially uncharged and therefore has zero potential difference between its plates. This potential difference cannot change instantaneously, since

$$v_C = \frac{q}{C} \quad \text{V} \tag{2.5}$$

where q is the charge on each plate. For v_C to change instantaneously, q must change instantaneously, and this would call for an infinite current. Thus immediately after the switch is closed at $t = 0^+$, $v_C = 0$, and from equation 2.2

$$V = v_R = Ri \quad \text{V} \tag{2.6}$$

so that at $t = 0^+$,

$$i = \frac{V}{R} \quad \text{A} \tag{2.7}$$

substitution of $t = 0$ and $i = V/R$ in equation (2.4) yields

$$A = \frac{V}{R} \quad \text{A} \tag{2.8}$$

so that

$$i = \frac{V}{R} \epsilon^{-t/RC} \quad \text{A} \tag{2.9}$$

and this relationship is shown in Fig. 2.3b. As v_R falls, v_C rises, until in infinite time (in practical circuits often merely a fraction of a second) the capacitor is fully charged, so that

$$i = 0 \quad \text{A} : v_C = V \quad \text{V} \tag{2.10}$$

If the switch were opened at $t = t_1$ before the capacitor was fully charged, then the voltage across the switch would be

$$v_S = V - v_C \quad \text{V} \tag{2.11}$$

From equation 2.9 and from the curves of Fig. 2.3b it may be seen that if the resistance in the circuit is very low, then the initial current may be extremely high, and the flow of current will form a pulse of very short duration. If the switch were a power semiconductor, it would be liable to be destroyed by this high current.

2.2.3 RL Load Circuit   In the circuit of Fig. 2.4a, when SW is closed at $t = 0$

$$V = v_L + v_R = L\frac{di}{dt} + Ri \quad \text{V} \tag{2.12}$$

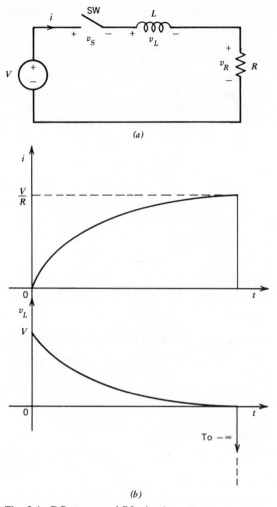

(a)

(b)

Fig. 2.4   DC source and RL circuit.

or

$$\frac{di}{dt} + \frac{R}{L}i = \frac{V}{L} \quad \text{A/s} \tag{2.13}$$

The current in the circuit may be divided into two components. The first is the forced or steady-state component of the current, and this represents the condition of operation of the circuit reached after SW has been closed for an infinitely long time. It is determined by the applied excitation and is the particular integral solution of the differential equation describing the circuit. The second component is the natural or transient component of the current, and this represents a condition of operation of the circuit that has disappeared after an infinite time. It is determined by the circuit parameters and the initial conditions existing in the circuit at $t=0$ and is the complementary function of the solution to the differential equation.

When the steady-state has been reached, the derivative in equation 2.13 is by definition equal to zero, so that from equation 2.13 the forced component of the current is

$$i_F = \frac{V}{R} \quad \text{A} \tag{2.14}$$

The natural component is obtained by solution of the homogeneous equation formed from equation 2.13 which is

$$\frac{di_N}{dt} + \frac{R}{L} i_N = 0 \quad \text{A/s} \tag{2.15}$$

of which the solution is

$$i_N = A\epsilon^{-(R/L)t} \quad \text{A} \tag{2.16}$$

where A is a constant of integration that is to be determined. The complete solution of equation 2.13 is thus

$$i = i_F + i_N = \frac{V}{R} + A\epsilon^{-(R/L)t} \quad \text{A} \tag{2.17}$$

At $t=0$, $i=0$, and substitution in equation 2.17 yields

$$A = -\frac{V}{R} \quad \text{A} \tag{2.18}$$

thus

$$i = \frac{V}{R}(1 - \epsilon^{-(R/L)t}) \quad \text{A} \tag{2.19}$$

and this function is shown in Fig. 2.4b. The voltage across the inductance is

$$v_L = L\frac{di}{dt} = V\epsilon^{-(R/L)t} \quad \text{V} \tag{2.20}$$

and this function also is shown in Fig. 2.4b.

If the switch is reopened, the stored energy in the inductance is released, inducing a voltage at the terminals of the inductance which tends to maintain the current that was flowing while the switch was closed. The opening of the switch tends to reduce the current instantaneously to zero, so that di/dt approaches a value of minus infinity. Since the voltage across the terminals of the inductance is $v_L = L\,di/dt$, this voltage also approaches infinity as indicated in Fig. 2.4. In the case of a simple mechanical switch operating in atmosphere, the air between the opening contacts is ionized by the field due to the high voltage v_S and momentarily conducts, forming a high-resistance arc in which much of the energy formerly stored in the inductor is dissipated as heat. A power transistor operating as a switch would be destroyed in such a situation.

2.2.4 Inductive Load Circuit   In practical circuits, resistance may be very small and inductance large, so that the result obtained by neglecting resistance in analyzing the circuit approximates closely to its actual behaviour. For such an approximate circuit shown in Fig. 2.5a, equation 2.13 becomes

$$\frac{di}{dt} = \frac{V}{L} \quad \text{A/s} \tag{2.21}$$

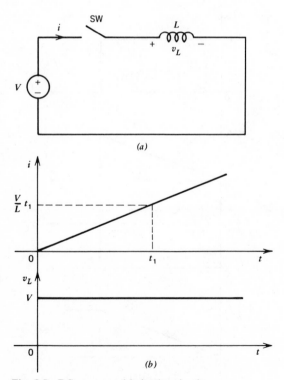

(a)

(b)

Fig. 2.5   DC source and inductive circuit.

and the time variation of current is that shown in Fig. 2.5b, where if SW is closed at $t=0$, then at $t=t_1$

$$i = \frac{V}{L}t_1 \quad \text{A} \tag{2.22}$$

The problem of how to open the switch without the appearance of an infinite voltage across its contacts remains, and one solution is shown in Fig. 2.6, where an ideal diode is connected in parallel with the inductance.

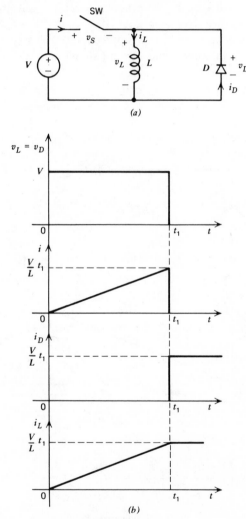

Fig. 2.6 Circuit of Fig. 2.5 with free-wheeling diode.

The diode D shown in Fig. 2.6a earns its name "free-wheeling diode" by its ability to permit current i_L to continue to flow when the energy source has been removed by the opening of the switch at $t = t_1$. The operation of the circuit for the interval $0 < t < t_1$ is not changed by the addition of the ideal diode and is described by equations 2.21 and 2.22.

For negative values of v_L, the inductor is now short-circuited by the diode, so that immediately after SW is reopened

$$v_L = L \frac{di_L}{dt} = 0 \quad \text{V}: \qquad t > t_1 \quad \text{s} \tag{2.23}$$

Thus di_L/dt is zero, and $i_L = i_D$ continues to flow at the value given by equation 2.22. The voltage v_S immediately after the switch is reopened is simply equal to the source voltage V.

In practice, the energy stored in the inductance at time $t = t_1$ would be dissipated in whatever resistance existed in the inductance-diode mesh of the circuit, and i_D would decay exponentially to zero. If the resistance of this mesh were extremely low, then energy would be trapped in the inductance for an appreciable period and might be further increased by a subsequent closing of the switch. This may be clearly seen from a consideration of the current curve in Fig. 2.6b, since if at some instant $t_2 > t_1$ switch SW were again closed, then the current would once more begin to increase at a constant rate V/L A/s, starting from the value $i = (V/L)t_1$ A.

2.3 RECOVERY OF TRAPPED ENERGY

In the ideal circuit of Fig. 2.6a, the energy stored in the inductance is trapped there and is not dissipated by current i_D, since this flows in a circuit without resistance. In a practical circuit, it is desirable not only to avoid the production of heat by the dissipation of energy stored in inductors, but also to improve the efficiency of the system by returning such stored energy to the source. This may be achieved by the circuit shown in Fig. 2.7, provided that the source is such as to be able to accept reverse current.

In the circuit of Fig. 2.7, a second winding has been added to the inductor of Fig. 2.5, thus forming a transformer. The secondary winding of the transformer is connected to the voltage source V via the diode D. The transformer may be assumed to be ideal apart from the magnetizing current, which is in fact the current flowing in the original inductor coil, now represented by the winding N_1. The energy stored in this transformer by current i_1 may be released from winding N_2 by current i_2.

Figure 2.8a shows a circuit equivalent to that of Fig. 2.7, which has been obtained by replacing the transformer by an ideal transformer and a

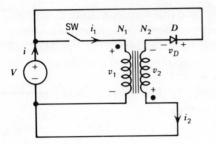

Fig. 2.7 Energy-recovery circuit.

magnetizing inductance equal to that of the inductor of Fig. 2.5. The turns ratio of the ideal transformer is

$$n = \frac{N_1}{N_2} = \frac{v_1}{v_2} \tag{2.24}$$

If now the diode and source V are referred to the primary side of the ideal transformer, then the resulting equivalent circuit is that shown in Fig. 2.8b. In this last circuit, any energy released by the magnetizing inductance L and supplied to the voltage source nV corresponds to that released by the transformer in the circuit of Fig. 2.7 and returned to the source V. The circuit of Fig. 2.8b is a convenient way of showing an energy-recovery

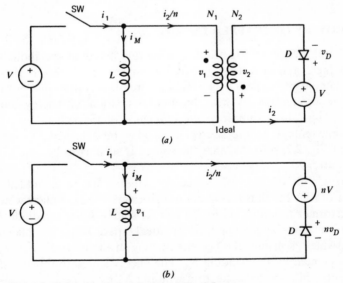

Fig. 2.8 Circuits equivalent to that of Fig. 2.7.

circuit in a complicated circuit diagram without crowding such a diagram with unnecessary detail.

Switch SW in Fig. 2.8b is closed at $t = 0$, and for the interval $0 < t < t_1$, at the end of which the switch is reopened,

$$v_1 = V = n(v_D - V) \quad \text{V} \tag{2.25}$$

from which

$$v_D = V\left(1 + \frac{1}{n}\right) \quad \text{V} \tag{2.26}$$

Owing to the presence of the diode, $i_2/n = 0$, and

$$i_1 = i_M \quad \text{A} \tag{2.27}$$

$$v_1 = V = L\frac{di_1}{dt} \quad \text{V} \tag{2.28}$$

and since at $t = 0$, $i_1 = 0$, the solution of equation 2.28 is

$$i_1 = \frac{V}{L}t \quad \text{A:} \qquad 0 < t < t_1 \quad \text{s} \tag{2.29}$$

and the circuit functions exactly as did that of Fig. 2.5a.

For $t > t_1$, after SW has been reopened, the energy stored in the inductance L during the interval $0 < t < t_1$ is released, so that v_1 and i_2/n are both negative, and current flows through the diode. Thus

$$i_M = \frac{-i_2}{n} \quad \text{A} \tag{2.30}$$

and

$$nv_D = 0 \quad \text{V} \tag{2.31}$$

Also

$$v_1 = -nV = L\frac{di_M}{dt'} = -\frac{L}{n}\frac{di_2}{dt'} \quad \text{V} \tag{2.32}$$

where

$$t' = t - t_1 \quad \text{s} \tag{2.33}$$

that is, when $t = t_1$, $t' = 0$. Since the current in the inductor cannot change instantaneously when SW is reopened at $t' = 0$, it follows that

$$i_M\big|_{t'=0^-} = \frac{-i_2}{n}\bigg|_{t'=0^+} = i_1\big|_{t'=0^-} \quad \text{A} \tag{2.34}$$

where $t'=0^-$ and $t'=0^+$ indicate respectively the instants immediately before and after the opening of the switch. But from equations 2.29 and 2.34

$$\left.\frac{i_2}{n}\right|_{t'=0^+} = -\frac{V}{L}t_1 \quad \text{A} \tag{2.35}$$

and from equation 2.32

$$\frac{di_2}{dt'} = \frac{n^2}{L}V \quad \text{A/s} \tag{2.36}$$

By integration of equation 2.36 and substitution of the initial conditions of equation 2.35,

$$i_2 = \frac{n^2 V}{L}\left(t' - \frac{t_1}{n}\right) \quad \text{A} \tag{2.37}$$

and i_2 becomes zero when

$$t' = \frac{t_1}{n} = t_2 - t_1 \quad \text{s} \tag{2.38}$$

or from equation 2.33

$$t = \frac{n+1}{n}t_1 = t_2 \quad \text{s} \tag{2.39}$$

At $t=t_2$, $v_1=0$, forward current through the diode ceases, and nv_D rises instantaneously to the value nV, that is,

$$v_D = V \quad \text{V}: \qquad t > t_2 \quad \text{s} \tag{2.40}$$

all other variables in the circuit becoming zero. All of the energy stored in the transformer magnetizing inductance L at $t=t_1$ has now been returned to the source.

The time variations of the variables for the complete cycle of operation are illustrated in Fig. 2.9, in which it is assumed that $n=0.5$.

Example 2.1 For the energy-recovery circuit of Fig. 2.7, the magnetizing inductance of the transformer is 200 μH and the leakage inductances and winding resistances are negligible; $N_1 = 20$ and $N_2 = 60$. The source voltage $V = 200$ V, and switch SW has been open for a long time, so that all currents in the circuit are zero.

If switch SW is closed for 200 μs and is then reopened:

(a) Calculate the peak value of the reverse voltage v_D across the diode.
(b) Determine the time for which the diode conducts.

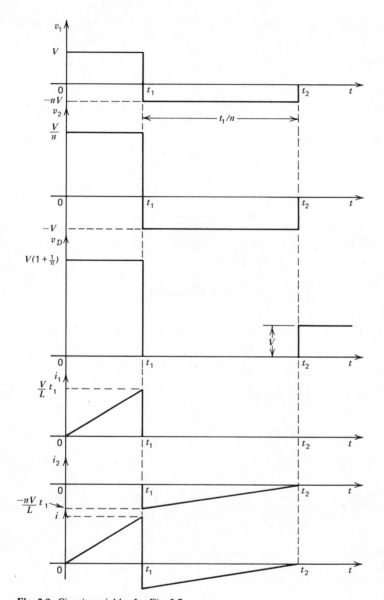

Fig. 2.9 Circuit variables for Fig. 2.7.

(c) Show that the whole of the energy taken from the source while SW is closed is returned to the source after SW is reopened.

Solution

(a) From equation 2.26, the diode voltage is

$$v_D = V\left(1 + \frac{1}{n}\right) \quad \text{V}$$

where

$$V = 200 \quad \text{V}, \qquad n = N_1/N_2 = \frac{1}{3}$$

so that

$$v_D = 800 \, \text{V}$$

(b) From equation 2.38 the conduction interval of the diode is t_1/n s, where $t_1 = 200$ μs; thus the conduction interval is

$$t_2 - t_1 = \frac{200}{1/3} = 600 \, \mu\text{s}$$

(c) The input energy provided by the source while SW is closed is

$$W_I = \int_0^{t_1} Vi_1 \, dt \quad \text{J:} \qquad 0 < t < t_1 \quad \text{s} \tag{2.41}$$

From equations 2.29 and 2.41

$$W_I = \frac{V^2}{L} \int_0^{t_1} t \, dt = \frac{1}{2} \frac{V^2}{L} t_1^2 \quad \text{J}$$

$$= 4 \quad \text{J} \tag{2.42}$$

During the interval $t_1 < t < t_2$, or $0 < t' < t_1/n$, when SW is opened and the diode conducts, the energy provided by the source is

$$W_R = \int_0^{t_1/n} Vi_2 \, dt' \quad \text{J} \tag{2.43}$$

From equations 2.37 and 2.43,

$$W_R = \frac{n^2 V^2}{L} \int_0^{t_1/n} \left(t' - \frac{t_1}{n}\right) dt' = \frac{n^2 V^2}{L} \left[\frac{(t')^2}{2} - \frac{t_1}{n} t'\right]_0^{t_1/n}$$

$$= \frac{n^2 V^2}{L} \left[\frac{t_1^2}{2n^2} - \frac{t_1^2}{n^2}\right] = -\frac{1}{2} \frac{V^2}{L} t_1^2 \quad \text{J} \tag{2.44}$$

in which the negative sign shows that energy is in fact returned to the source. From equations 2.42 and 2.44,

$$W_I + W_R = 0 \quad J$$

so that the net energy provided by the source is zero.

2.4 DC SOURCE AND RLC CIRCUIT

In the circuit of Fig. 2.10 when SW is closed at $t = 0$

$$V = v_L + v_C + v_R = L\frac{di}{dt} + \frac{1}{C}\int_0^t i\,dt + v_C(0) + Ri \quad V \qquad (2.45)$$

where $v_C(0)$ is the initial charge on the capacitor. From equation 2.45

$$\frac{d^2i}{dt^2} + \frac{R}{L}\frac{di}{dt} + \frac{1}{LC}i = 0 \quad A/s^2 \qquad (2.46)$$

The forced component of the current in the second-order system described by equation 2.46 is zero, since when the steady state is reached, and the derivatives are both zero, then i must be zero. This is obvious also on the basis of physical reasoning from the circuit of Fig. 2.10, since when the switch is closed, the capacitor will charge up until $v_C = V$, when current will cease to flow.

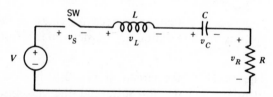

Fig. 2.10 DC source and RLC circuit.

The natural component of the current is in this case the total current and may be determined by solving equation 2.46. The solution is

$$i_N = i = A_1 \epsilon^{s_1 t} + A_2 \epsilon^{s_2 t} \quad A \qquad (2.47)$$

where s_1 and s_2 are the roots of the characteristic equation

$$s^2 + \frac{R}{L}s + \frac{1}{LC} = 0 \qquad (2.48)$$

while A_1 and A_2 are constants of integration to be determined from the initial conditions. It is convenient to define two properties of the circuit as

$$\zeta = \frac{R}{2L} = \text{damping factor} \tag{2.49}$$

$$\omega_o = \frac{1}{\sqrt{LC}} = \text{resonant frequency} \tag{2.50}$$

Equation 2.46 may now be written

$$\frac{d^2i}{dt^2} + 2\zeta\frac{di}{dt} + \omega_o{}^2i = 0 \quad \text{A/s}^2 \tag{2.51}$$

of which the solution is equation 2.47 in another form, namely,

$$i = \epsilon^{-\zeta t}\left[A_1\epsilon^{\sqrt{\zeta^2-\omega_o{}^2}\,t} + A_2\epsilon^{-\sqrt{\zeta^2-\omega_o{}^2}\,t} \right] \quad \text{A} \tag{2.52}$$

If $\zeta^2 > \omega_o{}^2$, then the current is made up of two decaying exponential components. If $\zeta^2 < \omega_o{}^2$, then equation 2.52 becomes

$$i = \epsilon^{-\zeta t}[B_1\cos\omega_r t + B_2\sin\omega_r t] \quad \text{A} \tag{2.53}$$

where

$$\omega_r = \sqrt{\omega_o{}^2 - \zeta^2} = \text{ringing frequency in rad/s} \tag{2.54}$$

and the current consists of a damped or decaying sinusoid. Once again B_1 and B_2 are constants of integration to be determined from the initial conditions.

Example 2.2   The switch in the circuit of Fig. 2.10 is closed at $t=0$. If $V = 100\text{V}$, $L = 10\text{mH}$, $C = 1\,\mu\text{F}$, $R = 80\,\Omega$, and the capacitor is initially charged to a voltage $v_C(0-) = 25\text{V}$, determine: (a) an expression for the current in the circuit; (b) the initial rate of change of current at $t = 0^+$; and (c) the time taken for the current to complete the first positive half cycle (see Fig. E2.2).

Solution   From equations 2.49 and 2.50

$$\zeta = \frac{80}{20 \times 10^{-3}} = 4 \times 10^3 \quad \text{rad/s}$$

$$\omega_o = \frac{1}{\sqrt{10^{-2} \times 10^{-6}}} = 10^4 \quad \text{rad/s}$$

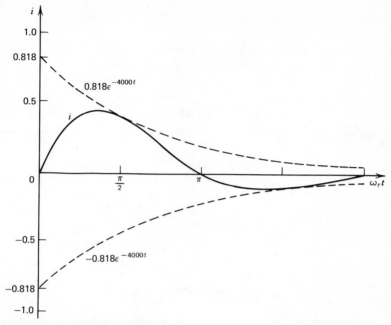

Fig. E2.2 Current in circuit of Fig. 2.10.

since $\zeta^2 < \omega_o^2$, the response is oscillatory, so that from equations 2.54 and 2.53

$$\omega_r = 10^3 \sqrt{100 - 16} = 9170 \quad \text{rad/s}$$

$$i = \epsilon^{-4000t}[B_1 \cos 9170t + B_2 \sin 9170t] \quad \text{A} \tag{1}$$

The circuit possesses inductance; therefore the current cannot change instantaneously, so that

$$i(0^+) = i(0^-) = 0 \quad \text{A}$$

and substitution in (1) yields

$$B_1 = 0 \quad \text{A}$$

consequently,

$$i = \epsilon^{-4000t} B_2 \sin 9170t \quad \text{A} \tag{2}$$

When SW is closed

$$V = v_L + v_C + v_R \quad \text{V}$$

But $v_R(0^+) = Ri(0^+) = 0$, so that

$$v_L = L\frac{di}{dt} = V - v_C = 100 - 25 = 75 \quad \text{V}$$

therefore

$$\left.\frac{di}{dt}\right|_{t=0^+} = \frac{75}{10 \times 10^{-3}} = 7500 \quad \text{A/s}$$

From equation 2

$$\frac{di}{dt} = -4000\epsilon^{-4000t}B_2\sin 9170t + 9170\epsilon^{-4000t}B_2\cos 9170t \quad \text{A/s}$$

thus at $t = 0^+$

$$9170B_2 = 7500$$

$$B_2 = 0.818 \quad \text{A}$$

and

$$i = 0.818\epsilon^{-4000t}\sin 9170t \quad \text{A}$$

The time to complete the first positive half cycle may be obtained from the relationship

$$f = \frac{\omega_r}{2\pi} = \frac{1}{T} \quad \text{Hz}$$

Thus

$$\frac{T}{2} = \frac{1}{2}\frac{2\pi}{9170} = 0.347 \quad \text{ms}$$

2.5 AC SOURCE AND RLC CIRCUIT

In the circuit of Fig. 2.11

$$v(t) = \sqrt{2}\, V\sin\omega t \quad \text{V} \tag{2.55}$$

so that after switch SW is closed at $t = 0$,

$$L\frac{di}{dt} + \frac{1}{C}\int_0^t i\,dt + v_C(0) + Ri = \sqrt{2}\, V\sin\omega t \quad \text{V} \tag{2.56}$$

and differentiation of equation 2.56 yields

$$\frac{d^2i}{dt^2} + \frac{R}{L}\frac{di}{dt} + \frac{1}{LC}i = \frac{\sqrt{2}\, V\omega}{L}\cos\omega t \quad \text{A/s}^2 \tag{2.57}$$

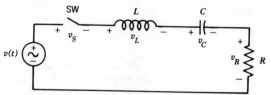

Fig. 2.11   AC source and RLC circuit.

The natural component of the current does not depend in any way on the source voltage, since it is obtained from solution of the homogeneous equation

$$\frac{d^2i}{dt^2} + \frac{R}{L}\frac{di}{dt} + \frac{1}{LC}i = 0 \quad \text{A/s}^2 \tag{2.58}$$

and this is exactly the same as equation 2.46 for the RLC circuit with a dc source. Thus as in equation 2.47

$$i_N = A_1\epsilon^{s_1 t} + A_2\epsilon^{s_2 t} \quad \text{A} \tag{2.59}$$

and this may be an exponential or an oscillatory component of the total current.

The forced component of the response is obtained by steady-state ac circuit analysis and is

$$i_F = \frac{\sqrt{2}\, V\sin(\omega t - \phi)}{\left[R^2 + \left(\omega L - \dfrac{1}{\omega C}\right)^2\right]^{1/2}} \quad \text{A} \tag{2.60}$$

where

$$\phi = \tan^{-1}\frac{\omega L - \dfrac{1}{\omega C}}{R} \quad \text{rad} \tag{2.61}$$

The constants of integration A_1 and A_2 are now obtained by writing

$$i = i_F + i_N = \frac{\sqrt{2}\, V\sin(\omega t - \phi)}{\left[R^2 + \left(\omega L - \dfrac{1}{\omega C}\right)^2\right]^{1/2}} + A_1\epsilon^{s_1 t} + A_2\epsilon^{s_2 t} \quad \text{A} \tag{2.62}$$

and applying the initial conditions.

Example 2.3   In the circuit of Fig. E2.3, switch SW1 is closed at $t = 1.5\,\text{ms}$ and switch SW2 is closed at $t = 15\,\text{ms}$. Determine $i(t)$ for $t > 0$ if $v_C(0) = 0$.

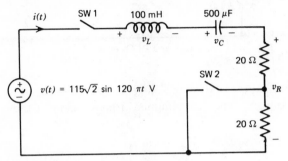

Fig. E2.3 RLC circuit.

Solution

Interval $0 < t < 1.5\,\text{ms}$:

$$i = 0 \quad \text{A}$$

Interval $1.5 < t < 15\,\text{ms}$:   At $t = 1.5\,\text{ms}$, $v(t) = 115\sqrt{2}\,\sin 32.4°$ V. Let $t' = t - 1.5 \times 10^{-3}$ s, so that the interval becomes $0 < t' < 13.5\,\text{ms}$, then

$$v(t') = 115\sqrt{2}\,\sin(120\pi t' + 32.4°) \quad \text{V}$$

For the circuit during this interval, from equation 2.61

$$\phi = \tan^{-1}\frac{32.4}{40} = 39.0°$$

From equation 2.60, the forced component of the current is

$$i_F(t') = \frac{115\sqrt{2}}{\sqrt{40^2 + 32.4^2}}\,\sin(120\pi t' + 32.4° - 39.0°)$$

$$= 3.16\sin(120\pi t' - 6.6°) \quad \text{A}$$

From equation 2.48

$$s^2 + 400s + 20 \times 10^3 = 0$$

so that

$$s_1 = -342, \qquad s_2 = -58.5$$

The natural component of the current is thus

$$i_N(t') = A_1 \epsilon^{-342t'} + A_2 \epsilon^{-58.5t'} \quad \text{A}$$

and

$$i(t') = i_F(t') + i_N(t')$$

$$= 3.16 \sin(120\pi t' - 6.6°) + A_1\epsilon^{-342t'} + A_2\epsilon^{-58.5t'} \quad \text{A} \qquad (1)$$

At $t' = 0$, $i(t') = 0$, and from equation 1

$$A_1 + A_2 = 0.363 \quad \text{A} \qquad (2)$$

At $t' = 0$, the entire source voltage is applied to the inductance, since the initial charge on the capacitance is zero, and the current in the resistance is zero. Thus

$$L\frac{di}{dt'}\bigg|_{t'=0} = 115\sqrt{2} \, \sin 32.4° \quad \text{V}$$

and

$$\frac{di}{dt'}\bigg|_{t'=0} = \frac{115\sqrt{2} \, \sin 32.4°}{100 \times 10^{-3}} = 872 \quad \text{A/s}$$

From (1)

$$\frac{di}{dt'} = 3.16 \times 120\pi \cos(120\pi t' - 6.6°) - 342A_1\epsilon^{-342t'} - 58.5A_2\epsilon^{-58.5t'} \quad \text{A/s}$$

so that at $t' = 0$

$$872 = 3.16 \times 120\pi \cos(-6.6°) - 342A_1 - 58.5A_2 \quad \text{A/s}$$

or

$$342A_1 + 58.5A_2 = 308 \qquad (3)$$

Solution of equations 2 and 3 yields

$$A_1 = 1.01, \qquad A_2 = -0.648 \quad \text{A}$$

and

$$i(t') = 3.16 \sin(120\pi t' - 6.6°) + 1.01\epsilon^{-342t'} - 0.648\epsilon^{-58.5t'} \quad \text{A} \qquad (4)$$

so that for this interval, substitution in (4) for t' yields

$$i(t) = 3.16 \sin(120\pi t - 39°) + 1.69\epsilon^{-342t} - 0.707\epsilon^{-58.5t} \quad \text{A} \qquad (5)$$

Interval $t > 15$ ms: At $t = 15$ ms, $v(t) = 115\sqrt{2}\,\sin(-36°)$ V. Let t'' $= t - 15 \times 10^{-3}$ s, so that the interval becomes $t'' > 0$, then $v(t'')$ $= 115\sqrt{2}\,\sin(120\pi t'' - 36°)$ V. During this interval

$$\phi = \tan^{-1}\frac{32.4}{20} = 58.3°$$

and

$$i_F(t'') = \frac{115\sqrt{2}}{\sqrt{20^2 + 32.4^2}}\,\sin(120\pi t'' - 36.0° - 58.3°)$$

$$= 4.27\sin(120\pi t'' - 94.3°)\quad\text{A}$$

From equation 2.48

$$s^2 + 200s + 20 \times 10^3 = 0$$

so that

$$s_1 = -100 + j100,\qquad s_2 = -100 - j100$$

and

$$i_N(t'') = \epsilon^{-100t''}\left[A_3\epsilon^{j100t''} + A_4\epsilon^{-j100t''}\right]\quad\text{A}$$

Thus

$i(t'') = i_F(t'') + i_N(t'')$

$$= 4.27\sin(120\pi t'' - 94.3°) + \epsilon^{-100t''}\left[A_5\cos 100t'' + A_6\sin 100t''\right]\quad\text{A}\quad(6)$$

Constants A_5 and A_6 must now be determined by applying the initial conditions when $t'' = 0$, or $t = 15 \times 10^{-3}$ s. At this instant, from (5), $i(t) = -3.34$ A. Substitution in (6) then yields

$$A_5 = 0.918\quad\text{A}$$

A second initial condition is required to determine A_6, and this may be obtained by first calculating v_L for $t'' = 0$. From the circuit of Fig. E2.3

$$v_L = v(t) - v_R - v_C\quad\text{V}$$

At $t = 15$ ms $v(t) = 115\sqrt{2}\,\sin(-36°) = -95.6\quad\text{V}$

$$v_R = Ri(t) = 20(-3.34) = -66.8\quad\text{V}$$

$$v_C = \frac{1}{500 \times 10^{-6}}\int_0^{t=15\times10^{-3}} i(t)\,dt\quad\text{V}$$

Substitution for $i(t)$ in this expression and integration gives

$$v_C = 6.1 \quad V$$

Thus

$$v_L = -95.6 + 66.8 - 6.1 = -34.9 \quad V$$

$$\left. \frac{di}{dt''} \right|_{t''=0} = \frac{-34.9}{100 \times 10^{-3}} = -349 \quad A/s$$

Differentiation of equation 6 and substitution for di/dt'', A_5 and t'' at $t'' = 0$ then yields

$$A_6 = -1.37 \quad A$$

$$i(t'') = 4.27 \sin(120\pi t'' - 94.3°)$$

$$+ \epsilon^{-100t''}[0.918 \cos 100t'' - 1.37 \sin 100t''] \quad A$$

Substitution for t'' in this expression then gives

$$i(t) = 4.27 \sin(120\pi t - 58.3°) + 1.92\epsilon^{-100t} \cos(100t + 47.6°) \quad A$$

The current for the three intervals of time is now known.

2.6 HALF-WAVE RECTIFIER

The excitations applied to the circuits so far considered in this chapter have been either sinusoidally varying functions of time or direct voltage excitations, which may be considered as sinusoidal excitations of zero frequency. When a diode is introduced between a sinusoidal voltage source and a linear load circuit, the excitation is no longer a pure sinusoid, and the forced response to such an excitation may no longer be determined simply by the method of sinusoidal steady-state circuit analysis at source frequency.

The excitation provided by an alternating voltage source in series with a diode is however periodic, and any such excitation can be represented as the sum of an infinite series of sinusoidal functions of related frequencies. The forced response to each significant term of this series may then be determined by sinusoidal steady-state analysis of the appropriate frequency, and since the load circuit is linear, the forced responses to all significant terms may be combined by superposition. It naturally follows also that the current supplied by the source to a half-wave rectifier circuit is nonsinusoidal, but periodic.

From the series of functions described in the preceding paragraphs, certain figures of merit for a converter system may be calculated, and these provide a measure of the suitability of the converter both for connection to the supply system and for excitation of the load circuit. These figures of merit may be calculated by means of Fourier analysis, which is briefly reviewed in Section 2.6.1, where it is also shown that, in half-wave rectifier circuits, the current includes a direct component. This means that the alternating sources discussed in the sections which follow must be considered to be ideal voltage sources with zero internal impedance. If a nonideal source in the form of a single-phase transformer were employed to excite such a circuit, the direct component of the current would saturate the core of the transformer unless the core were extremely large. While it may appear that on this account half-wave rectifier circuits are not of practical importance, nevertheless their analysis may be extended to that of full-wave single-phase and polyphase circuits that are not subject to unrealistic limitations.

2.6.1 Fourier Analysis   In the circuit of Fig. 2.12 current will flow when

$$v - v_O > 0: \qquad v_{AK} = 0 \quad \text{V} \tag{2.63}$$

The rectified voltage may be described by the Fourier series

$$v_O = V_O + \sum_{n=1}^{\infty} a_n \sin n\omega t + \sum_{n=1}^{\infty} b_n \cos n\omega t \quad \text{V} \tag{2.64}$$

If the angle at which the diode commences to conduct (the firing angle) is α, and that at which it ceases to conduct (the extinction angle) is β, then the conduction angle is

$$\gamma = \beta - \alpha \quad \text{rad} \tag{2.65}$$

For a diode supplying a passive load circuit, $\alpha = 0$ and $\gamma = \beta$. The magnitude of β depends on the nature of the load circuit.

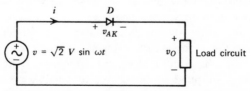

Fig. 2.12 Half-wave rectifier circuit.

The coefficients in equation 2.64 may be expressed for the general case as

$$V_O = \frac{1}{2\pi} \int_0^{2\pi} v_O \, d(\omega t) = \frac{1}{2\pi} \int_\alpha^\beta v_O \, d(\omega t) \quad \text{V} \tag{2.66}$$

$$a_n = \frac{1}{\pi} \int_0^{2\pi} v_O \sin n\omega t \, d(\omega t) = \frac{1}{\pi} \int_\alpha^\beta v_O \sin n\omega t \, d(\omega t) \quad \text{V} \tag{2.67}$$

$$b_n = \frac{1}{\pi} \int_0^{2\pi} v_O \cos n\omega t \, d(\omega t) = \frac{1}{\pi} \int_\alpha^\beta v_O \cos n\omega t \, d(\omega t) \quad \text{V} \tag{2.68}$$

and in particular V_O, which is the average value of the rectified voltage, may be called the "dc output voltage." The rms value of the nth harmonic voltage is given by

$$V_{nR} = \frac{1}{\sqrt{2}} [a_n^2 + b_n^2]^{1/2} \quad \text{V} \tag{2.69}$$

The rms value of the rectified voltage is

$$V_R = \left[\frac{1}{2\pi} \int_\alpha^\beta v_O^2 \, d(\omega t) \right]^{1/2} = \left[V_O^2 + \sum V_{nR}^2 \right]^{1/2} \quad \text{V} \tag{2.70}$$

so that the rms value of the harmonic components or "ripple voltage" is

$$V_{RI} = \left[\sum V_{nR}^2 \right]^{1/2} = [V_R^2 - V_O^2]^{1/2} \cdot \text{V} \tag{2.71}$$

The voltage ripple factor may then be defined as

$$K_v = \frac{V_{RI}}{V_O} \tag{2.72}$$

A group of equations exactly analogous to equations 2.64 to 2.72 may be written for the line or load current in the circuit of Fig. 2.12. The Fourier series describing the current is

$$i = I_O + \sum_{n=1}^{\infty} c_n \sin(n\omega t - \phi_n) + \sum_{n=1}^{\infty} d_n \cos(n\omega t - \phi_n) \quad \text{A} \tag{2.73}$$

where

$$c_n = \frac{a_n}{Z_n} \quad \text{A:} \quad d_n = \frac{b_n}{Z_n} \quad \text{A:} \quad \phi_n = \tan^{-1} \frac{n\omega L}{R} \quad \text{rad} \tag{2.74}$$

and

$$I_O = \frac{V_O}{R} \quad \text{A}$$

In equation 2.74, Z_n is the impedance of the load circuit of Fig. 2.12 to currents of angular frequency $n\omega$, and it is assumed that this impedance consists of resistance and inductance. The rms value of the nth harmonic current is

$$I_{nR} = \frac{1}{\sqrt{2}} [c_n^2 + d_n^2]^{1/2} \quad \text{A} \tag{2.75}$$

and the rms value of the harmonic components of current is

$$I_{RI} = \left[\sum I_{nR}^2 \right]^{1/2} = [I_R^2 - I_O^2]^{1/2} \quad \text{A} \tag{2.76}$$

where I_R is the rms value of the rectified current. Finally the current ripple factor may be defined as

$$K_i = \frac{I_{RI}}{I_O} \tag{2.77}$$

2.6.2 Resistive Load Circuit   For the circuit of Fig. 2.13a with switch SW closed, during the negative half cycle, $\alpha = 0$, and $\beta = \pi$, so that $\gamma = \pi$. Consequently

$$i = \frac{v}{R} \quad \text{A:} \qquad 0 < \omega t < \pi \quad \text{rad} \tag{2.78}$$

$$i = 0: \qquad \pi < \omega t < 2\pi \quad \text{rad} \tag{2.79}$$

From equation 2.66 the dc output voltage is

$$V_O = \frac{1}{2\pi} \int_0^\pi \sqrt{2}\, V \sin \omega t\, d(\omega t) = \frac{\sqrt{2}\, V}{\pi} \quad \text{V} \tag{2.80}$$

From equation 2.70 the rms rectified voltage is

$$V_R = \left[\frac{1}{2\pi} \int_0^\pi 2V^2 \sin^2 \omega t\, d(\omega t) \right]^{1/2} = \frac{V}{\sqrt{2}} \quad \text{V} \tag{2.81}$$

From equation 2.71 the ripple voltage is

$$V_{RI} = [V_R^2 - V_O^2]^{1/2} = V_O \left[\frac{\pi^2}{4} - 1 \right]^{1/2} = 1.211 V_O \quad \text{V} \tag{2.82}$$

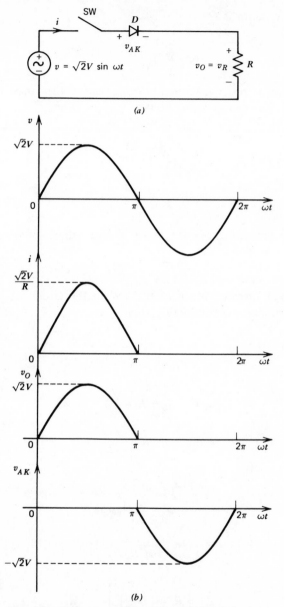

(a)

(b)

Fig. 2.13 Half-wave rectifier with resistive load circuit.

and the ripple factor is

$$K_v = \frac{V_{RI}}{V_O} = 1.211 \tag{2.83}$$

Since this is a resistive circuit, the current values may be obtained directly from the corresponding voltage values. Thus

$$I_O = \frac{V_O}{R} = \frac{\sqrt{2}\,V}{\pi R} \quad \text{A} \tag{2.84}$$

$$I_R = \frac{V_R}{R} = \frac{V}{\sqrt{2}\,R} \quad \text{A} \tag{2.85}$$

and since the waveforms of output current and output voltage are identical

$$K_i = K_v = 1.211 \tag{2.86}$$

2.6.3 RC Load Circuit When switch SW in the circuit of Fig. 2.14a is closed, and the diode is conducting, then

$$v_C + v_R = v_O = v \quad \text{V} \tag{2.87}$$

or

$$\frac{1}{C}\int_0^t i\,dt + v_C(0) + Ri = \sqrt{2}\,V \sin \omega t \quad \text{V} \tag{2.88}$$

The forced component of the current is

$$i_F = \frac{\sqrt{2}\,V}{Z}\sin(\omega t + \phi) \quad \text{A} \tag{2.89}$$

where

$$\phi = \tan^{-1}\frac{1}{\omega CR} \quad \text{rad} \tag{2.90}$$

$$Z = \left[R^2 + \left(\frac{1}{\omega C}\right)^2 \right]^{1/2} \quad \Omega \tag{2.91}$$

The natural component of the current is of the form

$$i_N = A\epsilon^{-t/RC} \quad \text{A} \tag{2.92}$$

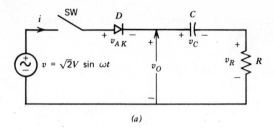

(a)

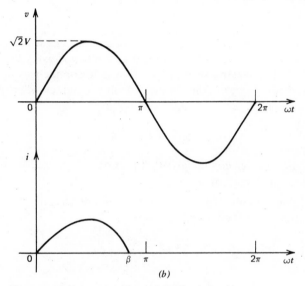

(b)

Fig. 2.14 Half-wave rectifier with RC load circuit.

so that

$$i = \frac{\sqrt{2}\,V}{Z} \sin(\omega t + \phi) + A\epsilon^{-t/RC} \quad \text{A} \tag{2.93}$$

If switch SW is closed during the negative half cycle of v and the capacitance has no initial charge, then at $t = 0$, the beginning of the first conducting period, $v_C = 0$, and $i = 0$, so that substitution of these initial conditions in equation 2.93 gives

$$i = \frac{\sqrt{2}\,V}{Z}[\sin(\omega t + \phi) - \epsilon^{-t/RC}\sin\phi] \quad \text{A} \tag{2.94}$$

and the time variation of current is indicated in Fig. 2.14*b*. The voltage across the capacitance is

$$v_C = \frac{1}{C} \int_0^t i \, dt$$

$$= \sqrt{2} \, V \sin \phi [\cos \phi \epsilon^{-t/RC} - \cos(\omega t + \phi)] \quad \text{V} \qquad (2.95)$$

At the end of the pulse of current, when

$$\omega t = \beta > \frac{\pi}{2} \quad \text{rad} \qquad (2.96)$$

v_C is necessarily positive, so that the capacitor is positively charged at the beginning of the next pulse of current, which commences when $v = v_C$. During succeeding cycles, the diode can only conduct when $v > v_C$, and eventually

$$v_C = \sqrt{2} \, V \quad \text{V} \qquad (2.97)$$

and conduction ceases completely. If R is zero, this condition is reached at the end of the first current pulse.

2.6.4 RL Load Circuit When switch SW in the circuit of Fig. 2.15*a* is closed and the diode is conducting

$$v_L + v_R = v_O = v \quad \text{V} \qquad (2.98)$$

or

$$L \frac{di}{dt} + Ri = \sqrt{2} \, V \sin \omega t \quad \text{V} \qquad (2.99)$$

If SW is closed during the negative half-cycle of v, then by sinusoidal steady-state circuit analysis the forced component of the current is

$$i_F = \frac{\sqrt{2} \, V \sin (\omega t - \phi)}{\left[R^2 + (\omega L)^2 \right]^{1/2}} \quad \text{A} \qquad (2.100)$$

where

$$\phi = \tan^{-1} \frac{\omega L}{R} \quad \text{rad} \qquad (2.101)$$

The natural response of such a circuit has already been determined in Section 2.2.3 and is

$$i_N = A \epsilon^{-(R/L)t} \quad \text{A} \qquad (2.12)$$

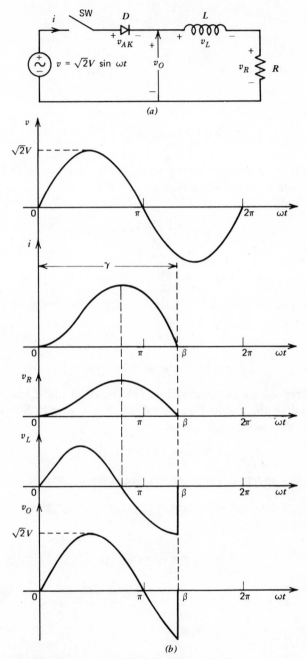

Fig. 2.15 Half-wave rectifier with RL load circuit.

Thus

$$i = i_F + i_N = \frac{\sqrt{2}\ V \sin(\omega t - \phi)}{Z} + A\epsilon^{-(R/L)t} \quad \text{A} \qquad (2.102)$$

where

$$Z = \left[R^2 + (\omega L)^2 \right]^{1/2} \quad \Omega \qquad (2.103)$$

The constant A is determined by substitution in equation 2.102 of the initial condition $i = 0$ at $t = 0$, giving

$$A = \frac{\sqrt{2}\ V \sin\phi}{Z} \quad \text{A} \qquad (2.104)$$

thus

$$i = \frac{\sqrt{2}\ V}{Z} \left[\sin(\omega t - \phi) + \epsilon^{-(R/L)t} \sin\phi \right] \quad \text{A:} \qquad 0 < \omega t < \beta \quad \text{rad} \quad (2.105)$$

Also

$$i = 0: \qquad \beta < \omega t < 2\pi \quad \text{rad} \qquad (2.106)$$

The waveforms of the circuit variables are shown in Fig. 2.15b.

At the end of the conduction period, $i = 0$, and $\omega t = \beta$. Substitution of these values in equation 2.105 gives

$$\sin(\beta - \phi) + \epsilon^{-R\beta/\omega L} \sin\phi = 0 \qquad (2.107)$$

and this transcendental equation may be solved numerically for given values of ω, L, and R.

The average value of the rectified current is conveniently obtained by starting from the equation

$$v - L\frac{di}{dt} - Ri = 0 \quad \text{V} \qquad (2.108)$$

from which

$$i = \frac{v}{R} - \frac{L}{R}\frac{di}{dt}$$

$$= \frac{\sqrt{2}\ V}{R} \sin\omega t - \frac{\omega L}{R}\frac{di}{d(\omega t)} \quad \text{A} \qquad (2.109)$$

Then

$$I_O = \frac{1}{2\pi} \int_0^\beta i \, d(\omega t)$$

$$= \frac{1}{2\pi} \int_0^\beta \left[\frac{\sqrt{2}\, V}{R} \sin \omega t - \frac{\omega L}{R} \frac{di}{d(\omega t)} \right] d(\omega t) \quad \text{A} \qquad (2.110)$$

The second term under this integral vanishes, and

$$I_O = \frac{\sqrt{2}\, V}{2\pi R} (1 - \cos \beta) \quad \text{A} \qquad (2.111)$$

Since no average value of rectified voltage can appear across the inductance, it follows that

$$V_O = R I_O = \frac{\sqrt{2}\, V}{2\pi} (1 - \cos \beta) \quad \text{V} \qquad (2.112)$$

At this point, several quantities required to determine whether or not a circuit design is satisfactory are unknown. These include the ripple voltage V_{RI}, the ripple current I_{RI}, the rms value of the rectified current I_R, and the voltage and current ripple factors. These may be obtained by Fourier analysis as explained in Section 2.6.1; however, a great deal of computation may be saved by the use of design curves showing normalized values of the average and rms current as functions of ϕ. As a first step in this procedure, a curve of ϕ versus β may be obtained from equation 2.107, and this is shown in Fig. 2.16.

If the normalized value of the current is defined as

$$i_N = \frac{Z}{\sqrt{2}\, V} i \qquad (2.113)$$

then from equation 2.105, the normalized value of the average rectified current is given by

$$I_N = \frac{1}{2\pi} \int_0^\beta \left[\sin(\omega t - \phi) + \epsilon^{-(R/L)t} \sin \phi \right] d(\omega t) \qquad (2.114)$$

and by entering Fig. 2.16 with a series of values of ϕ, the corresponding values of β may be determined and I_N calculated for each. The resulting curve of I_N versus ϕ is shown in Fig. 2.17.

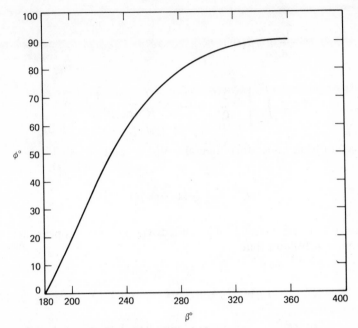

Fig. 2.16 ϕ versus β for circuit of Fig. 2.15a.

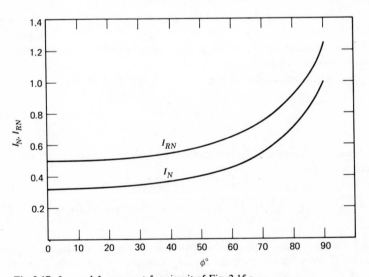

Fig. 2.17 I_{RN} and I_N versus ϕ for circuit of Fig. 2.15a.

44

Also from equation 2.105 the normalized value of the rms current is given by

$$I_{RN} = \left[\frac{1}{2\pi} \int_0^\beta \left[\sin(\omega t - \phi) + \epsilon^{-(R/L)t} \sin\phi \right]^2 d(\omega t) \right]^{1/2} \quad (2.115)$$

and again by means of Fig. 2.16, the curve of I_{RN} versus ϕ shown in Fig. 2.17 is obtained.

Inspection of the curve of v_O versus ωt in Fig. 2.15b shows also that the rms value of the output voltage is given by

$$V_R = \left[\frac{1}{2\pi} \int_0^\beta \left[\sqrt{2} \, V \sin \omega t \right]^2 d(\omega t) \right]^{1/2} \quad V \qquad (2.116)$$

If the resistance in the circuit is negligibly small, so that

$$\omega L \gg R \quad \Omega \qquad (2.117)$$

then from equations 2.101, 2.103, and 2.105,

$$i = \frac{\sqrt{2} \, V}{\omega L} (1 - \cos \omega t) \quad A \qquad (2.118)$$

and the waveform of the current is as shown in Fig. 2.18.

From equation 2.118 or Fig. 2.18 it may be seen that

$$I_O = \frac{\sqrt{2} \, V}{\omega L} \quad A \qquad (2.119)$$

and that the only harmonic present in this current is the first or fundamental, of which the rms value is

$$I_{1R} = \frac{V}{\omega L} = \frac{I_O}{\sqrt{2}} \quad A \qquad (2.120)$$

The rms value of the rectified current is then

$$I_R = \left[I_O{}^2 + I_{1R}{}^2 \right]^{1/2} = 1.225 I_O \quad A \qquad (2.121)$$

The voltage across the inductance is $v_L = v$ for the entire cycle, so that

$$V_O = 0 \quad V \qquad (2.122)$$

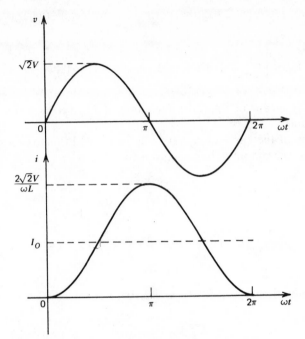

Fig. 2.18 Current in circuit of Fig. 2.15a when $\omega L \gg R$.

and the ripple factors are

$$K_v = \frac{V_{1R}}{V_O} = \infty \tag{2.123}$$

$$K_i = \frac{I_{1R}}{I_O} = 0.707 \tag{2.124}$$

2.6.5 RL Load Circuit with Free-wheeling Diode Half-wave rectifiers with a load circuit consisting of resistance and inductance in series are characterized by discontinuous current and high ripple content. The first of these characteristics can be eliminated and the second much reduced by means of a free-wheeling diode shown as D_2 in Fig. 2.19a.

At this point it is necessary to make a distinction that has not been needed in any of the previous analyses. Care has always been taken to stipulate that switch SW in the circuits shall be closed during the negative half-cycle of the voltage source. This was done to avoid the need to analyze the special case of switching on the circuit at a positive value of source voltage which would give a "transient" response applying to the

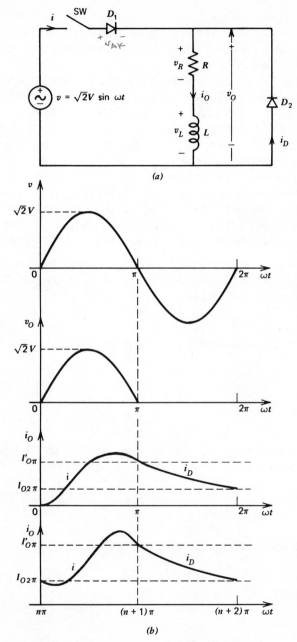

Fig. 2.19 Half-wave rectifier with free-wheeling diode.

first cycle of operation only. In this way, the response which was in fact
determined for the first cycle of operation was that which was repeated in
each subsequent cycle, and could therefore be called the "steady-state"
response of the circuit.

Even when the stipulation is repeated that the switch is to be closed
during the negative half-cycle of the voltage source, the steady-state
response of the circuit of Fig. 2.19a is not determined during the first cycle
of operation. The reason for this is that the current in the load circuit is no
longer discontinuous, but has a finite value at the commencement of each
subsequent cycle. Only after several cycles of response have been completed does this continuous current settle down to a steady state of cyclical
variation. It is therefore necessary to divide the analysis which follows into
two parts: the "transient analysis," which gives the response in the first
cycle only, and the "steady-state analysis," which gives the response when
the circuit has settled down into a repeated cycle of operation.

Transient Analysis If switch SW is closed during the negative half cycle of
v, then during the first positive half cycle of v, D_2 has no effect on the
operation of the circuit, and from equation 2.105

$$i_O = i = \frac{\sqrt{2}\,V}{Z}\left[\sin(\omega t - \phi) + \epsilon^{-(R/L)t}\sin\phi\right] \quad \text{A:} \quad 0 < \omega t < \pi \quad \text{rad}$$

$$(2.125)$$

also

$$v_{AK} = 0, \quad v_O = v \quad \text{V:} \quad 0 < \omega t < \pi \quad \text{rad} \quad (2.126)$$

At $\omega t = \pi$, v goes negative, and the stored energy in the inductance is
dissipated by current i_D flowing in the RLD_2 mesh. Thus

$$i_O = i_D = I_{O\pi}\epsilon^{-(R/L)t'} \quad \text{A:} \quad 0 < \omega t' < \pi \quad \text{rad} \quad (2.127)$$

where

$$\omega t' = \omega t - \pi \quad \text{rad} \quad (2.128)$$

$$I_{O\pi} = i|_{\omega t = \pi} = i|_{\omega t' = 0} \quad \text{A} \quad (2.129)$$

Also

$$v_{AK} = v, \quad v_O = 0 \quad \text{V:} \quad 0 < \omega t' < \pi \quad \text{rad} \quad (2.130)$$

At $\omega t = 2\pi$, or $\omega t' = \pi$, v goes positive, and since $v_O = 0$, D_1 once more commences to conduct. At this instant, the current in the load circuit is

$$I_{O2\pi} = I_{O\pi} \epsilon^{-(R/L)(\pi/\omega)} \quad \text{A} \qquad (2.131)$$

The variation of i_O during this first cycle is shown in the upper current curve of Fig. 2.19b.

At the beginning of the next cycle of v, i_O is no longer zero, and in succeeding cycles $I_{O\pi}$ and $I_{O2\pi}$ approach steady-state values such that the current at beginning and end of the cycle is a constant value $I'_{O2\pi}$ shown in the lower current curve of Fig. 2.19b.

Steady-State Analysis Let

$$\omega t'' = \omega t - n\pi \quad \text{rad} \qquad (2.132)$$

where $n/2$ is the number of cycles of response that have been completed since switch SW was closed and is very large. From equation 2.102

$$i_O = i = \frac{\sqrt{2}\,V}{Z}\sin(\omega t'' - \phi) + A\epsilon^{-(R/L)t''} \quad \text{A} \qquad (2.133)$$

Also

$$i_O|_{t''=0} = I'_{O2\pi} \quad \text{A} \qquad (2.134)$$

Substitution of the initial conditions of equation 2.134 in equation 2.133 yields

$$i_O = \frac{\sqrt{2}\,V}{Z}\sin(\omega t'' - \phi) + \left(I'_{O2\pi} + \frac{\sqrt{2}\,V}{Z}\sin\phi\right)\epsilon^{-(R/L)t''} \quad \text{A} \quad (2.135)$$

At $\omega t'' = \pi$, diode D_2 begins to conduct; i falls instantaneously to zero, and from equation 2.135

$$i_O|_{t''=\frac{\pi}{\omega}} = I'_{O\pi} = \frac{\sqrt{2}\,V}{Z}\sin\phi + \left(I'_{O2\pi} + \frac{\sqrt{2}\,V}{Z}\sin\phi\right)\epsilon^{-(R/L)(\pi/\omega)} \quad \text{A} \quad (2.136)$$

During the succeeding half-cycle v_O is zero, and from equation 2.127

$$i_O = i_D = I'_{O\pi}\epsilon^{-(R/L)(t''-\pi/\omega)} \quad \text{A} \qquad (2.137)$$

At $\omega t'' = 2\pi$, v and hence v_O become positive, and

$$i_O|_{t''=\frac{2\pi}{\omega}} = I'_{O\pi}\epsilon^{-(R/L)(\pi/\omega)} = I'_{O2\pi} \quad \text{A} \qquad (2.138)$$

Thus from equations 2.136 and 2.138,

$$\frac{\sqrt{2}\,V}{Z}\sin\phi+\left(I'_{O2\pi}+\frac{\sqrt{2}\,V}{Z}\sin\phi\right)\epsilon^{-(R/L)(\pi/\omega)}=I'_{O2\pi}\epsilon^{(R/L)(\pi/\omega)}\quad \text{A}$$

$$(2.139)$$

so that

$$I'_{O2\pi}=\frac{\dfrac{\sqrt{2}\,V}{Z}\sin\phi[1+\epsilon^{-(R/L)(\pi/\omega)}]}{\epsilon^{(R/L)(\pi/\omega)}-\epsilon^{-(R/L)(\pi/\omega)}}\quad \text{A}\qquad (2.140)$$

and from equation 2.138

$$I'_{O\pi}=I'_{O2\pi}\epsilon^{(R/L)(\pi/\omega)}\quad \text{A}\qquad (2.141)$$

The steady-state variation of i_O is shown in the lower current curve of Fig. 2.19b. It should be noted that while this circuit ensures a continuous and fairly constant load current i_O, the line current i is discontinuous and has a very high harmonic content. It should also be remarked that the current i during the interval $n\pi<\omega t<(n+1)\pi$ may exceed $I'_{O\pi}$.

If the resistance R is negligibly small, then it may be seen from equation 2.127 that the current will not decrease during the interval $0<\omega t'<\pi$, so that as already explained in Section 2.2.4, each successive positive half-wave of the source voltage will increase the current and the energy stored in the inductance.

An alternative method of steady-state analysis of the circuit of Fig. 2.19a is provided by Fourier analysis. This provides the essential information for an engineering design in addition to determining the figures of merit defined in equations 2.72 and 2.77.

As shown in Fig. 2.19b, the time variation of the load-circuit voltage v_O consists simply of the positive half cycles of the source voltage v. This voltage wave may be analyzed by means of equations 2.66 to 2.68, in which $\alpha=0$ and $\beta=\pi$, into the following series:

$$v_O=\frac{2\sqrt{2}\,V}{\pi}\left[\frac{1}{2}+\frac{\pi}{4}\sin\omega t-\frac{1}{3}\cos 2\omega t\right.$$

$$\left.-\frac{1}{15}\cos 4\omega t-\frac{1}{70}\cos 6\omega t-\cdots\right]\quad \text{V}\qquad (2.142)$$

It may be seen from equation 2.142 that the amplitudes of the harmonics decrease very rapidly as frequency increases.

From equation 2.142, the average value of v_O is

$$V_O = \frac{\sqrt{2}\, V}{\pi} \quad \text{V} \tag{2.143}$$

and this expression could equally well have been obtained by simply calculating the average value of the curve of v_O in Fig. 2.19b.

The rms value of the load-circuit voltage v_O is from equation 2.70

$$V_R = \left[\frac{1}{2\pi} \int_0^\pi \left[\sqrt{2}\, V \sin \omega t \right]^2 d(\omega t) \right]^{1/2} = \frac{V}{\sqrt{2}} \quad \text{V} \tag{2.144}$$

so that from equation 2.71 the ripple voltage is

$$V_{RI} = \left[\frac{V^2}{2} - \frac{2V^2}{\pi^2} \right]^{1/2} = 0.545\, V \quad \text{V} \tag{2.145}$$

The voltage ripple factor is then

$$K_v = \frac{V_{RI}}{V_O} = \frac{0.545\pi}{\sqrt{2}} = 1.21 \tag{2.146}$$

A series describing the load current i_O may be obtained from equation 2.142 and the parameters of the load circuit, giving

$$i_O = \frac{2\sqrt{2}\, V}{\pi} \left[\frac{1}{2R} + \frac{\pi}{4Z_1} \sin(\omega t - \phi_1) - \frac{1}{3Z_2} \cos(2\omega t - \phi_2) \right.$$

$$\left. - \frac{1}{15Z_4} \cos(4\omega t - \phi_4) - \cdots \right] \quad \text{A} \tag{2.147}$$

where

$$Z_n = \left[R^2 + (n\omega L)^2 \right]^{1/2} \quad \Omega \tag{2.148}$$

and Z_1 is the impedance at fundamental frequency normally signified by the symbol Z. Also

$$\phi_n = \tan^{-1} \frac{n\omega L}{R} \quad \text{rad} \tag{2.149}$$

From equations 2.138 and 2.139 it may be seen that the amplitudes of the current harmonics decrease more rapidly than do those of the voltage harmonics due to the presence of the inductive reactance $n\omega L$ in the load circuit.

The average value of the rectified current is

$$I_O = \frac{V_O}{R} = \frac{\sqrt{2}\,V}{\pi R} \quad \text{A} \tag{2.150}$$

so that the series in equation 2.147 might equally well be written

$$i_O = I_O + \sqrt{2}\,I_{1R}\sin(\omega t - \phi_1) - \sqrt{2}\,I_{2R}\cos(2\omega t - \phi_2)$$

$$- \sqrt{2}\,I_{4R}\cos(4\omega t - \phi_4) - \cdots \quad \text{A} \tag{2.151}$$

in which the rms values of the harmonic currents are

$$I_{1R} = \frac{1}{2}\frac{V}{Z_1}: \qquad I_{2R} = \frac{2}{3\pi}\frac{V}{Z_2}: \qquad I_{4R} = \frac{2}{15\pi}\frac{V}{Z_4}: \qquad \cdots \quad \text{A} \tag{2.152}$$

The ripple current may be determined from equation 2.151 as

$$I_{RI} = \sqrt{\sum I_{nR}^2} \quad \text{A} \tag{2.153}$$

but the degree of accuracy to which it is determined will depend on the number of terms calculated for the series in equation 2.147. From the rapidity with which the amplitudes of these terms decrease with increase of frequency, however, it may be seen that only a small number will be needed to give a very accurate value of I_{RI}. The current ripple factor may then be determined from

$$K_i = \frac{I_{RI}}{I_O} \tag{2.77}$$

and the rms value of the rectified current may be obtained by rearranging equation 2.76 to give

$$I_R = \left[I_O^2 + I_{RI}^2 \right]^{1/2} \quad \text{A} \tag{2.154}$$

Example 2.4 In the circuit of Fig. 2.19a, the source voltage $v = 110\sqrt{2}\ \sin 120\pi t$ V, $R = 5\ \Omega$, and $L = 30$ mH. Calculate the following:

(a) The average value of the load current i_O.
(b) The rms magnitudes of the fundamental, second, and fourth harmonics in i_O.
(c) The rms value of i_O.
(d) The boundary values of i_O in the steady state ($I'_{O\pi}$ and $I'_{O2\pi}$ in Fig. 2.19b) using Fourier analysis.
(e) Also verify the results of part (d), using the solution obtained from the differential equation.

Solution

(a) From equation 2.150, the average value of i_O is given by

$$I_O = \frac{\sqrt{2}\ V}{\pi R} = \frac{\sqrt{2} \times 110}{\pi \times 5} = 9.9 \quad \text{A}$$

(b) From equation 2.152,

$$I_{1R} = \text{rms fundamental current} = \frac{V}{2Z_1} = \frac{0.5V}{Z_1}$$

$$I_{2R} = \text{rms second harmonic current} = \frac{2}{3\pi}\frac{V}{Z_2} = \frac{0.212V}{Z_2}$$

$$I_{4R} = \text{rms fourth harmonic current} = \frac{2}{15\pi}\frac{V}{Z_4} = \frac{0.043V}{Z_4}$$

where

$$Z_1 = \sqrt{R^2 + (\omega L)^2} = 12.3 \quad \Omega$$

$$Z_2 = \sqrt{R^2 + (2\omega L)^2} = 23.2 \quad \Omega$$

$$Z_4 = \sqrt{R^2 + (4\omega L)^2} = 45.5 \quad \Omega$$

Hence

$$I_{1R} = 4.47 \quad \text{A}$$

$$I_{2R} = 1 \quad \text{A}$$

$$I_{4R} = 0.104 \quad \text{A}$$

(c)   From equation 2.154,

$$I_R = \text{rms current} \cong \sqrt{I_0^2 + I_{1R}^2 + I_{2R}^2 + I_{4R}^2} = 10.9 \quad \text{A}$$

(d)   The minimum value of $i_O = I'_{O2\pi}$ occurs at $\omega t = 0$, while the value $i_O = I'_{O\pi}$ occurs at $\omega t = \pi$. Hence from equation 2.151,

$I_{R1} = 1.56$

$K_i = 0.460$

$$I'_{O2\pi} \cong I_O - \sqrt{2} \left(I_{1R} \sin\phi_1 + I_{2R} \cos\phi_2 + I_{4R} \cos\phi_4 \right)$$

$$I'_{O\pi} \cong I_O + \sqrt{2} \left(I_{1R} \sin\phi_1 - I_{2R} \cos\phi_2 - I_{4R} \cos\phi_4 \right)$$

where

$$\phi_1 = \tan^{-1} \frac{\omega L}{R} = 66°, \qquad \phi_2 = \tan^{-1} \frac{2\omega L}{R} = 77.5°$$

$$\phi_4 = \tan^{-1} \frac{4\omega L}{R} = 83.7°$$

Hence

$$I'_{O2\pi} \cong 9.9 - \sqrt{2} \left(4.07 + 0.216 + 0.012 \right) = 3.82 \quad \text{A}$$

$$I'_{O\pi} \cong 9.9 + \sqrt{2} \left(4.07 - 0.216 - 0.012 \right) = 15.35 \quad \text{A}$$

(e)   The steady-state values of $I'_{O2\pi}$ and $I'_{O\pi}$ as obtained from the differential equation are shown in equations 2.140 and 2.141 respectively, and substitution of the values

$$V = 110, \qquad Z = Z_1 = 12.3, \qquad \phi = \phi_1 = 66°$$

yields

$$I'_{O2\pi} = 3.86 \quad \text{A}$$

and

$$I'_{O\pi} = 15.4 \quad \text{A}$$

2.6.6 Load Circuit with Electromotive Force   In this section, the effect of introducing a direct electromotive force into the load circuit of a half-wave rectifier is investigated. This is the situation that would arise if such a circuit were employed to charge a battery or to excite the armature circuit of a dc motor. In general the load circuit may also be expected to possess both resistance and inductance.

The analysis of the complete circuit is of importance because it shows the voltage and current waveforms that arise in the circuit. However, it does not yield the quantities and figures of merit needed to evaluate an engineering design without a good deal of computation. For this reason it is followed by an analysis of two limiting cases, since in practice circuits may approximate closely to a condition of zero inductance, or alternatively of zero resistance.

In a small battery-charging installation, the circuit inductance may be virtually zero, although in a larger installation it is probable that some inductance would be deliberately added to the circuit, first to reduce the harmonic content of the current, and second to regulate the current. Conversely the armature-circuit resistance of an integral horsepower dc motor is small and may often be neglected, while the inductance must be taken into consideration when designing a power semiconductor system to excite the armature. It is therefore of practical interest to consider these two limiting cases for which all of the quantities required to evaluate a design may be readily obtained.

For the circuit of Fig. 2.20a, if SW is closed during the negative half cycle of v, then when the diode is conducting, the current may be considered to consist of two components; one due to the ac source, the other due to the direct emf. Each of these components has in turn a forced component. The component due to the ac source is

$$i_{SF} = \frac{\sqrt{2}\, V}{Z} \sin(\omega t - \phi) \quad \text{A} \tag{2.155}$$

where

$$Z = \left[R^2 + (\omega L)^2 \right]^{1/2} \quad \Omega \tag{2.156}$$

$$\phi = \tan^{-1} \frac{\omega L}{R} \quad \text{rad} \tag{2.157}$$

The component due to the direct emf is

$$i_{CF} = -\frac{V_C}{R} \quad \text{A} \tag{2.158}$$

The two natural components will be single exponential terms of the same time constant, and may therefore be combined to give

$$i_N = A\epsilon^{-(R/L)t} \quad \text{A} \tag{2.159}$$

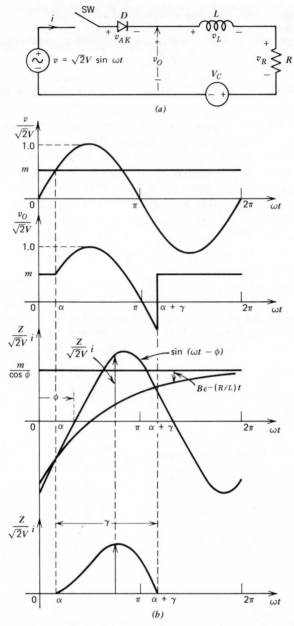

Fig. 2.20 Half-wave rectifier with EMF.

The total current in the circuit is the sum of these three components

$$i = \frac{\sqrt{2}\,V}{Z} \sin(\omega t - \phi) - \frac{V_C}{R} + A\epsilon^{-(R/L)t} \quad A: \qquad \alpha < \omega t < \alpha + \gamma \quad \text{rad}$$

(2.160)

where α is the angle at which conduction begins, and γ is the conduction angle. Necessarily, as may be seen from the voltage curve in Fig. 2.20b

$$\sin \alpha = \frac{V_C}{\sqrt{2}\,V} = m$$

(2.161)

At $\omega t = \alpha$, $i = 0$, so that from equation 2.160

$$A = \left[\frac{V_C}{R} - \frac{\sqrt{2}\,V}{Z} \sin(\alpha - \phi) \right] \epsilon^{(R/L)(\alpha/\omega)} \quad A$$

(2.162)

Also

$$R = Z \cos \phi \quad \Omega$$

(2.163)

and substitution from equations 2.161, 2.162, and 2.163 in equation 2.160 yields

$$\frac{Z}{\sqrt{2}\,V} i = \sin(\omega t - \phi) - \left[\frac{m}{\cos \phi} - B\epsilon^{-(R/L)t} \right] : \qquad \alpha < \omega t < \alpha + \gamma \quad \text{rad}$$

(2.164)

where

$$B = \left[\frac{m}{\cos \phi} - \sin(\alpha - \phi) \right] \epsilon^{(R/L)(\alpha/\omega)}$$

(2.165)

The terms on the right-hand side of equation 2.164 may be represented separately as shown in Fig. 2.20b, and from these the waveform of the normalized current $Zi/\sqrt{2}\,V$, also shown there, may be obtained.

At the end of the conduction period $i = 0$, and

$$\omega t = \alpha + \gamma \quad \text{rad}$$

(2.166)

Substitution in equation 2.164 then yields

$$\frac{(m/\cos \phi) - \sin(\alpha + \gamma - \phi)}{(m/\cos \phi) - \sin(\alpha - \phi)} = \epsilon^{-\gamma/\tan \phi}$$

(2.167)

For any given circuit, m and hence α, as well as ϕ, will be known. The transcendental equation 2.167 could then be solved numerically to determine γ. On the other hand, if γ and ϕ are specified, then equation 2.167 may readily be solved to determine $\sin\alpha = m$, and this procedure may be employed to calculate a family of curves such as are shown in Fig. 2.21. It should be noted, however, that when $R = 0$, $\phi = \pi/2$, and $\cos\phi = 0$. In this case equation 2.167 reduces to the trivial statement $m = m$. Equation 2.167 cannot therefore be employed to determine the curve in Fig. 2.21 marked $\phi = 90°$. This curve is obtained when the limiting case of negligible load-circuit resistance is considered.

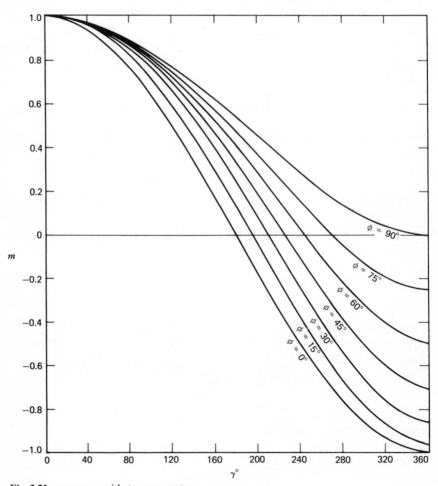

Fig. 2.21 m versus γ with ϕ as parameter.

The voltage and current waveforms for the circuit are now known, and it is necessary to obtain from them the quantities needed to evaluate its performance.

The voltage across the inductance is

$$v_L = \frac{d\lambda}{dt} \quad \text{V} \tag{2.168}$$

where λ is the flux linkage of the circuit and is directly proportional to the current. Since this flux linkage varies cyclically with i, it follows that

$$\int_0^{2\pi/\omega} d\lambda = \int_0^{2\pi/\omega} v_L \, dt = 0 \quad \text{Wb} \tag{2.169}$$

and in particular

$$\int_0^{2\pi} v_L \, d(\omega t) = 0 \quad \text{Wb} \tag{2.170}$$

so that the average value of v_L during one cycle must be zero. Thus the average value of v_R is

$$V_{RES} = \frac{1}{2\pi} \int_\alpha^{\alpha+\gamma} \left[\sqrt{2} \, V \sin \omega t - V_C \right] d(\omega t)$$

$$= \frac{\sqrt{2} \, V}{2\pi} \left[\sqrt{1-m^2} \, (1-\cos\gamma) - m(\gamma - \sin\gamma) \right] \quad \text{V} \tag{2.171}$$

The average value of the rectified current is

$$I_O = \frac{V_{RES}}{R} = \frac{V_{RES}}{Z \cos\phi} \quad \text{A} \tag{2.172}$$

so that from equations 2.171 and 2.172, the normalized value of the output current is

$$I_N = \frac{Z}{\sqrt{2} \, V} I_O = \frac{1}{2\pi \cos\phi} \left[\sqrt{1-m^2} \, (1-\cos\gamma) - m(\gamma - \sin\gamma) \right] \tag{2.173}$$

From equation 2.173 with the aid of equation 2.167 or of the corresponding curves in Fig. 2.21, a family of curves representing the relationship between m and I_N may be produced, as shown in Fig. 2.22.

The dc output voltage is

$$V_O = RI_O + V_C \quad \text{V} \tag{2.174}$$

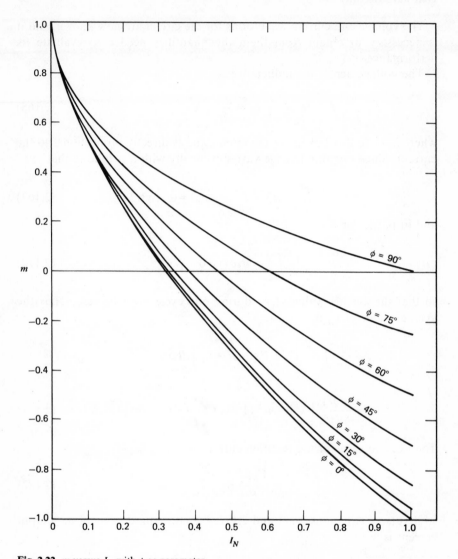

Fig. 2.22 m versus I_N with ϕ as parameter.

To complete the information needed to evaluate the design it is still necessary to determine the rms value of the current as well as the voltage and current ripple factors. The rms output voltage may be obtained by determining γ from Fig. 2.21 or equation 2.167 and then evaluating

$$V_R = \left\{ \frac{1}{2\pi} \left[\int_\alpha^{\alpha+\gamma} [\sqrt{2}\ V \sin \omega t]^2 \, d(\omega t) + \int_{\alpha+\gamma}^{2\pi+\alpha} V_C^2 \, d(\omega t) \right] \right\}^{1/2} \qquad (2.175)$$

The ripple voltage is then calculated from

$$V_{RI} = [V_R{}^2 - V_O{}^2]^{1/2} \quad \text{V} \qquad (2.71)$$

and the voltage ripple factor is

$$K_v = \frac{V_{RI}}{V_O} \qquad (2.72)$$

The normalized value of the rms current I_{RN} may be obtained from

$$I_{RN} = \left[\frac{1}{2\pi} \int_\alpha^{\alpha+\gamma} \left[\frac{Z}{\sqrt{2}\ V} i \right]^2 d(\omega t) \right]^{1/2} \qquad (2.176)$$

in which substitution is made for $Zi/\sqrt{2}\ V$ from equation 2.164. Thus for any set of values of ϕ, m, and γ obtained from Fig. 2.21, I_{RN} may be calculated with the aid of equations 2.157 and 2.161, and this procedure may be employed to calculate a family of curves such as are shown in Fig. 2.23. The actual rms current is then obtained from

$$I_R = \frac{\sqrt{2}\ V}{Z} I_{RN} \quad \text{A} \qquad (2.177)$$

and the ripple current I_{RI}, from

$$I_{RI} = \left[\sum I_{nR}{}^2 \right]^{1/2} = [I_R{}^2 - I_O{}^2]^{1/2} \quad \text{A} \qquad (2.76)$$

The current ripple factor is then

$$K_i = \frac{I_{RI}}{I_O} \qquad (2.77)$$

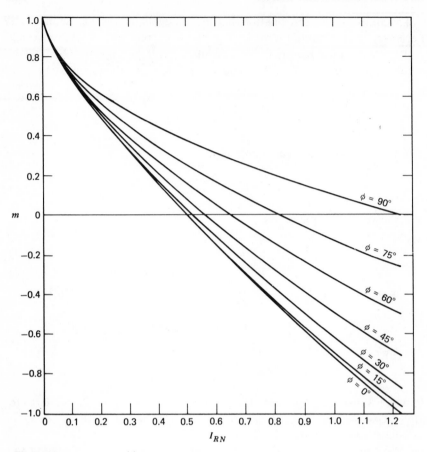

Fig. 2.23 m versus I_{RN} with ϕ as parameter.

The negative values of m occurring at $\gamma = 360°$ in Fig. 2.21 represent limiting values that may be employed in circuits with the corresponding values of ϕ. If a greater negative value of m were to be employed than is shown there, the current would build up in successive cycles until limited only by the circuit resistance. This may readily be seen if the simple case is considered where $m = -1$, and the circuit has some inductance, no matter how small. Under these conditions, v_L in the circuit of Fig. 2.20a would be positive throughout the cycle except at the instant when $v = -\sqrt{2}\ V$, when v_L would momentarily be zero. Thus di/dt in the circuit is positive throughout the cycle, except at this instant, and therefore the current increases continuously.

The two limiting cases mentioned earlier may now be analyzed.

Load-Circuit Inductance Negligible From equations 2.156, 2.157, and 2.160 with L set to zero

$$i = \frac{\sqrt{2}\,V}{R}\sin\omega t - \frac{V_C}{R} \quad \text{A:} \quad \alpha < \omega t < \alpha + \gamma \quad \text{rad} \qquad (2.178)$$

or

$$\frac{R}{\sqrt{2}\,V}i = \sin\omega t - m \qquad (2.179)$$

and this current is shown in Fig. 2.24, from which it may be seen that

$$\gamma = \pi - 2\alpha \quad \text{rad} \qquad (2.180)$$

In this case, I_O and V_O may most conveniently be obtained from equation 2.178. Thus

$$I_O = \frac{1}{2\pi}\int_{\alpha}^{\pi-\alpha} \frac{\sqrt{2}\,V}{R}(\sin\omega t - m)\,d(\omega t) \quad \text{A} \qquad (2.181)$$

from which by integration and substitution from equation 2.161, the normalized current is

$$I_N = \frac{R}{\sqrt{2}\,V}I_O = \sqrt{1-m^2} - m\cos^{-1}m \qquad (2.182)$$

Also

$$V_O = RI_O + V_C \quad \text{V} \qquad (2.183)$$

From Fig. 2.24 the output voltage is seen to be

$$v_O = V_C \quad \text{V:} \quad 0 < \omega t < \alpha \quad \text{rad}$$

$$v_O = \sqrt{2}\,V\sin\omega t \quad \text{V:} \quad \alpha < \omega t < \pi - \alpha \quad \text{rad} \qquad (2.184)$$

$$v_O = V_C \quad \text{V:} \quad \pi - \alpha < \omega t < 2\pi \quad \text{rad}$$

The rms value of the output voltage is therefore

$$V_R = \left\{ V_C^2 + \frac{1}{2\pi}\int_{\alpha}^{\pi-\alpha}\left[(\sqrt{2}\,V\sin\omega t)^2 - V_C^2\right]d(\omega t)\right\}^{1/2} \quad \text{V} \qquad (2.185)$$

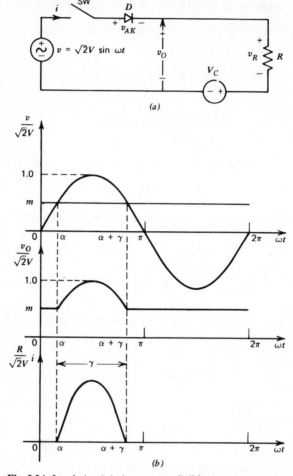

Fig. 2.24 Load-circuit inductance negligible.

The ripple voltage V_{RI} may now be determined from equation 2.71 and the voltage ripple factor from equation 2.72.

From equation 2.178, the rms value of the output current is

$$I_R = \left[\frac{1}{2\pi R^2} \int_{\alpha}^{\pi - \alpha} \left(\sqrt{2}\, V \sin \omega t - V_C \right)^2 d(\omega t) \right]^{1/2} \quad \text{A} \qquad (2.186)$$

The ripple current I_{RI} may now be determined from equation 2.76 and the current ripple factor from equation 2.77.

Example 2.5 In the circuit of Fig. 2.24a, the source voltage $v = 110\sqrt{2}\,\sin 120\pi t$ V, $R = 1\ \Omega$, and the load-circuit emf $V_C = 100$ V. If switch SW is closed during the negative half cycle of the source voltage, calculate;

  (a) The angle α at which diode D starts to conduct.
  (b) The conduction angle γ.
  (c) The average value of the current i.
  (d) The rms value of the current i.
  (e) The power delivered by the ac source.
  (f) The power factor at the ac source.

Solution

  (a) The angle α at which diode D starts to conduct is, from equation 2.161 or Fig. 2.24b

$$\alpha = \sin^{-1} m = \sin^{-1} \frac{V_C}{\sqrt{2}\,V}$$

$$= \sin^{-1} \frac{100}{110\sqrt{2}} = 0.697^c \text{ or } 40°$$

  (b) From equation 2.180 or Fig. 2.24b, the conduction angle is

$$\gamma = \pi - 2\alpha = \pi - 2 \times 0.697 = 1.75^c \text{ or } 100°$$

  (c) From equation 2.181, the average value of the rectified current is

$$I_O = \frac{1}{2\pi} \int_{0.697}^{\pi - 0.697} 110\sqrt{2} \left(\sin 120\pi t - \frac{100}{110\sqrt{2}} \right) d(\omega t)$$

$$= 10.2 \quad \text{A}$$

  (d) From equation 2.186, the rms value of the output current is

$$I_R = \left[\frac{1}{2\pi} \int_{0.697}^{\pi - 0.697} (110\sqrt{2})^2 \left(\sin 120\pi t - \frac{100}{110\sqrt{2}} \right)^2 d(\omega t) \right]^{1/2}$$

$$= 21.2 \quad \text{A}$$

(e)   The power delivered by the ac source is

$$P = RI_R^2 + V_C I_O = 1 \times 21.2^2 + 100 \times 10.2$$

$$= 1469 \quad W$$

(f)   The power factor is

$$\frac{\text{power delivered}}{VI_R} = \frac{1469}{110 \times 21.2} = 0.630$$

Load-Circuit Resistance Negligible   For this limiting case it is not possible to obtain an expression for i from equation 2.160, since the term V_C/R is unbounded. However, it is possible once again to consider this current as being made up of two components, one due to the ac source, the other, to the direct emf. For the first of these

$$\sqrt{2}\, V \sin \omega t = L \frac{di_S}{dt} = \omega L \frac{di_S}{d(\omega t)} \quad V \tag{2.187}$$

from which

$$i_S = \frac{\sqrt{2}\, V}{\omega L} \int_\alpha^{\omega t} \sin \omega t\, d(\omega t) = \frac{\sqrt{2}\, V}{\omega L}[\cos \alpha - \cos \omega t] \quad A \tag{2.188}$$

and this component of current is zero at $\omega t = \alpha$ and at $\omega t = 2\pi - \alpha$.
    For the component due to the direct emf

$$-V_C = L \frac{di_C}{dt} = \omega L \frac{di_C}{d(\omega t)} \quad V \tag{2.189}$$

from which

$$i_C = -\frac{V_C}{\omega L} \int_\alpha^{\omega t} d(\omega t) = -\frac{V_C}{\omega L}[\omega t - \alpha] \quad A \tag{2.190}$$

Thus

$$i = i_S + i_C = \frac{\sqrt{2}\, V}{\omega L}[\cos \alpha - \cos \omega t - m(\omega t - \alpha)] \quad A \tag{2.191}$$

At instant $\omega t = \alpha$, from equation 2.187

$$\frac{di_S}{dt} = \frac{\sqrt{2}\, V}{L} \sin \alpha = \frac{V_C}{L} \quad A/s \tag{2.192}$$

and from equation 2.189

$$\frac{di_C}{dt} = -\frac{V_C}{L} \quad \text{A/s} \qquad (2.193)$$

Thus at $\omega t = \alpha$

$$\frac{di_S}{dt} = -\frac{di_C}{dt} \quad \text{A/s} \qquad (2.194)$$

Equation 2.194 shows that the curve of $-i_C(t)$ is tangential to the curve of $i_S(t)$ at $\omega t = \alpha$, and this may be seen from the curves of i_S and $-i_C$ normalized with respect to $\sqrt{2}\ V/\omega L$ and shown in Fig. 2.25b. The vertical intercept between these two curves gives the normalized value of i, and the waveform of that current also is shown in Fig. 2.25; i has its maximum value when $v = V_C$ and $v_L = L\,di/dt = 0$.

The average value of the voltage across the inductance is zero, so that, as in the general case,

$$\int_0^{2\pi} v_L\, d(\omega t) = 0 \quad \text{Wb} \qquad (2.170)$$

This means that the shaded areas on the waveform of $v/\sqrt{2}\,V$ in Fig. 2.25b must be equal.

From Fig. 2.25b it is seen that the current falls to zero at instant $\omega t = \alpha + \gamma$, so that from equation 2.191,

$$\cos\alpha - \cos(\alpha + \gamma) - m\gamma = 0 \qquad (2.195)$$

and this transcendental equation may be solved numerically to determine γ. On the other hand, if γ is specified, then equation 2.195 may readily be solved to determine $\sin\alpha = m$, and this procedure may be employed to calculate the curve marked $\phi = 90°$ in Fig. 2.21.

For zero circuit resistance, $\phi = \pi/2$ and $\cos\phi = 0$, thus equation 2.173 cannot be employed to determine I_N. However, from equation 2.191

$$I_O = \frac{1}{2\pi}\int_\alpha^{\alpha+\gamma} \frac{\sqrt{2}\,V}{\omega L}\left[\cos\alpha - \cos\omega t - m(\omega t - \alpha)\right]d(\omega t) \quad \text{A} \qquad (2.196)$$

from which by integration and substitution from equation 2.161

$$I_N = \frac{\omega L}{\sqrt{2}\,V}I_O = \frac{1}{2\pi}\left[\sqrt{1-m^2}\ (\gamma - \sin\gamma) + m(1 - \cos\gamma) - \frac{m\gamma^2}{2}\right] \quad \text{A}$$

$$(2.197)$$

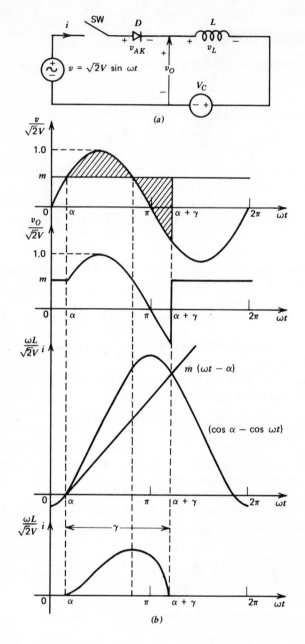

(a)

(b)

FIG. 2.25

From Fig. 2.25, the output voltage is seen to be

$$v_O = v_C \quad \text{V}: \qquad 0 < \omega t < \alpha \cdot \text{ rad}$$

$$v_O = \sqrt{2}\, V \sin \omega t \quad \text{V}: \qquad \alpha < \omega t < \alpha + \gamma \quad \text{rad} \qquad (2.198)$$

$$v_O = V_C \quad \text{V}: \qquad \alpha + \gamma < \omega t < 2\pi \quad \text{rad}$$

and since the average voltage across the inductance must be zero, it follows that

$$V_O = V_C \quad \text{V} \qquad (2.199)$$

The rms value of the output voltage is

$$V_R = \left\{ V_C^2 + \frac{1}{2\pi} \int_\alpha^{\alpha+\gamma} \left[(\sqrt{2}\, V \sin \omega t)^2 - V_C^2 \right] d(\omega t) \right\}^{1/2} \quad \text{V} \qquad (2.200)$$

The ripple voltage V_{RI} may now be determined from equation 2.71 and the voltage ripple factor from equation 2.72.

From equation 2.191 the rms value of the output current is

$$I_R = \frac{\sqrt{2}\, V}{\omega L} \left[\frac{1}{2\pi} \int_\alpha^{\alpha+\gamma} [\cos \alpha - \cos \omega t - m(\omega t - \alpha)]^2 d(\omega t) \right]^{1/2} \quad \text{A} \qquad (2.201)$$

The ripple current I_{RI} may now be determined from equation 2.76 and the current ripple factor from equation 2.77.

Example 2.6 Five 12-V batteries are connected in series and are to be charged from a 110-V single phase 60-Hz ac voltage supply. Using the circuit in Fig. 2.25 with $L = 30$ mH, calculate the following:

(a) The average current and the power delivered to the batteries if each battery emf is 6 V.

(b) The average current and the power delivered to the batteries if each battery emf is 13 V.

(c) The average current and the power delivered to the batteries if each battery emf is 0 V.

Solution

(a) Angle α at which diode D starts to conduct is given by equation 2.161 or Fig. 2.25

$$\alpha = \sin^{-1} m, \qquad m = \frac{V_C}{\sqrt{2}\, V}$$

with

$$V_C = 30 \text{ V}, \qquad V = 110 \text{ V}, \qquad m = \frac{30}{\sqrt{2} \times 110} = 0.192$$

The conduction interval γ of the diode D is calculated from the solution of the transcendental equation 2.195. Knowing γ, the average current I_O is calculated from equation 2.196. Such results are shown graphically in Figs. 2.21 and 2.22. For $\phi = 90°$, $m = 0.192$, $\gamma = 264°$, $I_N = 0.57$

$$\text{average current} = \frac{\sqrt{2} \, V}{\omega L} I_N = \frac{\sqrt{2} \times 110}{0.377 \times 30} \times 0.57 = 7.8 \quad \text{A}$$

$$\text{power delivered to batteries} = 30 \times 7.8 = 234 \text{ W}$$

(b) $\quad V_C = 65 \text{ V}, \qquad V = 110 \text{ V}, \qquad m = \frac{V_C}{\sqrt{2} \, V} = 0.416$

From Figs. 2.21 and 2.22, $\gamma = 212°$, $I_N = 0.26$

$$\text{average current} = \frac{\sqrt{2} \, V}{\omega L} \times 0.26 = 3.56 \quad \text{A}$$

$$\text{power delivered to batteries} = 65 \times 3.56 = 231 \text{ W}$$

(c) $\quad V_C = 0, \qquad m = 0, \qquad \gamma = 360°, \qquad I_N = 1$

$$\text{average current} = \frac{\sqrt{2} \, V}{\omega L} = 13.7 \quad \text{A}$$

PROBLEMS

2.1   In the circuit of Fig. P2.1, switch SW is closed at $t = 0$ and opened again at $t = 100$ μs. At $t = 0$, $v_C = 0$. Sketch approximately to scale the time variations of i and v_C and determine expressions describing these variations.

2.2   In the circuit of Fig. P2.2, switch SW is closed at $t = 0$ and opened again at $t = 10$ ms. At $t = 0$, $i_L = 0$. Sketch approximately to scale the time variations of i, i_L, i_D, and v_O and determine expressions describing these variations.

2.3   In the circuits of Fig. P2.3a to g, determine the value of di/dt at the instant $t = 0$ when switch SW is closed and the initial conditions are:

(a) $i = 0$
(b) $i = 0$, $v_C = 0$
(c) $i = 0$, $v_C = -75$ V
(d) $i = 0$, $i_L = 0$, $v_C = 0$
(e) $i = 0$, $v_C = 50$ V
(f) $i = 0$
(g) $i = 0$, $v_C = -50$ V

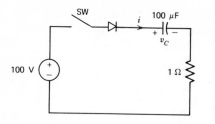

Fig. P2.1

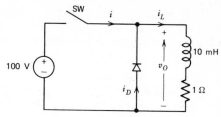

Fig. P2.2

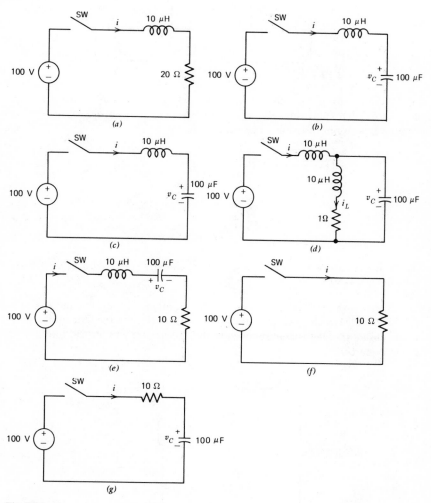

Fig. P2.3

2.4 In the circuit of Fig. P2.4, $N_1 = 100$, $N_2 = 300$. Switch SW is opened at $t = 0$ after being closed for a long time. Sketch approximately to scale the time variations of i_1, i_2, v_1, and v_2 and determine expressions describing these variations.

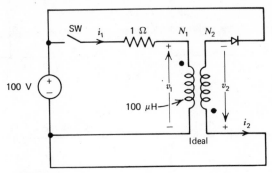

Fig. P2.4

2.5 In the circuit of Fig. P2.5, switch SW is closed at $t = 0$. Also at $t = 0$, $i = 0$, and $v_C = 0$. Sketch approximately to scale the time variations of i, v_L, and v_C and determine expressions describing these variations.

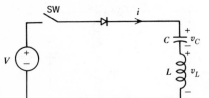

Fig. P2.5

2.6 In the circuit of Fig. P2.6, switch SW is closed at $t = 0$. Also at $t = 0$, $i = 0$, $i_D = 0$ and $v_C = -50$ V.

(a) Sketch approximately to scale the time variations of i, v_L, v_C, and i_D.

(b) Show the circuit modification which, without affecting the time variation of i, would ensure that the energy supplied to the 75-V source was instead returned to the 100-V source.

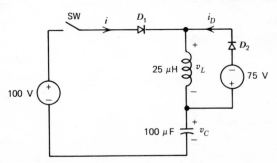

Fig. P2.6

2.7 In the circuit of Fig. P2.7, $\omega = 377$ rad/s. Switch SW is closed during the negative half-cycle of the source voltage.

(a) Sketch approximately to scale the time variations of v, i, v_O, and v_{AK}.
(b) Calculate the average and rms values of i and v_O.

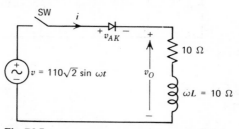

Fig. P2.7

2.8 The Fourier series describing waveform a in Fig. P2.8 is

$$v(t) = \sqrt{2}\, V\left[\frac{1}{\pi} + \frac{1}{2}\sin\omega t - \frac{2}{3\pi}\cos 2\omega t - \frac{2}{15\pi}\cos 4\omega t \cdots\right]\ \ V$$

Determine the Fourier series describing waveform b, in which the cusps have the same shape as those of waveform a.

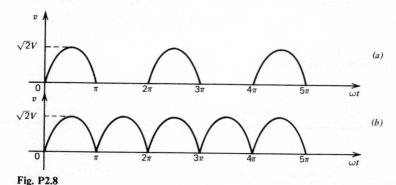

Fig. P2.8

2.9 The circuit of Fig. P2.9 is operating under steady-state conditions.

(a) Calculate the average and rms values of i_O and v_O (consider the dc component and the lowest harmonic only).
(b) Sketch approximately to scale the time variations of v, v_D, i_D, v_O, and i_O.
(c) Calculate the average and rms values of i_D and i_1.
(d) Calculate the power factor at the source.

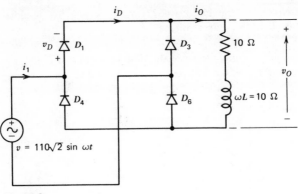

Fig. P2.9

2.10   In the circuit of Fig. P2.10, $V = 67$ V, and

$$v_{an} = \sqrt{2}\, V \sin \omega t \quad \text{V}$$

$$v_{bn} = \sqrt{2}\, V \sin (\omega t - 120°) \quad \text{V}$$

$$v_{cn} = \sqrt{2}\, V \sin (\omega t + 120°) \quad \text{V}$$

(a)   Sketch approximately to scale the time variations of v_{an}, v_{bn}, v_{cn}, v_O, and i_1, if i_O is a constant current of 100 A.

(b)   If the time variation of v_O may be described by the Fourier series

$$v_O = 1.17V(1 - 0.25 \cos 3\omega t) \quad \text{V}$$

calculate the average and rms values of i_1 and v_O.

(c)   Calculate the power factor at the source.

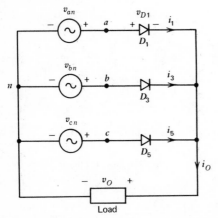

Fig. P2.10

2.11 The Fourier-series representation of the output voltage for the half-wave rectifier in Fig. 2.13a is given in problem 2.8. Using the first three terms of this series, verify that the rms voltage obtained approaches the value $V/\sqrt{2}$ given in equation 2.81.

2.12 For the circuits shown in Fig. P2.11, in which $v_s = 110\sqrt{2} \sin 120\pi t$ V, sketch approximately to scale the time variations of v, i, i_O, i_D, v_O and v_{AK}, and calculate the average value of i_O.

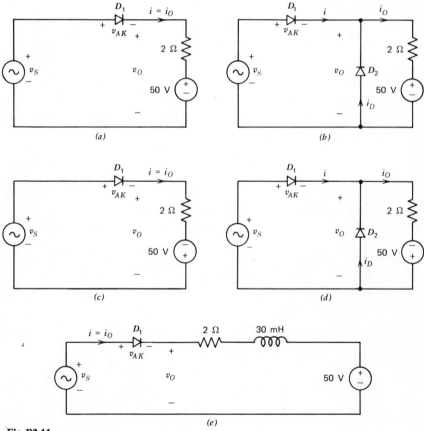

Fig. P2.11

2.13 In the circuit of Fig. P2.12, in which $v_s = 100\sqrt{2} \sin 120\pi t$ V, L is so large that the output current i_O may be considered to have a constant value I_O. Sketch

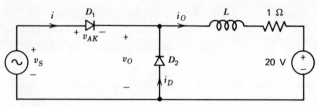

Fig. P2.12

approximately to scale the time variations of v_O, i, i_D, and v_{AK}, and calculate the average values of v_O and i_O.

THREE

POWER SEMICONDUCTOR SWITCHES

The preceding chapter concerned switches and diodes and the effect of their introduction into linear circuits. This chapter discusses the effect of introducing into such circuits devices that combine the properties of a switch and a diode; these are the power transistor and the thyristor or silicon controlled rectifier (SCR). While it is assumed that the reader understands the operation of a transistor, that of the more complicated thyristor is briefly described, so that the reasons for certain restrictions on the manner in which it may be operated are understood.

Once the need for operating restrictions has been appreciated and methods of applying them have been discussed, it is possible to employ manufacturers' published data to select devices for particular applications. The use of a typical manufacturer's data sheet is therefore explained. The basic methods of commutation, that is, the techniques by which the anode current of a thyristor may be reduced to zero to permit the "switch" to be opened again, are then described.

3.1 IDEAL MODELS OF POWER SWITCHES

Diodes, power transistors, and thyristors each possess internal resistances and capacitances, but these are in general so small compared with the parameters of the circuits in which the devices operate that they may be neglected for many purposes and the device considered to be ideal. The effect of some of them must be taken into account, however, when determining the rating and operating limitations of the devices, and they are therefore occasionally mentioned in later sections.

The ideal model of the power semiconductor diode has already been discussed in Section 2.1 and is illustrated in Fig. 2.1.

The circuit symbol for an n-p-n power transistor is shown marked T in Fig. 3.1a. An ideal power transistor has zero resistance to positive collector current i_C when the base-to-emitter voltage v_{BE} is positive. It has infinite resistance to positive collector current i_C if v_{BE} is zero. It also has infinite resistance to current in the reverse direction regardless of the magnitude of v_{BE}. If the switching time of the transistor is neglected, so that it has no active region, in the circuit of Fig. 3.1a there are two possible operating states, and these are illustrated in the diagram of i_C versus v_{CE} in Fig. 3.1b. The ideal transistor turns off the collector current i_C whenever v_{BE} is reduced to zero. In general power transistors may be employed only in relatively low-power circuits, since available voltage and current ratings are low. In addition, they possess no surge current capacity and are capable of withstanding only low rates of change of current, both increasing and decreasing. For relatively high-power circuits, they would therefore be replaced by thyristors in the same power-circuit configuration. The control circuitry would be different in the two cases, however.

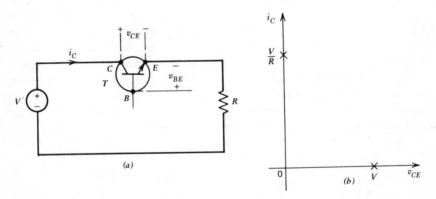

Fig. 3.1 Ideal transistor.

The circuit symbol for a thyristor is shown marked Q in Fig. 3.2a. An ideal thyristor has infinite resistance to positive anode current i_A unless the cathode-to-gate current i_G is momentarily given a positive value, upon which the anode-to-cathode resistance falls to zero and remains at that value until anode current ceases to flow. If i_G is then zero, after a brief interval the thyristor again assumes infinite anode-to-cathode resistance. The flow of anode current does not cease when i_G is reduced to zero, and this is the feature that distinguishes the thyristor from the transistor as a circuit element.

The ideal thyristor has infinite resistance to current from cathode to anode. If it is remembered that anode current i_A continues to flow after gate current i_G has ceased, in which state v_{AK} is also zero, then the diagram of Fig. 3.2b shows the range of the operating point of the ideal thyristor for the circuit of Fig. 3.2a.

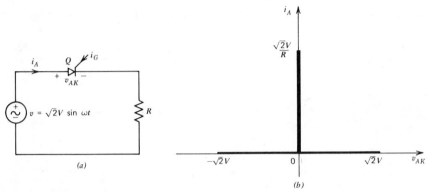

Fig. 3.2 Ideal thyristor.

At this point, mention should be made of composite devices that, in effect, enclose two or more devices in one case and thus reduce the number of terminals to be connected. Prominent among these is the bilateral triode switch (Triac), in which in effect two thyristors in inverse-parallel connection are incorporated and have a single gate terminal. Thus a single composite three-terminal device replaces two separate three-terminal devices. The main disadvantage of this and similar arrangements is that the internal losses of two devices must be dissipated as heat from a single case, and this tends to restrict the use of composite devices to relatively low power levels. However, composite devices, which may be considered as high-power integrated circuits, are being developed rapidly, and the design engineer would be well advised to keep in touch with manufacturers' literature describing their capabilities. The circuits embodying composite devices may be analyzed exactly as if the separate devices were employed.

3.2 OPERATION OF THE THYRISTOR

While it is frequently satisfactory to analyze a power circuit including thyristors on the assumption that they are ideal, to understand their operating limitations it is necessary to have some conception of their operating mechanism. A somewhat simplified description of that mechanism is therefore given in this section.

The thyristor is a four layer *p-n-p-n* structure which therefore possesses three junctions. It has external anode, cathode, and gate terminals connected to appropriate areas of this structure and is connected in series with an external voltage source and load as indicated in Fig. 3.3*a*. The auxiliary voltage source V_G and switch SW may be employed to turn on the thyristor when the voltage v_{AK} applied to its power terminals is positive.

A simple analogy that may be employed to explain many of the operating characteristics of a thyristor may be obtained by considering it to be made up of two interconnected transistors as shown in Fig. 3.3*b*. The circuit corresponding to this arrangement is shown in Fig. 3.3*c*. The *n*1-*p*2

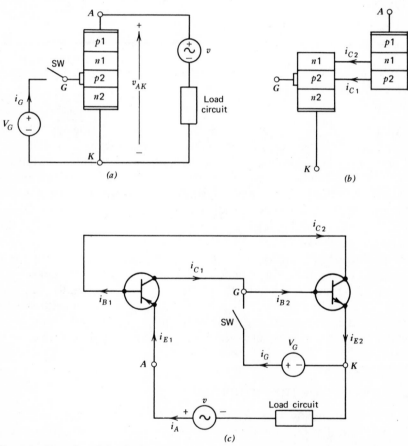

Fig. 3.3 Operation of the thyristor.

junction of the thyristor is common to the two transistors. The ratio of collector current i_C to emitter current i_E for each of the two transistors may be defined as

$$\alpha_1 = \frac{i_{C1}}{i_{E1}} : \qquad \alpha_2 = \frac{i_{C2}}{i_{E2}} \tag{3.1}$$

The two collector currents may be considered to pass from the *p-n-p* to the *n-p-n* transistor as indicated in Fig. 3.3*b* and *c*. In addition to these, there will be a leakage current I_{CO} crossing the common junction *n*1-*p*2. The entire current passing through the thyristor is thus

$$i_A = i_{C1} + i_{C2} + I_{CO} \quad \text{A} \tag{3.2}$$

$$= \alpha_1 i_{E1} + \alpha_2 i_{E2} + I_{CO} \quad \text{A} \tag{3.3}$$

But, as may be seen from Fig. 3.3*c*,

$$i_{E1} = i_{E2} = i_A \quad \text{A} \tag{3.4}$$

thus substitution in equation 3.3 yields

$$i_A = (\alpha_1 + \alpha_2) i_A + I_{CO} \quad \text{A} \tag{3.5}$$

or

$$i_A = \frac{I_{CO}}{1 - (\alpha_1 + \alpha_2)} \quad \text{A} \tag{3.6}$$

If the current-transfer ratios for the common-emitter connection of the two transistors are

$$\beta_1 = \frac{i_{C1}}{i_{B1}} : \qquad \beta_2 = \frac{i_{C2}}{i_{B2}} \tag{3.7}$$

where i_{B1} and i_{B2} are the base currents, then since

$$i_{E1} = i_{B1} + i_{C1} \quad \text{A:} \qquad i_{E2} = i_{B2} + i_{C2} \quad \text{A} \tag{3.8}$$

it can readily be shown by substitution for i_{E1} and i_{E2} in equations 3.1 that

$$\alpha_1 = \frac{\beta_1}{1 + \beta_1} : \qquad \alpha_2 = \frac{\beta_2}{1 + \beta_2} \tag{3.9}$$

and substitution from equations 3.9 in equation 3.6 then yields

$$i_A = \frac{(1+\beta_1)(1+\beta_2)I_{CO}}{1-\beta_1\beta_2} \quad \text{A} \tag{3.10}$$

If $\beta_1\beta_2 \ll 1$, then i_A will be small because I_{CO} is small, and this corresponds to the "turned-off" or forward blocking state of the thyristor. However, if $\beta_1\beta_2 \cong 1$, then i_A may be very large and will be limited only by the impedance of the load circuit. There are four chief factors which may cause $\beta_1\beta_2$ to approach unity. These are:

Voltage   As v_{AK} is increased, the collector-to-emitter voltages of the two transistors are increased. This increases the energy of the minority carriers in the reverse-biased $n1$-$p2$ junction, and a sufficient increase of their energy enables them to dislodge more carriers by collision. These dislodged carriers in their turn acquire high energy, and the result is an avalanche breakdown at the junction. The resultant large increase in the collector current of the two transistors causes $\beta_1\beta_2$ to approach unity.

Rate of Change of Voltage   The depletion area of the reverse-biased $n1$-$p2$ junction has the characteristics of a capacitor due to the space-charge field existing there. In any capacitor

$$i = C\frac{dv}{dt} \quad \text{A} \tag{3.11}$$

however, in a junction, the capacitance is a function of the junction potential difference. If, as an approximation, it is assumed that the entire forward voltage v_{AK} appears across the $n1$-$p2$ junction, then the charging current which will flow when v_{AK} is varied is given by

$$i = \frac{d(C_j v_{AK})}{dt} = C_j\frac{dv_{AK}}{dt} + v_{AK}\frac{dC_j}{dt} \quad \text{A} \tag{3.12}$$

where C_j is the junction capacitance. For increasing v_{AK}, the second term on the right-hand side of equation 3.12 is negative; however, if dv_{AK}/dt is great enough, a large charging current flows across the junction and again $\beta_1\beta_2$ approaches unity. This is commonly called the "dv/dt effect."

Temperature   At high temperatures I_{CO} is high. The resultant increase in collector currents and hence in the magnitudes of β_1 and β_2 may again be sufficient to cause $\beta_1\beta_2$ to approach unity.

Injection of Base Current   This is the normal method of turning on a thyristor. If current i_{B2} is increased by momentarily closing switch SW in

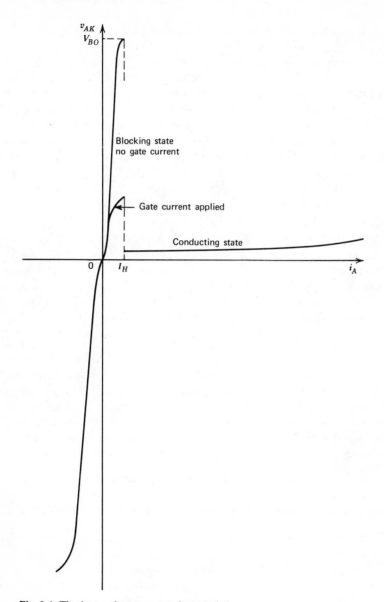

Fig. 3.4 Thyristor voltage-current characteristic.

Fig. 3.3c, then i_{C2} and β_2 increase. But since $i_{C2} = i_{B1}$, then i_{C1} and β_1 also increase. Thus $\beta_1 \beta_2$ approaches unity, and a large anode current i_A restricted only by the load-circuit impedance is able to flow.

If v_{AK} is negative, then the thyristor may be turned on by one of the first three factors and permit reverse current to flow. However in this case there are two reverse-biased n-p junctions to be brought to conduction, and this requires more extreme conditions than those resulting in forward current.

Figure 3.4 indicates the form of the steady-state voltage-current characteristic of a thyristor at normal operating temperature. However, this diagram is not drawn to scale, since differences in the magnitudes of voltages and currents in the different states are very great. Voltage V_{BO} is the breakdown voltage at which the thyristor will begin to conduct and the anode-to-cathode resistance fall to a very low value without the application of a gate current. The voltage required to start conduction when a gate current is applied is very much exaggerated in the diagram; I_H is the minimum anode current (holding current) which must be maintained if the thyristor is not to revert to the nonconducting state.

3.3 CONTROLLED HALF-WAVE RECTIFIER

As in the case of the diode half-wave rectifier discussed in Section 2.6, it must be assumed that the controlled half-wave rectifier, in which the diode is replaced by a thyristor, is supplied from an ideal alternating-voltage source. The discussion of Fourier analysis in Section 2.6.1 is applicable here also, but in the case of the thyristor the angle α of equations 2.65 to 2.68 is no longer zero, since the instant at which the thyristor commences to conduct is that in the positive half cycle of the source voltage v at which gate current i_G is given a positive value.

3.3.1 RL Load Circuit After each pulse of gate current i_G, while thyristor Q in Fig. 3.5a is conducting

$$v_L + v_R = v_O = v \quad \text{V} \tag{3.13}$$

or

$$L\frac{di}{dt} + Ri = \sqrt{2}\, V \sin \omega t \quad \text{V} \tag{3.14}$$

The solution of equation 3.14 is of the form

$$i = \frac{\sqrt{2}\,V}{Z} \sin(\omega t - \phi) + A\epsilon^{-(R/L)t} \quad \text{A} \tag{3.15}$$

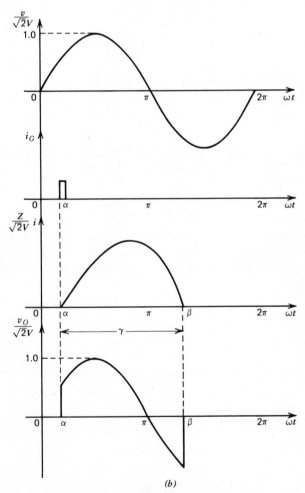

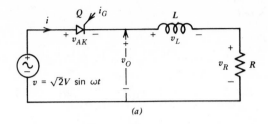

Fig. 3.5 Controlled half-wave rectifier with RL load circuit.

where, as before,

$$Z = \left[R^2 + (\omega L)^2 \right]^{1/2} \quad \Omega \tag{3.16}$$

$$\phi = \tan^{-1} \frac{\omega L}{R} \quad \text{rad} \tag{3.17}$$

Substitution in equation 3.15 of the initial conditions $i = 0$ at $\omega t = \alpha$ yields

$$i = \frac{\sqrt{2}\, V}{Z} \left[\sin(\omega t - \phi) - \sin(\alpha - \phi)\epsilon^{(R/L)(\alpha/\omega - t)} \right] \quad \text{A} \tag{3.18}$$

At $\omega t = \beta$, i is again zero, so that from equation 3.18

$$\sin(\beta - \phi) = \sin(\alpha - \phi)\epsilon^{(R/L)[(\alpha - \beta)/\omega]} \tag{3.19}$$

and β may be determined by the solution of this transcendental equation.

The time variations of the circuit variables are shown in Fig. 3.5b, in which the conduction angle is

$$\gamma = \beta - \alpha \quad \text{rad} \tag{3.20}$$

A family of curves of γ versus α for various values of ϕ have been obtained from equations 3.19 and 3.20 and are shown in Fig. 3.6.

From Equation 3.18 the normalized value of the average rectified current is given by

$$I_N = \frac{1}{2\pi} \int_\alpha^{\alpha + \gamma} \left[\sin(\omega t - \phi) - \sin(\alpha - \phi)\epsilon^{(R/L)(\alpha/\omega - t)} \right] d(\omega t) \tag{3.21}$$

and a family of curves of I_N versus α for various values of ϕ have been calculated from equation 3.21 and are shown in Fig. 3.7. The limiting curves for $\phi = 0$ and $\phi = 90°$, however, have been obtained from expressions derived in Sections 3.3.2 and 3.3.3.

Also from equation 3.18 the normalized value of the rms output current is given by

$$I_{RN} = \left[\frac{1}{2\pi} \int_\alpha^{\alpha + \gamma} \left[\sin(\omega t - \phi) - \sin(\alpha - \phi)\epsilon^{(R/L)(\alpha/\omega - t)} \right]^2 d(\omega t) \right]^{1/2} \tag{3.22}$$

and a corresponding family of curves of I_{RN} versus α are shown in Fig. 3.8. Here also the limiting curves have been obtained from expressions derived later.

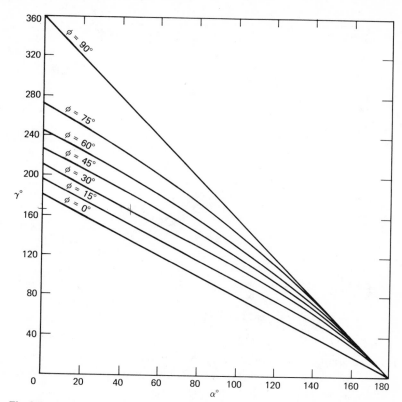

Fig. 3.6 $\gamma°$ versus $\alpha°$ for the circuit of Fig. 3.5a.

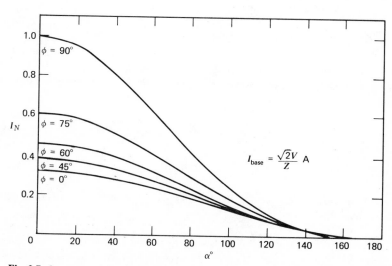

Fig. 3.7 I_N versus $\alpha°$ for the circuit of Fig. 3.5a.

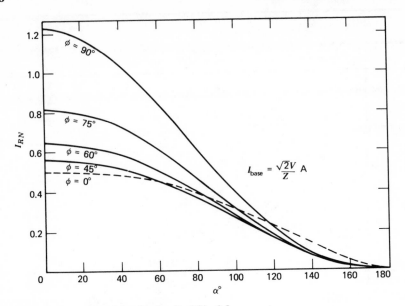

Fig. 3.8 I_{RN} versus $\alpha°$ for the circuit of Fig. 3.5a.

The two limiting conditions which are of particular interest and assistance in visualizing the operation of this circuit are (a) a purely resistive load circuit, and (b) a purely inductive load circuit.

3.3.2 Resistive Load Circuit For a resistive load circuit, $Z = R$, and $\phi = 0$, so that from equation 3.15

$$i = \frac{\sqrt{2}\,V}{R}\sin \omega t \quad \text{A:} \quad \alpha < \omega t < \pi \quad \text{rad} \tag{3.23}$$

and the time variations of the circuit variables are shown in Fig. 3.9.

From equation 3.23 the normalized value of the average rectified current is given by

$$I_N = \frac{1}{2\pi}\int_\alpha^\pi \sin \omega t\, d(\omega t) = \frac{1 + \cos \alpha}{2\pi} \tag{3.24}$$

and the corresponding curve of I_N versus α is shown marked $\phi = 0$ in Fig. 3.7.

Also from equation 3.23 the normalized value of the rms output current is given by

$$I_{RN} = \left[\frac{1}{2\pi}\int_\alpha^\pi \sin^2 \omega t\, d(\omega t)\right]^{1/2} = \left[\frac{1}{4\pi}\left(\pi - \alpha + \frac{\sin 2\alpha}{2}\right)\right]^{1/2} \tag{3.25}$$

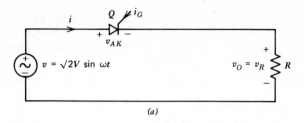

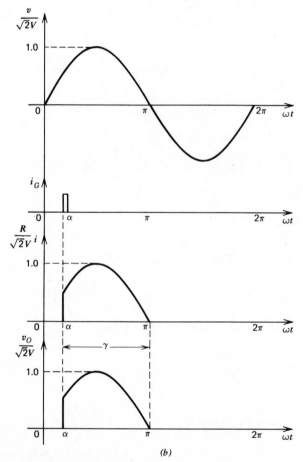

Fig. 3.9 Controlled half-wave rectifier with resistive load circuit.

and the corresponding curve of I_{RN} versus α is shown in Fig. 3.8.

Since this is a purely resistive circuit, and the voltage waveform shown in Fig. 3.9 is identical with that of the current, it follows that the curves marked $\phi = 0$ in Figs. 3.7 and 3.8 may also be employed as curves of normalized average rectified voltage and normalized rms output voltage.

3.3.3 Inductive Load Circuit

For an inductive load circuit, $Z = \omega L$, and $\phi = \pi/2$, so that from equation 3.18

$$i = \frac{\sqrt{2}\,V}{\omega L}(\cos\alpha - \cos\omega t) \quad \text{A} \tag{3.26}$$

and the time variations of the circuit variables are shown in Fig. 3.10, where it is seen that

$$\beta = 2\pi - \alpha \quad \text{rad} \tag{3.27}$$

and this may be confirmed by substitution in equation 3.19. The limiting case for $\alpha = 0$ has already been illustrated in Fig. 2.16.

From equation 3.26 the normalized value of the average rectified current is given by

$$I_N = \frac{1}{2\pi}\int_{\alpha}^{2\pi-\alpha}(\cos\alpha - \cos\omega t)\,d(\omega t)$$

$$= \frac{1}{\pi}[(\pi-\alpha)\cos\alpha + \sin\alpha] \tag{3.28}$$

and the corresponding curve of I_N versus α is shown marked $\phi = 90°$ in Fig. 3.7.

Also from equation 3.26 the normalized value of the rms output current is given by

$$I_{RN} = \left[\frac{1}{2\pi}\int_{\alpha}^{2\pi-\alpha}(\cos\alpha - \cos\omega t)^2\,d(\omega t)\right]^{1/2} \tag{3.29}$$

and the corresponding curve of I_{RN} versus α is shown in Fig. 3.8.

Since for this circuit $v_O = v_L$, and the average value of the voltage across an inductance carrying a cyclical current is zero, then the two shaded areas under the curve of v_O versus ωt in Fig. 3.10 must be equal. This also means that the average rectified output voltage is always zero. The normalized rms output voltage is

$$V_{RN} = \left[\frac{1}{2\pi}\int_{\alpha}^{2\pi-\alpha}\sin^2\omega t\,d(\omega t)\right]^{1/2} = \left[\frac{1}{2} - \frac{\alpha}{2\pi} + \frac{1}{4\pi}\sin 2\alpha\right]^{1/2} \tag{3.30}$$

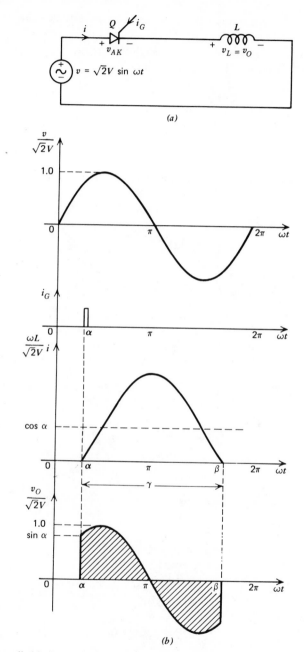

(a)

(b)

Fig. 3.10 Controlled half-wave rectifier with inductive load circuit.

Example 3.1 The rms value of the source voltage in the circuit of Fig. 3.5 is 100 V. Calculate the values of the average rectified current I_O and the rms output current I_R, and draw the waveforms of the thyristor voltage v_{AK} for $\alpha = 45°$ and $\alpha = 135°$ if

(a) $R = 10\,\Omega$, $L = 0$
(b) $R = 10\,\Omega$, $\omega L = 10 \quad \Omega$

Solution

(a)   For $\alpha = 45°$, $\phi = 0$, from Fig. 3.7, $I_N = 0.27$, and from Fig. 3.8, $I_{RN} = 0.48$.
    Base current is

$$\frac{\sqrt{2}\,V}{R} = \frac{100\sqrt{2}}{10} = 14.14 \quad A$$

$$\therefore \quad I_O = 14.14 \times 0.27 = 3.82 \quad A$$

$$I_R = 14.14 \times 0.48 = 6.80 \quad A$$

For $\alpha = 135°$, $\phi = 0$, from Fig. 3.7, $I_N = 0.05$, and from Fig. 3.8, $I_{RN} = 0.10$.

$$\therefore \quad I_O = 14.14 \times 0.05 = 0.71 \quad A$$

$$I_R = 14.14 \times 0.10 = 1.41 \quad A$$

The waveforms of v_{AK} for $\phi = 0$ are shown in Fig. E3.1a.
(b)   For $\alpha = 45°$, $\phi = 45°$, from Fig. 3.6, $\gamma = 180°$; from Fig. 3.7, $I_N = 0.32$; and from Fig. 3.8, $I_{RN} = 0.50$.
    Base current is

$$\frac{\sqrt{2}\,V}{Z} = \frac{100\sqrt{2}}{10\sqrt{2}} = 10 \quad A$$

$$\therefore \quad I_O = 10 \times 0.32 = 3.2 \quad A$$

$$I_R = 10 \times 0.50 = 5.0 \quad A$$

For $\alpha = 135°$, $\phi = 45°$, from Fig. 3.6, $\gamma = 74°$; from Fig. 3.7, $I_N = 0.05$;

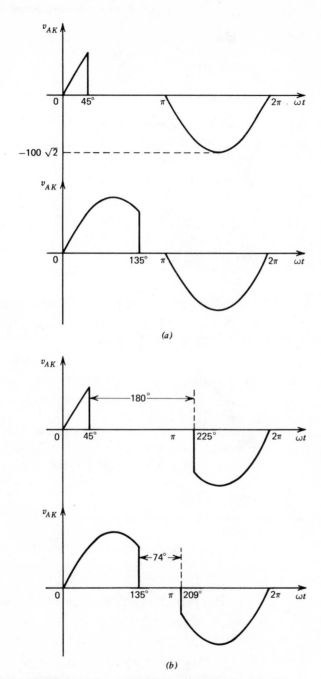

Fig. E3.1 (*a*) Waveforms of V_{AK}; $\phi = 0$. (*b*) Waveforms of V_{AK}; $\phi = 45°$.

and from Fig. 3.8, $I_{RN} = 0.10$.

$$\therefore \quad I_O = 10 \times 0.05 = 0.5 \quad A$$

$$I_R = 10 \times 0.10 = 1.0 \quad A$$

The waveforms of v_{AK} for $\phi = 45°$ are shown in Fig. E3.1b.

3.3.4 RC Load Circuit After the first pulse of gate current i_G has been applied, and thyristor Q in Fig. 3.11a is conducting, then

$$v_C + v_R = v_O = v \quad V \tag{3.31}$$

and from equation 2.92 the current is

$$i = \frac{\sqrt{2} \, V}{Z} \sin(\omega t + \phi) + A\epsilon^{-t/RC} \quad A \tag{3.32}$$

If the capacitor has no initial charge, then at $\omega t = \alpha$, the beginning of the first conducting period, $v_C = 0$, and $i = (\sqrt{2} \, V \sin \alpha)/R$ so that substitution of these initial conditions in equation (3.32) yields

$$\frac{Z}{\sqrt{2} \, V} i = \sin(\omega t + \phi) + \left[\frac{Z}{R} \sin \alpha - \sin(\alpha + \phi) \right] \epsilon^{(1/RC)(\alpha/\omega - t)} \quad V \tag{3.33}$$

As in the case of the uncontrolled rectifier, v_C has a positive value at the end of the conduction period, and this value is increased with each successive pulse of rectified current until conduction ceases when

$$v_C = \sqrt{2} \, V \quad V: \qquad \alpha < \frac{\pi}{2} \quad \text{rad}$$

$$\tag{3.34}$$

$$v_C = \sqrt{2} \, V \sin \alpha \quad V: \qquad \alpha > \frac{\pi}{2} \quad \text{rad}$$

If R is zero, the condition described in equation 3.34 is reached at the end of the first current pulse. However, if $\alpha \neq 0$, then an infinite pulse of current will flow at the instant $\omega t = \alpha$ to make $v_C = \sqrt{2} \, V \sin \alpha$ and such a pulse would destroy a thyristor.

3.3.5 Load Circuit with Electromotive Force In analyzing this circuit, shown in Fig. 3.12, use may be made of some of the conclusions reached in analyzing the uncontrolled half-wave rectifier in Section 2.6.6. In partic-

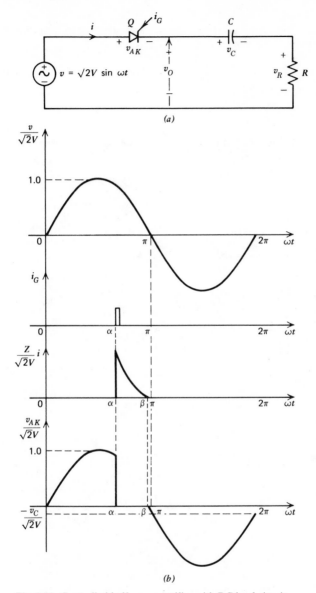

$v = \sqrt{2}V \sin \omega t$

(a)

(b)

Fig. 3.11 Controlled half-wave rectifier with RC load circuit.

ular, the earliest point in the cycle of the ac source voltage at which conduction can begin is given from equation 2.161 by the angle

$$\eta = \sin^{-1} \frac{V_C}{\sqrt{2}\, V} = \sin^{-1} m \quad \text{rad} \tag{3.35}$$

If a positive pulse of i_G is applied earlier in the cycle than $\omega t = \eta$, conduction will not commence, and if the pulse is not repeated, no conduction will take place during the entire cycle. If the pulse is applied at $\omega t = \alpha$, where $\alpha \geqslant \eta$, then conduction will begin, and the resulting current

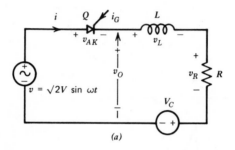

(a)

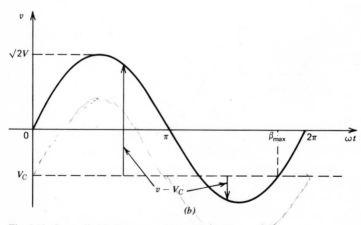

(b)

Fig. 3.12 Controlled half-wave rectifier with EMF.

will be that described by equations 2.164 and 2.165, so that

$$\frac{Z}{\sqrt{2}\,V}\,i = \sin(\omega t - \phi) - \left[\frac{m}{\cos\phi} - B\epsilon^{(\alpha-\omega t)/\tan\phi}\right]: \qquad \alpha < \omega t < \alpha + \gamma \quad \text{rad}$$

(3.36)

where

$$B = \frac{m}{\cos\phi} - \sin(\alpha - \phi) \tag{3.37}$$

Similarly, from equation 2.167,

$$\frac{(m/\cos\phi) - \sin(\alpha + \gamma - \phi)}{(m/\cos\phi) - \sin(\alpha - \phi)} = \epsilon^{-\gamma/\tan\phi} \tag{3.38}$$

When any particular circuit is to be analyzed, ϕ will have a definite numerical value. Equation 3.38 may then be employed to give a family of curves showing the relationship between m and γ for various values of α. Such a family of curves for $\phi = \pi/6$ is shown in Fig. 3.13.

The boundary shown in broken line in the first quadrant of Fig. 3.13 is identical with the curve marked $\phi = 30°$ in Fig. 2.21. To the right of this boundary, $\eta > \alpha$, and the thyristor does not turn on, since the gating signal i_G has been removed before $v > V_C$.

The boundary shown in broken line in the fourth quadrant of Fig. 3.13 gives the theoretical operating limit of the system below which control would be lost, and the current would build up to a very large value. The physical interpretation of this boundary may be seen from Fig. 3.12b, where a negative value of V_C is shown. If the inductance of the circuit is such that the current does not fall to zero before $\omega t = \beta_{\max}$, where

$$\beta_{\max} = 2\pi - \sin^{-1}\frac{|V_C|}{\sqrt{2}\,V} \quad \text{rad} \tag{3.39}$$

then the thyristor does not turn off, because the net source emf acting in a positive direction in the circuit is $v - V_C$, and this again becomes positive for $\omega t > \beta_{\max}$. In practice, the boundary should be somewhat higher, since the thyristor requires a finite time to turn off, and therefore conduction must cease at some angle $\omega t < \beta_{\max}$.

By interpolation within the boundaries of Fig. 3.13, γ may be determined for any given values of m and α.

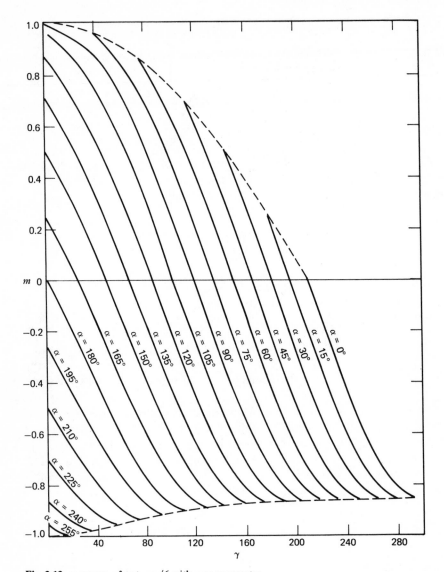

Fig. 3.13 m versus γ for $\phi = \pi/6$ with α as parameter.

The normalized average value of the rectified current is given by

$$I_N = \frac{1}{2\pi} \int_\alpha^{\alpha+\gamma} \frac{Z}{\sqrt{2}\,V} i\, d(\omega t) \qquad (3.40)$$

For given values of m, ϕ, and α, the value of γ may be determined from equation 3.38 and these four values may be substituted in equations 3.36, 3.37, and 3.40 to give I_N. By this procedure, a family of curves of m versus I_N for a given value of ϕ and a series of values of α may be produced. Such a family of curves for $\phi = \pi/6$ is shown in Fig. 3.14. The boundary shown in broken line in the first quadrant is identical with that marked $\phi = 30°$ in

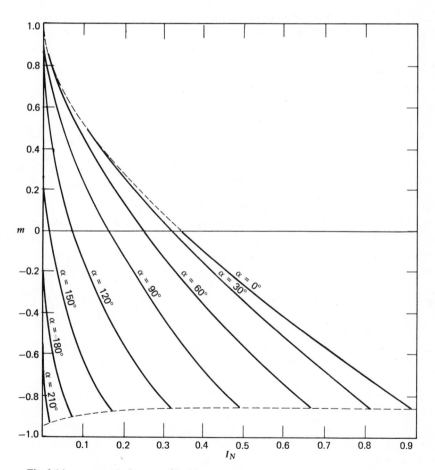

Fig. 3.14 m versus I_N for $\phi = \pi/6$ with α as parameter.

Fig. 2.22, while that shown in the fourth quadrant is defined by the fourth-quadrant boundary of Fig. 3.13. The operating point of the circuit must lie within these boundaries.

The normalized value of the rms output current is given by

$$I_{RN} = \left[\frac{1}{2\pi} \int_{\alpha}^{\alpha + \gamma} \left[\frac{Z}{\sqrt{2} \ V} i \right]^2 d(\omega t) \right]^{1/2} \tag{3.41}$$

and a family of curves of m versus I_{RN} for $\phi = \pi/6$ have been derived from this expression and are shown in Fig. 3.15. The envelope curve shown in broken line is identical with that marked $\phi = 30°$ in Fig. 2.23.

Load-Circuit Inductance Negligible   Once again, the limiting cases where $L = 0$ or $R = 0$ are of interest, and in Fig. 3.16a is shown the circuit for zero inductance in the load. The current may be determined as for the uncontrolled case, and from equation 2.179 it is seen that

$$\frac{R}{\sqrt{2} \ V} i = \sin \omega t - m \tag{3.42}$$

The time variations of currents and voltages are shown in Fig. 3.16b, where the conduction angle is seen to be

$$\gamma = (\pi - \eta) - \alpha \quad \text{rad} \tag{3.43}$$

A family of curves of m versus γ obtained from equation 3.43 is shown in Fig. 3.17, where the boundary shown in broken line in the first quadrant is identical with that marked $\phi = 0$ in Fig. 2.21.

Due to the fact that the circuit is purely resistive, the fourth-quadrant boundary at $m = -1$ simply indicates the dividing line between controlled operation with discontinuous current and uncontrolled operation with continuous current.

From equations 3.42 and 3.43 it may be seen that the normalized value of the average rectified current is

$$I_N = \frac{1}{2\pi} \int_{\alpha}^{\pi - \eta} (\sin \omega t - m) d(\omega t)$$

$$= \frac{1}{2\pi} \left[\cos \alpha + \sqrt{1 - m^2} - m(\pi - \eta - \alpha) \right] \tag{3.44}$$

Curves of $m = \sin \eta$ versus I_N for a series of values of α may be obtained from equation 3.44 and are shown in Fig. 3.18. The first-quadrant boundary is identical with that marked $\phi = 0$ in Fig. 2.22, and the operating

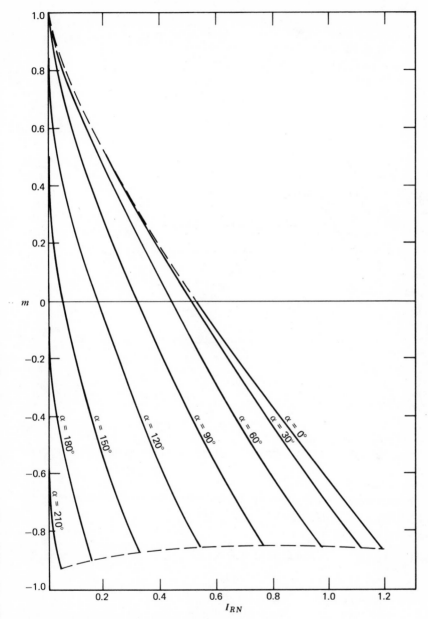

Fig. 3.15 m versus I_{RN} for $\phi = \pi/6$ with α as parameter.

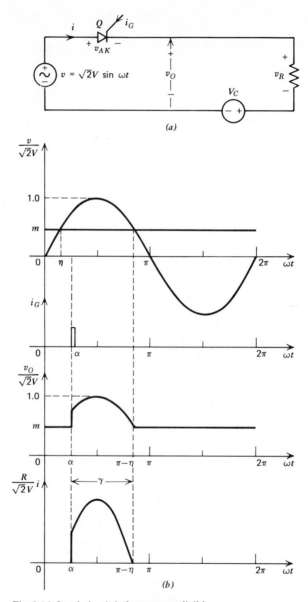

Fig. 3.16 Load-circuit inductance negligible.

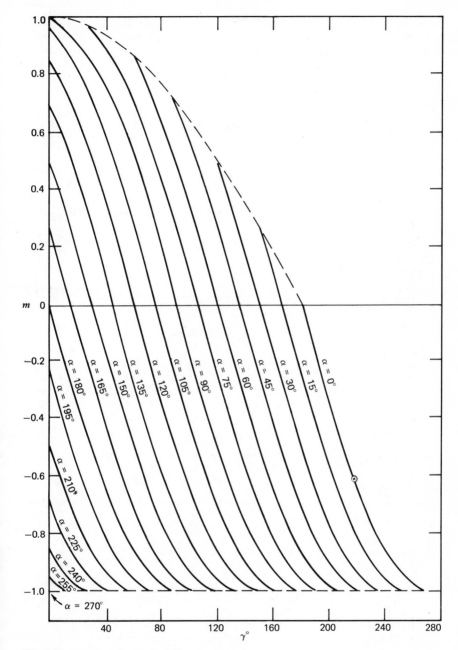

Fig. 3.17 m versus γ for $\phi = 0$ with α as parameter.

point of the circuit must lie to the left of this boundary for the reasons already given in the discussion of the LR circuit with electromotive force.

The normalized value of the rms output current is given by

$$I_{RN} = \left[\frac{1}{2\pi} \int_{\alpha}^{\pi - \eta} (\sin \omega t - m)^2 \, d(\omega t) \right]^{1/2} \tag{3.45}$$

Curves of m versus I_{RN} obtained from equation 3.45 are shown in Fig. 3.19, where the first-quadrant boundary is from Fig. 2.23.

Load-Circuit Resistance Negligible In Fig. 3.20a is shown the circuit for the case in which resistance is negligibly small. The current is again

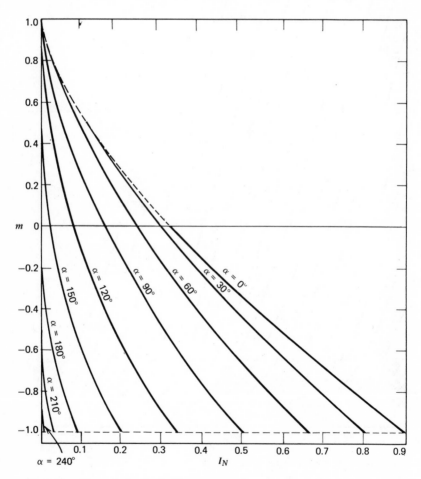

Fig. 3.18 m versus I_N for $\phi = 0$ with α as parameter.

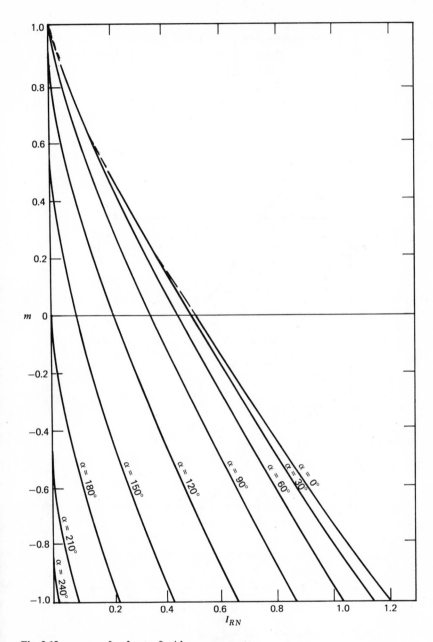

Fig. 3.19 m versus I_{RN} for $\phi = 0$ with α as parameter.

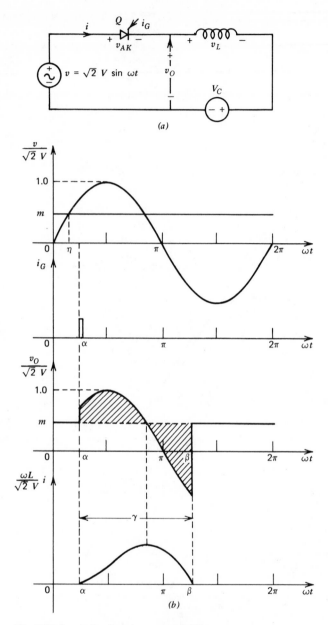

Fig. 3.20 Load-circuit resistance negligible.

determined as for the uncontrolled rectifier, so that from equation 2.191.

$$\frac{\omega L}{\sqrt{2}\ V}i = \cos\alpha - \cos\omega t - m(\omega t - \alpha) \tag{3.46}$$

Conduction ceases at $\omega t = \alpha + \gamma$, so that substitution in equation 3.46 gives

$$m\gamma = \cos\alpha - \cos(\alpha + \gamma) \tag{3.47}$$

The two shaded areas in the waveform of $v_0/\sqrt{2}\ V$ must again be equal for the reasons explained in Section 2.6.6.

A family of curves of m versus γ obtained from equation 3.47 is shown in Fig. 3.21, where the first-quadrant boundary is identical with that marked $\phi = 90°$ in Fig. 2.21, while the fourth-quadrant boundary is defined by equation 3.39. Beyond this boundary, control would be lost, and the current would build up to a magnitude, limited only by the saturated value of the circuit inductance.

From equation 3.46, the normalized value of the average rectified current is

$$I_N = \frac{1}{2\pi}\int_{\alpha}^{\alpha+\gamma}\left[\cos\alpha - \cos\omega t - m(\omega t - \alpha)\right]d(\omega t)$$

$$= \frac{1}{2\pi}\left[\gamma\cos\alpha + \sin\alpha - \sin(\alpha+\gamma) - \frac{m\gamma^2}{2}\right] \tag{3.48}$$

Curves of m versus I_N may be obtained from equations 3.47 and 3.48, and are shown in Fig. 3.22, where the first-quadrant boundary is from Fig. 2.22.

The normalized value of the rms output current is given by

$$I_{RN} = \left[\frac{1}{2\pi}\int_{\alpha}^{\alpha+\gamma}\left[\cos\alpha - \cos\omega t - m(\omega t - \alpha)\right]^2 d(\omega t)\right]^{1/2} \tag{3.49}$$

Curves of m versus I_{RN} may be obtained from equations 3.47 and 3.49, and are shown in Fig. 3.23, where the first-quadrant boundary is from Fig. 2.23.

Example 3.2 The circuit shown in Fig. E3.2 is employed to charge a bank of batteries of which the nominal terminal voltage is $V_C = 72$ V. Calculate the average and rms line currents and the power factor at the source if

(a) $V_C = 48$ V, $\alpha = 60°$(i.e., batteries discharged)
(b) $V_C = 78$ V, $\alpha = 120°$ (i.e., batteries fully charged)

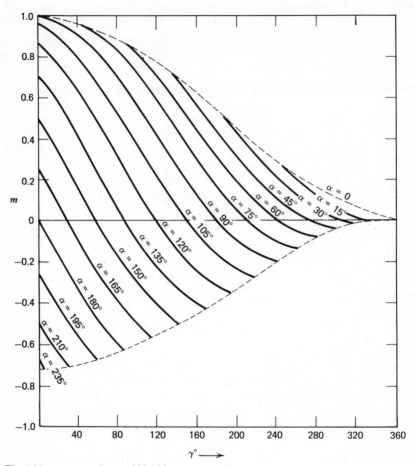

Fig. 3.21 m versus γ for $\phi = 90°$ with α as parameter.

Solution

(a) $\phi = 90°, \qquad \alpha = 60°$

$$m = \frac{V_C}{\sqrt{2}\,V} = \frac{48}{110\sqrt{2}} = 0.315$$

From Figs. 3.22 and 3.23, $I_N = 0.27$, $I_{RN} = 0.43$
    The base value of the current is

$$\frac{\sqrt{2}\,V}{\omega L} = \frac{110\sqrt{2}}{3} = 43 \text{ A}$$

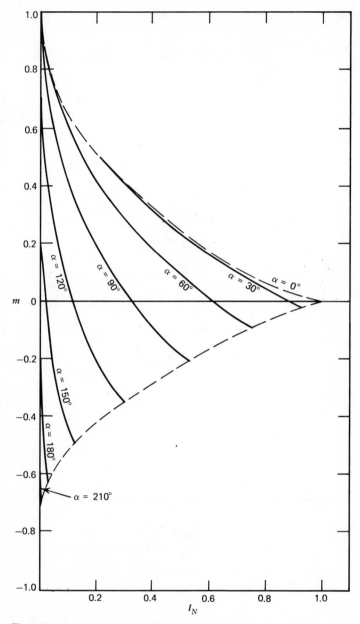

Fig. 3.22 m versus I_N for $\phi = 90°$ with α as parameter.

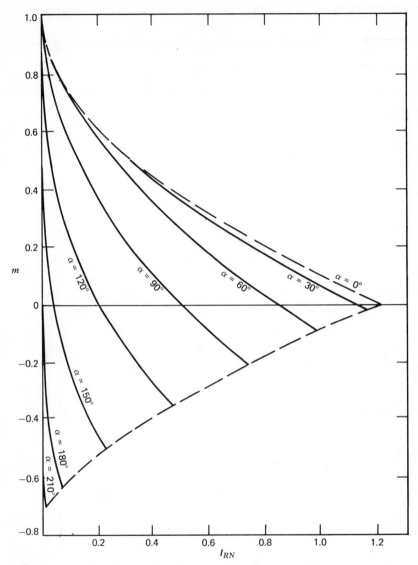

Fig. 3.23 m versus I_{RN} for $\phi = 90°$ with α as parameter.

110

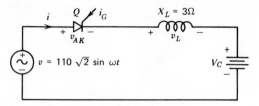

Fig. E3.2 Battery charging circuit.

thus

$$I_O = 0.27 \times 43 = 11.6 \quad A$$

$$I_R = 0.43 \times 43 = 18.5 \quad A$$

Power delivered to battery $= 11.6 \times 48 = 567 \quad W$

$$\text{Power factor} = \frac{567}{110 \times 18.5} = 0.280$$

(b) $\phi = 90°, \qquad \alpha = 120°$

$$m = \frac{78}{110\sqrt{2}} = 0.50$$

From Figs. 3.22 and 3.23, $I_N = 0.015$, $I_{RN} = 0.030$ thus

$$I_O = 0.015 \times 43 = 0.65 \text{ A}$$

$$I_R = 0.030 \times 43 = 1.29 \text{ A}$$

power delivered to battery $= 0.65 \times 78 = 50.7 \text{ W}$

$$\text{power factor} = \frac{50.7}{110 \times 1.29} = 0.357$$

3.4 THYRISTOR DATA SHEETS

A typical data sheet for a thyristor is shown in Fig. 3.24a and the reverse side of that same sheet is shown in Fig. 3.24b. The manner of presentation of the information given in this sheet varies with different manufacturers; nevertheless, all of the information which appears in Fig. 3.24 will be made available in some form. Apart from the dimensioned drawing of the device, its properties are specified in great detail in the table and graphs, and the correct application of the thyristor can only be made if the significance of these data is understood.

Westinghouse

Thyristor
Silicon Controlled Rectifiers
Westinghouse Type 261+

Forward Current 200 Amps RMS
125 Amperes Half-Wave Average
Forward Blocking Voltages to 1500 Volts

Dimensions in Inches

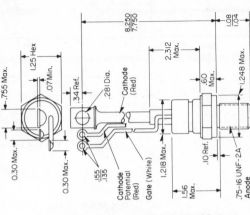

.755 Max.
1.25 Hex
.07 Min.
.34 Ref.
.281 Dia.
Cathode (Red)
0.30 Max.
0.30 Max.
.55 / .35
Cathode Potential (Red)
Gate (White)
1.218 Max.
1.56 Max.
.10 Ref.
.60 Max.
2.312 Max.
8.250 / 7.750
1.08 / 1.04
1.248 Max.
.75-16 UNF-2A Anode

TO 93 Case
Approximate Weight: 8 oz.

The Westinghouse Type 261 Series Features

• All diffused design
• Guaranteed dv/dt (300V/μsec)
• Low gate current
• Guaranteed value of di/dt
• Low thermal impedance
• High surge current capability

The exclusive Westinghouse CBE construction technique provides a thermal fatigue-free device by eliminating solder joints. In addition, the entire series carries the Westinghouse lifetime guarantee.

✦ Westinghouse Lifetime Guarantee

Westinghouse warrants to the original purchaser that it will correct any defects in workmanship, by repair or replacement f.o.b. factory, for any silicon power semiconductor bearing this symbol ✦ **TM** during the life of the equipment in which it is originally installed, provided said device is used within manufacturer's published ratings and applied in accordance with good engineering practice. The foregoing warranty is exclusive and in lieu of all other warranties of quality whether written, oral, or implied (including any warranty of merchantability or fitness for purpose). Westinghouse shall not be liable for any consequential damages.

Maximum Ratings and Characteristics

Block State (T_J=125°C)	Symbol	Westinghouse Type														
		261B	261D	261F	261H	261K	261M	261P	261S	261V	261Z	261ZB	261ZD	261ZF	261ZH	261ZK②
Repetitive Peak Forward and Reverse Voltage,② volts	V_{FB} V_{RB}	100	200	300	400	500	600	700	800	900	1000	1100	1200	1300	1400	1500
Non-repetitive Transient Peak Forward and Reverse Voltage, volts ≤5.0 msec	V_{FBT} V_{RBT}	200	300	400	500	600	700	850	950	1100	1200	1300	1450	1550	1700	1800
Peak Forward and Reverse Leakage Current, mA	I_{FB} I_{RB}															

← 15 →

Conducting State (T_J=125°C)	Symbol	All Types
RMS Forward Current, amps	I_{RMS}	200
Ave. Forward Current (180° Conduction), amps	I_{AVE}	125
Surge Current (at 60 Hz): ½ Cycle, amps	I_{FM}	3,300
3 Cycles, amps	I_{FM}	2,400
10 Cycles, amps	I_{FM}	2,000
I^2t for Fusing (at 60 Hz half-wave), amps²-sec	I^2t	45,000
Forward Voltage Drop at T_J=25°C: I_F=100 Adc, volts	V_F	1.40
I_F=625 Adc, volts	V_F	2.05

Thermal Characteristics	Symbol	All Types
Oper. Junction Temp. Range, °C	T_J	-40 to +125
Storage Temperature Range, °C	T_{stg}	-40 to +150
Max. Thermal Impedance, °C/Watt: Junction to Case	θ_{JC}	0.15
Case to Sink, Lubricated	θ_{CS}	0.075
Max. Thread Torque, Lubricated, in.-lbs		240

Gate Parameters (T_J=25°C)	Symbol	All Types
Gate Current to Trigger (V_{FB}=12V), ma	I_{GT}	150
Gate Voltage to Trigger (V_{FB}=12V), volts	V_{GT}	3
Non-Triggering Gate Voltage at T_J=125°C (Rated V_{FB}), volts	V_{GNT}	0.15
Peak Forward Gate Current, amps	I_{GFM}	4
Peak Reverse Gate Voltage, volts	V_{GRM}	5
Peak Gate Power, watts	P_{GM}	16
Average Gate Power, watts	$P_{G(AV)}$	3

Switching State

	Symbol	All Types
Typical Turn-On Time, I_F=100A, 10-90%, V_{FD}=10 volts②, T_J=25°C, μsec	t_{on}	5.0
Min. di/dt, Linear to 5.0 I_{AVE}.④ amps/μsec	di/dt	100
Typical Turn-Off Time, I_F=150A, T_J=125°C. dip/dt=50A/μsec., dv/dt=20V/μsec. Linear to .8 V_{FB}. μsec	t_{off}	60
Min. dv/dt, Exp. to Full V_{FB}. volts/μsec	dv/dt	300

② Applies for zero or negative gate voltage.
④ With recommended gate drive. See AD 54-560.

② For higher voltages refer to Westinghouse.

February, 1969
Supersedes TD 54-567, pages 1 and 2, dated August, 1966
E, D, C/2115/DB: E, D, C/2117

(a)

Thyristor
Silicon Controlled Rectifiers
Westinghouse Type 261

Forward Current 200 Amps RMS
125 Amperes Half-Wave Average
Forward Blocking Voltages to 1500 Volts

Electrical Characteristics

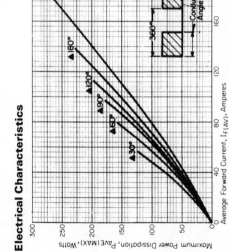

Figure 1. Max. power dissipation, full cycle average, rectangular wave.

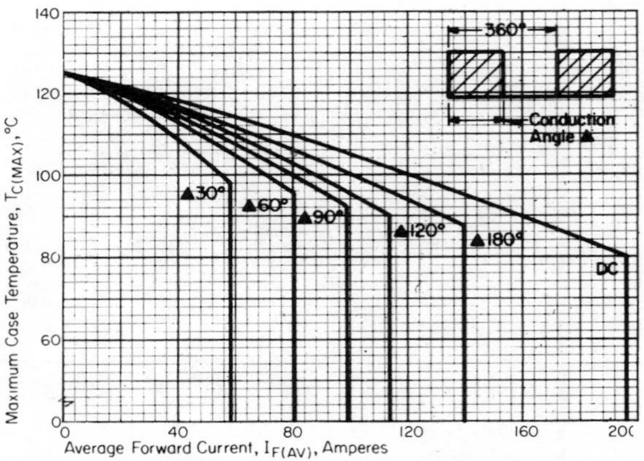

Figure 2. Max. allowable stud temperature, rectangular wave.

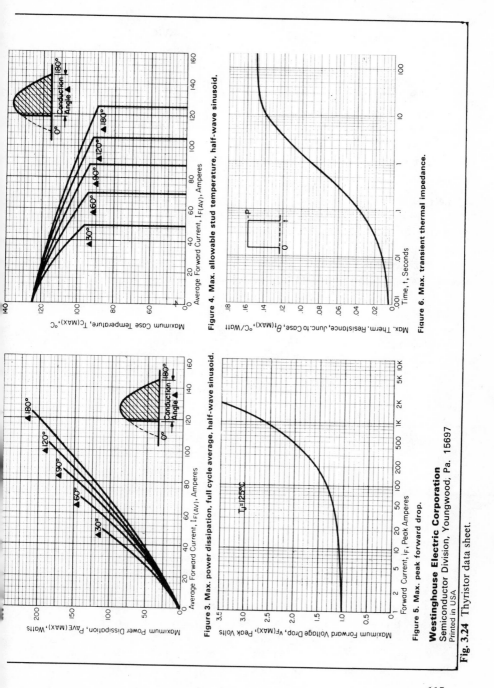

Figure 3. Max. power dissipation, full cycle average, half-wave sinusoid.

Figure 4. Max. allowable stud temperature, half-wave sinusoid.

Figure 5. Max. peak forward drop.

Figure 6. Max. transient thermal impedance.

Westinghouse Electric Corporation
Semiconductor Division, Youngwood, Pa. 15697
Printed in USA

Fig. 3.24 Thyristor data sheet.

115

As is the case with other power devices, a thyristor will only give satisfactory and continuous service if it is protected from deterioration due to excessive heating of its parts and in particular of the junction. The major source of junction heating at power frequencies is that of forward conduction losses. For the thyristor specified in Fig. 3.24, the permissible operating junction temperature range is $-40 < T_J < 125°C$. The lower temperature limit is required to keep stress within the silicon crystal to an acceptable value. Thyristors may be stored at temperatures somewhat higher than the maximum permissible junction operating temperature, and this is because under such conditions they are not subjected to electrical as well as to thermal stress.

3.4.1 Specified Voltages Many of the electrical quantities specified in Fig. 3.24 are given for the worst possible case, that is, when junction temperature T_J is at the permissible maximum value. This applies to the maximum repetitive peak forward blocking voltage V_{FB} and the maximum repetitive peak reverse voltage V_{RB}. The names of these quantities are self-explanatory, and their positions on the voltage-current characteristic of a thyristor are shown in Fig. 3.25. If the temperature of the junction is allowed to rise above the specified maximum, there is a danger of avalanche breakdown at the specified voltages. For very short periods, specified as 5 ms in Fig. 3.24a, the thyristor will safely block somewhat increased forward or reverse voltages V_{FBT} or V_{RBT}; V_{BO} is the breakover voltage at which the thyristor will begin to conduct in the forward direction at the specified maximum temperature. Provided that the external circuit is such as to limit the resulting current to the specified value, the thyristor is not damaged by this occurrence, and indeed this is one method of triggering the thyristor which is employed in practice. A large reverse current resulting from a reverse voltage exceeding V_{RBT} invariably destroys the thyristor. The fact that the voltage at which forward conduction commences is dependent upon the magnitude of the gate current is also indicated in Fig. 3.25, where $I_{G2} > I_{G1}$.

3.4.2 Anode Current and Heat Sink Specification Correct application of a thyristor, apart from observation of the voltage limitations discussed in the preceding section, depends almost entirely upon arriving at a current rating which will not result in excessive junction temperature, and the correct choice of current rating can only be made on the basis of some understanding of the heating process in the device. The maximum rms forward current I_R (I_{rms} in Fig. 3.24) is specified to prevent excessive heating in resistive elements of the thyristor, such as leads and joints.

The heat arising at the junction due to the forward conduction losses flows to the thyristor case, from there to the heatsink, and from the

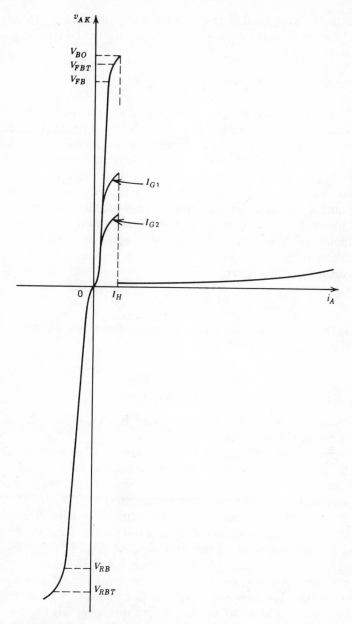

Fig. 3.25 Thyristor voltage-current characteristic.

heatsink to the surrounding atmosphere. The difference in temperature between junction and atmosphere under steady-state conditions is given by the equation

$$T_J - T_A = P_{\text{AVE}}(\theta_{JC} + \theta_{CS} + \theta_{SA}) \quad °C \tag{3.50}$$

where T_A is the ambient temperature, θ_{JC} and θ_{CS} are the thermal impedances specified in Fig. 3.24a, and P_{AVE} is the average rate of heat generation in watts; θ_{SA} is the sink-to-atmosphere thermal impedance and is not a property of the thyristor but of the heatsink. Moreover, it is not a constant quantity, but is a function of the sink material, the surface treatment, the size, and the temperature difference between the sink and the surrounding air. Data on heatsinks are available from manufacturers in various forms. Figure 3.26 refers to a series of extruded aluminum heatsinks similar to those shown in Fig. 1.4. These are cooled by natural convection, and the curves give the relationship between ΔT and P_{AVE}, the heat dissipated in watts, where

$$\Delta T = T_S - T_A \quad °C \tag{3.51}$$

and T_S is the sink temperature. Thus if for any point on a curve the values of ΔT and P_{AVE} are read off, then

$$\theta_{SA} = \frac{\Delta T}{P_{\text{AVE}}} \quad °C|W \tag{3.52}$$

Alternatively, if the power to be dissipated is known, then ΔT may be obtained for any chosen sink.

Equations 3.50 to 3.52 apply to a steady-state condition of operation of the device and are analogous to steady-state equations describing a purely resistive dc circuit. Temperature is the analogoue of potential. Power, or rate of heat transfer, is the analogue of current, or rate of charge transfer. Thermal impedance is analogous to resistance. An equivalent thermal circuit corresponding to equation 3.50 is shown in Fig. 3.27. In this circuit θ_{JC} and θ_{CS} are specified in the data sheet of Fig. 3.24a, while θ_{SA} is obtained from data equivalent to the curves in Fig. 3.26.

If the anode current i_A of the thyristor were a constant direct current, then the relationship between P_{AVE} and the magnitude of that current would be given by the curve marked DC in the upper left-hand graph of Fig. 3.24b. The permissible value of P_{AVE} could then be obtained from equation 3.50, where T_J was replaced by the specified maximum value of 125°C, and the permissible value of constant direct current read off from the curve. If the anode current of the thyristor consisted of a series of

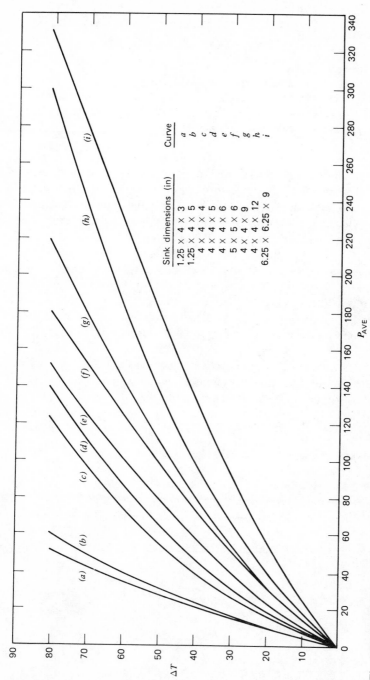

Fig. 3.26 Standard heat sink ratings for natural convection—aluminum extrusions.

Sink dimensions (in)	Curve
1.25 × 4 × 3	a
1.25 × 4 × 5	b
4 × 4 × 4	c
4 × 4 × 5	d
4 × 4 × 6	e
5 × 5 × 6	f
4 × 4 × 9	g
4 × 4 × 12	h
6.25 × 6.25 × 9	i

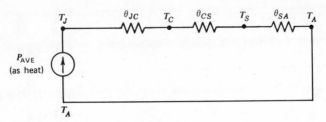

Fig. 3.27 Equivalent thermal circuit.

pulses, such as result from the operation of a controlled rectifier circuit, then substitution of the average forward current for the direct current in the foregoing procedure would not be acceptable. The reason for this is that the device has a very small heat capacity, and the temperature of the junction varies cyclically at power frequencies.

The variation of junction temperature resulting from a single rectangular pulse of anode current is as illustrated in Fig. 3.28, where T_J increases exponentially from beginning to end of the pulse and then decays exponentially to the value of T_A. A series of such pulses would result in a wave of T_J which, when steady-state conditions had been established, would be made up of sectors of exponential curves as illustrated in Fig. 3.29.

The lowest value of T_J in the cycle of variation shown in Fig. 3.29 would be greater than T_A, and the highest value should not exceed the specified maximum value of T_J. Thus due to the rapid increase of T_J when i_A begins to flow, the average value of i_A for a rectangular wave that will just cause T_J to reach 125°C is less than the value of the constant direct

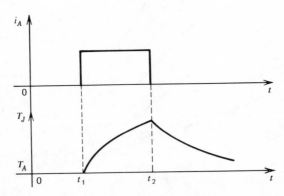

Fig. 3.28 Variation of junction temperature due to a pulse of i_A.

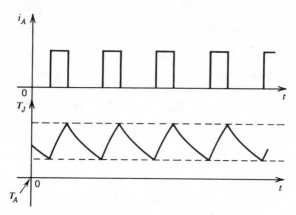

Fig. 3.29 Variation of junction temperature due to a rectangular wave of i_A.

current that also would cause T_J to rise to 125°C. It is therefore necessary that a thyristor must be rated at a lower value of average forward current $I_{F(AV)}$ when it is conducting a rectangular wave than when it is carrying a constant direct current. These considerations give rise to the remaining curves in the upper left-hand graph of Fig. 3.24b.

The permissible average forward current for a rectangular wave of a given conduction angle is obtained from the appropriate curve in Fig. 3.24b. Thus if the thyristor-heatsink combination is capable of dissipating 100 W at the given value of T_A with $T_J = 125°C$, then a constant direct current of 80 A can be carried, but if the current is a rectangular wave such that the conducting time and the nonconducting time are equal, giving a conduction angle of 180°, then $I_{F(AV)}$ must be reduced to approximately 70 A, corresponding to a current during the conduction period of 140 A.

It will be noted from the lower left-hand graph of Fig. 3.24b that the forward resistance of the thyristor is by no means constant, and this factor accounts for the approximately linear relationship between $P_{AVE(MAX)}$ and $I_{F(AV)}$.

The derating of the thyristor below the DC value is greater when the current pulses are parts of a sine wave than in the case of the rectangular wave. This is due to the higher form factor of the sinusoidal wave, where form factor is defined as the ratio of the rms to the average value of the waveform. For given values of $I_{F(AV)}$ and conduction angle, a wave of sinusoidal pulses has a higher peak value than does one of rectangular pulses, and allowance is made for this factor by further derating in the curves of the middle left-hand graph in Fig. 3.24b. In these latter curves, the value of $I_{F(AV)}$ for a given conduction angle and value of $P_{AVE(MAX)}$ is less than for a rectangular wave of the same average value and conduction angle.

Determination of the current rating of a thyristor by means of equations 3.50 to 3.52 and the relevant curve of $P_{AVE(MAX)}$ versus $I_{F(AV)}$ requires a knowledge of the ambient temperature and the sink-to-atmosphere thermal impedance. If these are yet to be specified, the current rating may be determined in terms of the case temperature T_C by means of the upper and middle right-hand graphs of Fig. 3.24b. Thus if $I_{F(AV)}$ is known, then the maximum case temperature $T_{C(MAX)}$ may be read off from the appropriate curve, and

$$P_{AVE(MAX)} = \frac{125 - T_{C(MAX)}}{\theta_{JC}} \quad W \qquad (3.53)$$

If the ambient temperature is specified, then the required value of θ_{SA} may be obtained from the curves of Fig. 3.26 or equivalent sink manufacturers' data.

It will be observed that the curves of $T_{C(MAX)}$ versus $I_{F(AV)}$ all reach limiting current values. Each limiting value for the particular waveform corresponds to an rms value of current equal to I_R specified (as I_{RMS}) in Fig. 3.24a. For the *DC* curve, this may readily be checked, since the limiting value is 200 A.

3.4.3 Surge Currents Equations 3.50 to 3.52 apply to a steady-state condition of operation of the device. However, the junction-case-heatsink combination possesses thermal capacity, so that only some time after the flow of heat commences do the parts of the combination, and in particular the junction, reach a steady temperature.

A simplified equivalent thermal circuit representing the transient behavior of the junction-case-heatsink combination is shown in Fig. 3.30. The source P_S represents the total power in the form of heat that is being produced at the junction. If this is applied as a step function, then the temperature of the junction will rise exponentially, as illustrated in Fig. 3.31.

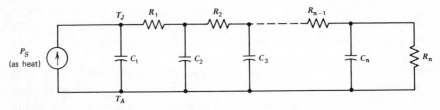

Fig. 3.30 Transient equivalent thermal circuit.

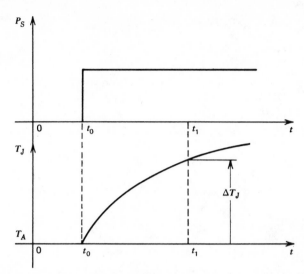

Fig. 3.31 Junction temperature variation due to a step function of power input.

If at any instant t_1 as shown in Fig. 3.31, the increase in junction temperature ΔT_J is divided by the power input P_S, then the result is a value of the transient thermal impedance θ_t of the combination at time $t = t_1 - t_0$ after the application of the step function of power. For a series of such instants a curve of θ_t versus t may be plotted and is similar to that shown in the bottom right-hand graph of Fig. 3.24*b*. That curve illustrates the variation with time of junction-to-case transient impedance, and as t becomes very large, θ_t approaches θ_{JC}.

The importance of the extremely low value of thermal impedance for a short time after the thyristor is switched on lies in the fact that it makes possible the acceptance of a large switching current surge without overheating of the junction. The longer this surge lasts, the smaller the current must be, and three different values of I_{FM} for three different surge durations are therefore specified on the data sheet.

If a fault on a system occurs when a thyristor is conducting, the device may be subjected to an extremely high overload current, and since it may already be operating at maximum junction temperature, this temperature will immediately commence to rise above the specified maximum. For overloads lasting for less than one half cycle an I^2t rating is specified, and this is employed in selecting a fuse or other equipment to protect the thyristor.

Example 3.3 For the circuit shown in Fig. E3.3, $V = 220$ V, $R = 1$ Ω, the delay angle α is zero, and the ambient temperature is 40°C. The thyristor Q is that specified in Fig. 3.24.

(a) Choose a suitable heat sink from the series illustrated in Fig. 3.26.

(b) Calculate the circuit efficiency.

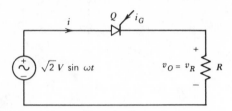

Fig. E3.3 Specification of heat sink.

Solution

(a) The conduction angle $\gamma = 180°$, and the average rectified current is

$$I_O = \frac{1}{2\pi} \int_0^\pi \frac{\sqrt{2}\, V}{R} \sin \omega t \, d(\omega t) = \frac{\sqrt{2}\, V}{R\pi} = \frac{220\sqrt{2}}{\pi} \simeq 100 \quad \text{A}$$

From the centre right-hand graph of Fig. 3.24b, $T_{C(MAX)}$ is 98°. From the centre left-hand graph of Fig. 3.24b P_{AVE} is 160 W. Thus the sink temperature is

$$T_S = T_C - P_{AVE} \times \theta_{CS} = 98 - 160 \times 0.075 = 86°C$$

$$\Delta T = T_S - T_A = 86 - 40 = 46°C$$

From Fig. 3.26, curve i represents a heat sink that will dissipate 160 W for this temperature difference.

(b) The power output is
$$P_O = RI_R^2 = \frac{V_R^2}{R}$$

$$V_R = \sqrt{2}\, V \left[\frac{1}{2\pi} \int_0^\pi \sin^2 \omega t \, d(\omega t) \right]^{1/2} = \frac{V}{\sqrt{2}}$$

$$P_O = \frac{V^2}{2R} = \frac{(220)^2}{2} = 24.2 \times 10^3 \quad \text{W}$$

Power dissipated in thyristor is 160 W; thus

$$\text{efficiency} = \frac{24.2 \times 10^3}{24.2 \times 10^3 + 160} = 0.997 \quad \text{pu}$$

Example 3.4 For the circuit of Example 3.3, employing the chosen heat sink, calculate the case and junction temperatures if the delay angle α is 120°.

Solution The conduction angle $\gamma = 60°$, and the average rectified current is

$$I_O = \frac{1}{2\pi} \int_\alpha^\pi \frac{\sqrt{2}\,V}{R} \sin \omega t\, d(\omega t) = \frac{V}{\sqrt{2}\,\pi R}(1 + \cos \alpha)$$

$$= \frac{220}{\sqrt{2}\,\pi \times 2} \cong 25 \quad \text{A}$$

From Fig. 3.24b, $P_{AVE} = 45$ W.
From Fig. 3.26, curve i, for $P_{AVE} = 45$ W, $\Delta T = 16°C$, so that the sink temperature is

$$T_S = T_A + \Delta T = 40 + 16 = 56° C$$

The case temperature is

$$T_C = T_S + P_{AVE}\theta_{CS} = 56 + 45 \times 0.075$$

$$= 59.4°C$$

The junction temperature is then

$$T_J = T_C + P_{AVE} \times \theta_{JC} = 59.4 + 45 \times 0.15 = 66.4°C$$

3.4.4 Limitation on di/dt When a forward voltage is applied to a thyristor, and it is turned on by means of a gate current, conduction of anode current across the junction commences in the immediate neighborhood of the gate connection and spreads from there across the whole area of the junction. Thyristors are so designed that the conduction area spreads as rapidly as possible; nevertheless if the rate of rise of anode current (di/dt) is great, a local hot spot will be formed in the neighborhood of the gate connection, due to the high current density in that part of the junction that has commenced to conduct. This localized heating may result in failure of the thyristor. It is for this reason that the minimum value of di/dt below, which the device will not suffer damage, is specified in Fig. 3.24a under the heading "Switching State." The di/dt may be held down to an acceptable value by including a small amount of inductance in the anode circuit, and when this is done, the time taken for the device to turn on to full conduction is that specified as t_{on}.

One way of ensuring a rapid spread of the conduction area is that of applying a gate current greater than the minimum value specified as I_{GT} under the heading "Gate Parameters." At the same time the peak value of the gate current pulse must not exceed that specified as i_{GFM}.

The circuit of Fig. 3.32 may be taken to include the thyristor Q specified in Fig. 3.24. The load circuit is a resistance R_L, and the output voltage v_O may be varied by the control resistor R_c in series with the diode D.

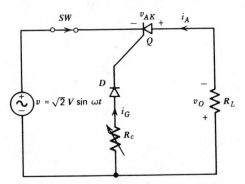

Fig. 3.32 Half-wave controlled rectifier circuit.

The peak forward gate current i_{GFM} to be employed in "triggering" or turning on this thyristor is 4 A. The value of gate current I_{GT} to trigger the thyristor at a forward voltage $V_{FB} = 12$ V is 150 mA, requiring a gate voltage $V_{GT} = 3$ V. The gate-to-cathode resistance of the thyristor may therefore be taken as

$$R_{GK} = \frac{3}{150 \times 10^{-3}} = 20 \quad \Omega$$

If the amplitude of the source voltage v is $\sqrt{2}\,V = 400$ V, then the maximum value of control resistance required $R_{c\,max}$ is given by

$$R_{c\,max} + 20 = \frac{400}{150 \times 10^{-3}} = 2.67 \times 10^3 \quad \Omega$$

The minimum permissible value of control resistance is given by

$$R_{c\,min} + 20 = \frac{400}{4} = 100 \quad \Omega$$

The range of control resistance which may be usefully and safely employed with this thyristor in the circuit of Fig. 3.32 is thus $80 < R_c < 2.65 \times 10^3 \ \Omega$.

If it is assumed for the purpose of making a simple illustration that an increase in V_{FB} does not reduce I_{GT} required to turn on the thyristor, then the waveforms of the circuit variables for $R_c = R_{c\,max}$ are shown in Fig. 3.33. The gate current only reaches the necessary value for turning on the thyristor at $\omega t = 3\pi/2$.

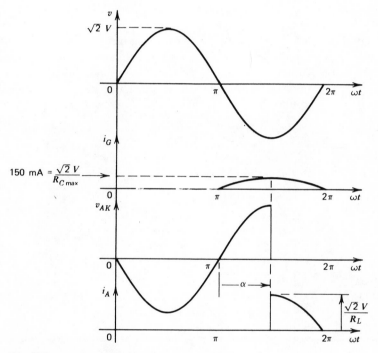

Fig. 3.33 Waveforms for Fig. 3.32. $R_c = 2.67 \, k\Omega$.

The delay angle α is defined as the interval in electrical angular measure by which the starting point of conduction is delayed by phase control in relation to the operation of the same circuit in which the thyristors were replaced by diodes.

Under the conditions of operation specified in this definition, positive current would commence to flow in R_L of Fig. 3.32 at $\omega t = \pi$. Thus for $R_c = R_{c\,max}$, $\alpha = \pi/2$.

The waveforms of the circuit variables for $R_c = R_{c\,min}$ are shown in Fig. 3.34. The delay angle is given by

$$\frac{400\sin\alpha}{80+20} = 150\times 10^{-3}\quad\text{A}$$

from which $\alpha \cong 2.2°$. Thus the range of delay angle for the circuit of Fig. 3.32 is $2.2° \leqslant \alpha \leqslant 90°$.

This method of control is manual and cannot give values of α greater than 90°. It is therefore impossible to reduce the output voltage v_O to zero. A better method of control is discussed in Chapter 4. However, what is much more important is that the circuit as it stands is not practical, since the load circuit is purely resistive and would permit a step increase of i_A. This would destroy the thyristor after a very few current pulses due to the production of a hot spot as described earlier. The next step is therefore that of calculating the amount of inductance which must be included in the load circuit to limit di_A/dt to a value below that specified on the data

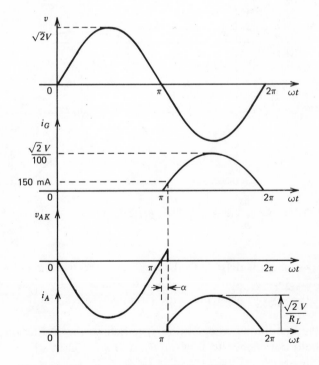

Fig. 3.34 Waveforms for Fig. 3.32. $R_c = 80\Omega$.

sheet. The modified circuit, in which R_L may be given the value of 1 Ω, is shown in Fig. 3.35a.

In the circuit of Fig. 3.35, when the thyristor is conducting,

$$-v = R_L i_A + L\frac{di_A}{dt'} \quad \text{V} \tag{3.54}$$

where t' is the time elapsed after the instant of turning on, and the negative sign appears because the thyristor conducts only during the negative half cycle of v. From equation 3.54

$$i_A = -\frac{v}{R_L}(1 - \epsilon^{-t'/\tau}) \quad \text{A} \tag{3.55}$$

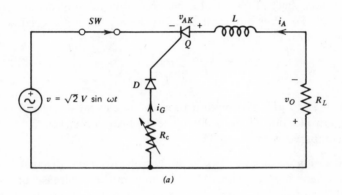

(a)

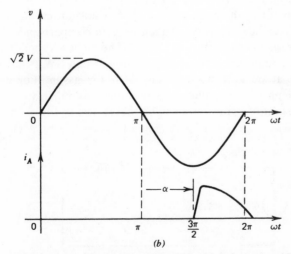

(b)

Fig. 3.35 Rectifier circuit of Fig. 3.32 with inductance added.

where

$$\tau = \frac{L}{R_L} \quad \text{s} \qquad (3.56)$$

Thus

$$\frac{di_A}{dt'} = \frac{v}{\tau R_L}\epsilon^{-t'/\tau} \quad \text{A/s} \qquad (3.57)$$

The maximum value of di_A/dt' will occur when $\alpha = 90°$, that is, when the maximum negative source voltage is applied to the circuit at $t' = 0$. Then

$$\frac{di_A}{dt'} = \frac{\sqrt{2}\,V}{L} \quad \text{A/s} \qquad (3.58)$$

From the data sheet of Fig. 3.24a, the allowable value of di/dt is seen to be 100 A/μs. To allow a margin of safety, let di_A/dt' not exceed 80 A/μs. Then substitution in equation 3.57 yields

$$80 \times 10^6 = \frac{400}{L} \quad \text{A/s}$$

from which $L = 5$ μH. The modification to the current waveform resulting from the introduction of this inductance is illustrated in Fig. 3.35b, which may be compared with Fig. 3.33.

3.4.5 Limitation on dv/dt   This topic has already been discussed in Section 3.2, where it is explained that a high rate of increase of forward voltage will turn on the thyristor, even when the gate current is zero.

When the thyristor in the circuit of Fig. 3.35 is not turned on, it may be considered as a very large resistive element R_{AK} in comparison with which the load-circuit resistance of 1 Ω is negligible. The highest value of dv_{AK}/dt will occur if the source is connected by closing switch SW (shown in the closed position) at peak forward voltage. On the assumption that v_{AK} builds up to its maximum value in a few microseconds, an equivalent circuit for the instant of connection will be that shown in Fig. 3.36.

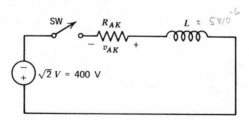

Fig. 3.36 Equivalent circuit corresponding to that of Fig. 3.35a.

The magnitude of R_{AK} is not specified as such in the data sheet; however, the peak forward leakage current is specified as 15 mA. Thus for a 400-V thyristor, R_{AK} must approach 30 kΩ. From Fig. 3.36

$$v_{AK} = \sqrt{2}\, V(1 - \epsilon^{-t/\tau}) \quad \text{V} \tag{3.59}$$

where

$$\tau = \frac{L}{R_{AK}} \quad \text{s} \tag{3.60}$$

and t is the time elapsed after the instant of closing switch SW in Fig. 3.36. Thus

$$\frac{dv_{AK}}{dt} = \frac{R_{AK}}{L} \sqrt{2}\, V \epsilon^{-t/\tau} \quad \text{V/s} \tag{3.61}$$

and this has its maximum value when $t = 0$, and

$$\left.\frac{dv_{AK}}{dt}\right|_{\max} = \frac{R_{AK}}{L} \sqrt{2}\, V \quad \text{V/s} \tag{3.62}$$

From equation 3.62 the maximum rate of increase of forward voltage in the equivalent circuit of Fig. 3.36 will thus be

$$\left.\frac{dv_{AK}}{dt}\right|_{\max} = \frac{30 \times 10^3}{5 \times 10^{-6}} \times 400 = 2.4 \times 10^6 \quad \text{V/}\mu\text{s}$$

which is far in excess of the specified value of 300 V/μs at which the circuit designer may be confident that the thyristor will not turn on upon connection of the source. The magnitude of dv_{AK}/dt may be reduced by increasing the inductance L, but to achieve the necessary reduction the inductance would need to be 40 mH. An inductor of this value capable of carrying the rated current of the thyristor would be both bulky and expensive, thus some other solution to this problem must be sought.

In this situation, the snubber circuit $R_s C_s$ shown in Fig. 3.37a is employed. Since the capacitor voltage, which is zero before switching on, cannot change instantaneously, the equivalent circuit for the worst possible instant of connection of the source is that shown in Fig. 3.37b, in which the load-circuit resistance is again negligible. By analogy with the circuit of Fig. 3.36.

$$\left.\frac{dv_{AK}}{dt}\right|_{\max} = \frac{R_s}{L} \sqrt{2}\, V \quad \text{V/s} \tag{3.63}$$

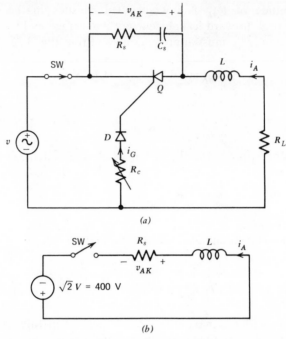

(a)

(b)

Fig. 3.37 Modification to circuit of Fig. 3.35a to limit dV_{AK}/dt.

In order that the amount of energy dissipated in R_s may be kept low, its value is made substantially greater than that which would be required merely to limit dv_{AK}/dt to the specified value. Typical values of R_s and C_s for the type of circuit considered here are 15 Ω and 0.1 μF, respectively. The adoption of this value of R_s demands an increase in the value of L to meet the specified rate of 300 V/μs. Thus

$$300 \times 10^6 = \frac{15}{L} \times 400 \quad \text{V/s}$$

so that $L = 20$ μH. Thus selection of appropriate snubber-circuit parameters amounts to a choice between the cost of reducing the circuit efficiency due to losses in R_s and the cost of providing a larger inductance than is needed to limit di_A/dt to an acceptable value.

Inspection of the circuit of Fig. 3.37a shows that, during the negative half cycle of the source voltage, the snubber capacitor C_s will be charged in a positive direction up to the instant at which the thyristor is turned on. Thus when the thyristor is turned on, there will be a step increase in current due to the discharge of C_s through a purely resistive circuit. This means that at the beginning of the conduction period di/dt approaches infinity, or at least some value well in excess of the specified limit. If the snubber circuit is correctly designed, however, this will not harm the thyristor. The purpose of limiting di/dt is to prevent the formation of a hot spot while the conduction area is spreading over the junction. Since capacitor C_s is small, it can only store a limited amount of energy, and the heat resulting from its partial dissipation at the thyristor junction is small enough to be absorbed by the thermal capacity of the junction without an excessive rise in junction temperature. Nevertheless, it is becoming the practice for manufacturers to specify a permissible limit to the energy in the snubber capacitor which may be dissipated at the junction of a given thyristor.

The cause of excessive dv/dt discussed in this section is the most direct and obvious one. It is also possible, however, for a thyristor to be turned on because of this when the power supply to the converter is switched on, or when the circuit configuration is changed by the turning on or off of a thyristor other than the one affected. Such possibilities depend on the circuit configuration of the converter, and the designer must therefore be on the look out for them at all times.

3.4.6 Turn-off Time The time which must elapse after forward current through the thyristor has ceased before forward voltage may again be applied without turn-on is called the "turn-off time." When conduction ceases, a high concentration of charge carriers still exists in the neighborhood of the center junction of the thyristor, and until this concentration has been sufficiently reduced by recombination, it is not possible to apply a forward voltage without conduction immediately taking place. A typical turn-off time t_{off} is therefore specified in the data sheet, and the time available for turn-off must always exceed this value.

However turn-off time is a function of a number of factors, and some of these are indicated in Fig. 3.24a under the heading "Switching State," where in each case the figure given is the maximum permissible value, or worst case. Thus if, immediately before turn off, the forward current I_F has not exceeded 150 A, the junction temperature T_J has not exceeded 125°C, and the rate of decrease of forward current has not exceeded 50 A/μs, then the thyristor will remain turned off after 60 μs, provided that at the end of that period the forward voltage is reapplied at a rate not exceeding 20 V/μs, increasing linearly up to a value of $0.8V_{FB}$. If one or more of these limits is exceeded, then a turn-off time greater than 60 μs will be necessary.

For the circuit of Fig. 3.35, the minimum value of the firing angle is approximately π. Forward blocking capability of the thyristor must therefore be regained during the positive half cycle of v. From the data sheet of Fig. 3.24a, $t_{off} = 60\mu s$, and this is the minimum half-cycle time of the source. Thus the maximum source frequency permissible in the circuit of Fig. 3.35 using this thyristor is

$$f_{max} = \frac{1}{120 \times 10^{-6}} = 8333 \quad \text{Hz}$$

3.5 COMMUTATION

An ideal transistor may be turned off; that is, its resistance to forward current restored to infinity, while forward current is flowing simply by reducing the base current to zero. This is not the case with a thyristor. Once turned on, a thyristor cannot recover its resistance to forward current unless that current is reduced to zero and held there for at least the turn-off time. A number of different methods may be employed to commutate or turn off a thyristor, and the method employed in a particular case depends on the function of the system embodying the thyristor. In this section, the three basic techniques of commutation are discussed and illustrated by means of simple circuits.

3.5.1 Line Commutation Line commutation is employed in circuits excited by a series ac source in which the current necessarily falls to zero at some point in the cycle. From that instant on, the forward voltage on the thyristor will be negative, and a series thyristor with zero gate current will turn off provided only that a forward voltage is not reapplied until the specified turn off time t_{off} has elapsed. This type of commutation has already been tacitly assumed in a number of earlier sections. It is also applied in ac voltage controllers and controlled rectifiers discussed in Chapters 4 and 5.

3.5.2 Load Commutation As its name implies, this type of commutation depends on the nature of the load circuit, that is, any natural tendency of the current in that circuit to fall to zero some time after the energy source has been applied to it by turning on a thyristor. Since in any ac-excited load circuit line commutation may be employed, it follows that load commutation is primarily of interest in dc-excited circuits.

In the circuit of Fig. 3.38a, it is clear that once thyristor Q is turned on, it will not turn off again. However the thyristor may be turned off in a finite time if some series capacitance is added to the circuit, as shown in Fig. 3.38b, provided that the relationship between R, L, and C is such that the current that flows when the thyristor is turned on is oscillatory. This means that a ringing frequency is established as explained in Section 2.4,

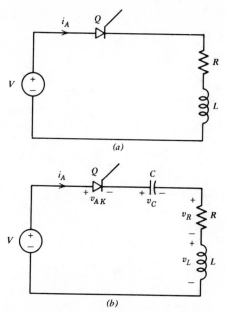

Fig. 3.38 Circuit employing load commutation.

so that from equations 2.49, 2.50, 2.53, and 2.54

$$\zeta = \frac{R}{2L} \quad \text{rad/s} \tag{3.64}$$

$$\omega_0 = \frac{1}{\sqrt{LC}} \quad \text{rad/s} \tag{3.65}$$

$$i_A = \epsilon^{-\zeta t}[B_1 \cos \omega_r t + B_2 \sin \omega_r t] \quad \text{A} \tag{3.66}$$

where

$$\omega_r = \sqrt{\omega_0{}^2 - \zeta^2} \quad \text{rad/s} \tag{3.67}$$

Example 3.5 The thyristor in the circuit of Fig. 3.38*b* is turned on at $t = 0$. If $V = 100$ V, $L = 10$ mH, $C = 1$ μF, $R = 80$ Ω, and the capacitor is initially charged to a voltage $v_C(0^-) = 25$ V, determine the time at which the thyristor is turned off and the final capacitor voltage.

Solution From Example 2.2

$$i_A = 0.818\epsilon^{-4000t} \sin 9170t \quad \text{A}$$

This current will fall to zero, and the thyristor will turn off when

$$9170t = \pi \quad \text{rad}$$

That is when

$$t = 343 \ \mu s$$

and this time is not affected by the initial charge on the capacitor. At the instant of turn-off, from the circuit of Fig. 3.38b, $v_R = 0$, $v_{AK} = 0$, and

$$v_C = V - L\frac{di_A}{dt} \quad \text{V}$$

Differentiation of the expression obtained for i_A and substitution in the above equation then yields

$$v_C = 119 \quad \text{V}$$

The time variations of i_A, v_C, and v_{AK} are shown in Fig. E3.5.

Example 3.5 shows that it is possible to bring the current in the load circuit to zero in finite time. However, the negative value of v_{AK} when the thyristor turns off shows that some method must be provided of at least partially discharging the capacitor before the thyristor can again conduct. This is discussed in connection with the series inverter in Chapter 7.

The limiting cases of negligible load-circuit inductance and resistance are once again of interest.

Load-Circuit Inductance Negligible It is immediately obvious that if the load-circuit inductance is negligible, then some series inductance must be added; otherwise the required oscillatory current would not be obtained.

Load-Circuit Resistance Negligible An ideal resistance-free circuit is shown in Fig. 3.39a. Thyristor Q is turned on by a pulse of gate current i_G at $t = 0$. The capacitor is charged to an initial voltage.

$$v_C(0^-) = XV \quad \text{V:} \quad X < 1 \tag{3.68}$$

so that at the instant of turn-on, the capacitor voltage aids that of the source in driving current round the circuit. Thus for $t > 0$,

$$V(1 - X) = L\frac{di_A}{dt} + \frac{1}{C}\int_0^t i_A \, dt \quad \text{V} \tag{3.69}$$

The solution of equation 3.69 is

$$i_A = I_m \sin \omega_r t \quad \text{A} \tag{3.70}$$

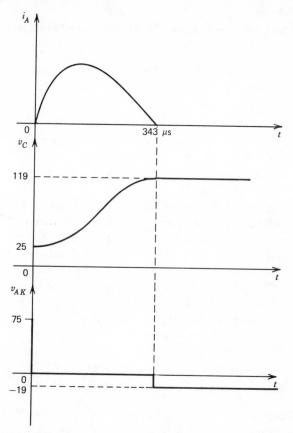

Fig. E3.5 Variables in the circuit of Fig. 3.38.

where

$$I_m = \frac{V(1-X)}{\omega_r L} = \frac{V(1-X)}{\sqrt{L/C}} \quad \text{A} \tag{3.71}$$

and from equation 2.50

$$\omega_r = \frac{1}{\sqrt{LC}} \quad \text{rad/s} \tag{3.72}$$

The quantity $\sqrt{L/C}$ in equation 3.71 has the dimensions of resistance and is called the "characteristic resistance" of the circuit. Thus

$$I_m = \frac{\text{initial driving voltage}}{\text{characteristic resistance}} \quad \text{A} \tag{3.73}$$

The thyristor turns off when the current returns to zero at $t = t_1$ or $\omega_r t_1 = \pi$, so that

$$t_1 = \frac{\pi}{\omega_r} = \pi\sqrt{LC} \quad \text{s} \tag{3.74}$$

Also

$$v_L = L\frac{di}{dt} = V(1-X)\cos\omega_r t \quad \text{V} \tag{3.75}$$

and

$$v_C = V - v_L = V - V(1-X)\cos\omega_r t \quad \text{V} \tag{3.76}$$

The time variations of the circuit variables are shown in Fig. 3.39b. Since the current in the circuit is zero at $t = 0$ and at $t = t_1$, then the energy stored in the inductor is also zero at both of these instants. Any energy supplied by the source during the interval $0 \leqslant t \leqslant t_1$ must therefore have been stored in the capacitor. The increase of stored energy in the capacitor is thus given by

$$W_C = \int_0^{t_1} Vi\,dt = VI_m \int_0^{\pi/\omega_r} \sin\omega_r t = \frac{2VI_m}{\omega_r}$$

$$= 2V\sqrt{LC}\,\frac{V(1-X)}{\sqrt{L/C}} = 2V^2(1-X)C \quad \text{J} \tag{3.77}$$

This same expression may be obtained by subtracting the initial energy stored in the capacitor at $t = 0$ from the final energy at $t = t_1$. Thus from equations 3.74 and 3.76 the final capacitor voltage is

$$v_C(t_1) = V(2-X) = 2V - v_C(0) \quad \text{V} \tag{3.78}$$

and

$$W_C = \tfrac{1}{2}C\left[V^2(2-X)^2 - V^2X^2\right] = 2V^2(1-X)C \quad \text{J} \tag{3.79}$$

In some systems it is possible, due to additional network elements not shown in Fig. 3.39a, that the current in the inductance may not be zero at the instant of turn-on. If

$$i_L(0^-) = I_{O1} \quad \text{A} \tag{3.80}$$

then the solution to equation 3.69 becomes

$$i_A = I_m \sin\omega_r t + I_{O1}\cos\omega_r t \quad \text{A} \tag{3.81}$$

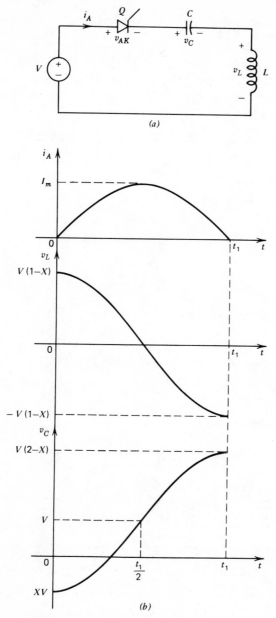

Fig. 3.39 Load-circuit resistance negligible ($X = -1$).

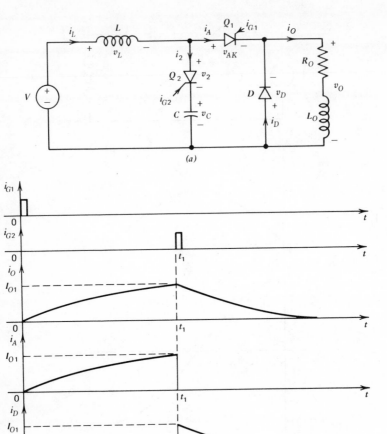

Fig. 3.40 Circuit employing forced commutation.

3.5.3 Forced Commutation It is not always convenient or economic to bring about load commutation of a typical RL load circuit by the introduction of a series capacitor large enough to carry the load current. Under such circumstances, some form of forced commutation is employed.

The principal additional circuit elements required for one method of forced commutation are shown in Fig. 3.40a. Their purpose is to force the thyristor voltage v_{AK} to a negative value, and so achieve "voltage commutation." A free-wheeling diode D is connected across the load circuit to release the energy stored in L_O when the main thyristor Q_1 is turned off.

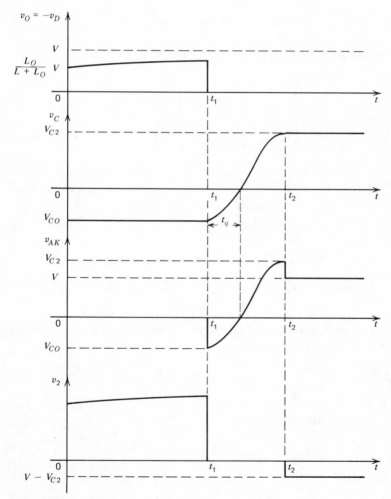

Fig. 3.40 (b) (continued).

The capacitor C is given an initial charge by means not shown in the diagram, such that $V_{CO} < 0$, and both Q_1 and the commutating thyristor Q_2 are initially turned off.

At $t = 0$, thyristor Q_1 is turned on by a pulse of gate current i_{G1}, and the current in the series circuit containing R_O, L_O, and L is

$$i_O = i_A = i_L = \frac{V}{R_O}(1 - \epsilon^{-R_O t/(L + L_O)}) \quad \text{A} \qquad (3.82)$$

Also

$$v_O = -v_D = R_O i_O + L_O \frac{di_O}{dt} \quad \text{V} \qquad (3.83)$$

and substitution from equation 3.82 in equation 3.83 yields

$$v_O = V(1 - \epsilon^{-R_O t/(L + L_O)}) + \frac{L_O}{L + L_O} V \epsilon^{-R_O t/(L + L_O)}$$

$$= V\left(1 - \frac{L}{L + L_O} \epsilon^{-R_O t/(L + L_O)}\right) \quad \text{V} \qquad (3.84)$$

The time variations of i_O and v_O are shown in Fig. 3.40b.

During this interval, i_2 and i_D are both zero, and for the center mesh of the circuit,

$$v_2 + v_C + v_D - v_{AK} = 0 \quad \text{V} \qquad (3.85)$$

But $v_C = V_{CO}$ and $v_{AK} = 0$, so that from equation 3.85

$$v_2 = -V_{CO} - v_D = -V_{CO} + v_O \quad \text{V} \qquad (3.86)$$

and the time variation of v_2 is also shown in Fig. 3.40b.

At $t = t_1$, thyristor Q_2 is turned on by a pulse of gate current i_{G2}, so that $v_2 = 0$, and from equation 3.84

$$v_{AK} = v_C + v_D \quad \text{V} \qquad (3.87)$$

But at $t = t_1$, $v_C = V_{CO} < 0$, and v_D can only be negative or zero. Thus it follows from equation 3.87 that

$$v_{AK} < 0 \quad \text{V:} \qquad t = t_1 \quad \text{s} \qquad (3.88)$$

and this negative voltage turns off thyristor Q_1. Let

$$t' = t - t_1 \quad \text{s} \qquad (3.89)$$

so that at $t' = 0^-$, from equations 3.82 and 3.89

$$i_O = i_A = i_L = \frac{V}{R_O}(1 - \epsilon^{-R_O t_1/(L+L_O)}) = I_{O1} \quad \text{A} \tag{3.90}$$

Since the load current in inductance L_O cannot change instantaneously, it is diverted through D, so that $v_D = 0$, and

$$i_O = i_D = I_{O1}\epsilon^{-(R_O/L_O)t'} \quad \text{A} \tag{3.91}$$

Since the line current i_L in inductance L cannot change instantaneously, it is diverted through the ringing circuit formed by L and C, so that $v_2 = 0$, $i_L = i_2$, and

$$L\frac{di_L}{dt'} + \frac{1}{C}\int_0^{t'} i_L \, dt' = V - V_{CO} \quad \text{V} \tag{3.92}$$

Differentiation of equation 3.92 yields

$$\frac{d^2 i_L}{(dt')^2} + \frac{1}{LC}i_L = 0 \quad \text{V/s} \tag{3.93}$$

of which the solution is of the form

$$i_L = A_1 \cos \omega_r t' + A_2 \sin \omega_r t' \quad \text{A} \tag{3.94}$$

where

$$\omega_r = \frac{1}{\sqrt{LC}} \quad \text{rad/s} \tag{3.95}$$

From equation 3.90 at $t = t_1$ or $t' = 0$, $i_L = I_{O1}$. Also at $t' = 0^+$ for the left-hand mesh of the circuit in Fig. 3.40a, $v_2 = 0$, and

$$v_L = L\frac{di_L}{dt} = V - V_{CO} \quad \text{V:} \qquad t' = 0^+ \quad \text{s} \tag{3.96}$$

Substitution of these initial conditions in equation 3.94 yields

$$i_L = \frac{V - V_{CO}}{\omega_r L}\sin \omega_r t' + I_{O1}\cos \omega_r t' \quad \text{A} \tag{3.97}$$

The peak value of i_L is given by the amplitude of the wave described by equation 3.97, and is

$$I_{Lm} = \left[\left(\frac{V - V_{CO}}{\omega_r L} \right)^2 + I_{O1}^2 \right]^{1/2} \quad \text{A} \tag{3.98}$$

At $t = t_2$, or $t' = t_2 - t_1$, i_L falls to zero, and thyristor Q_2 turns off, so that from equation 3.97

$$\omega_r (t_2 - t_1) = \tan^{-1} \frac{\omega_r L I_{O1}}{V_{CO} - V} = \psi \quad \text{rad} \tag{3.99}$$

Since $V_{CO} < 0$, it follows that the tangent of the angle ψ in equation 3.99 consists of a positive quantity divided by a negative quantity, and Fig. 3.41 shows that ψ must be an angle in the second quadrant. Inspection of the waveform of i_L in Fig. 3.40b confirms this fact, since the sinusoidal wave of i_L during the interval $t_1 < t < t_2$ has completed between one quarter and one half of a cycle. Thus

$$90° < \psi < 180° \tag{3.100}$$

$t_2 - t_1$ may now be determined from equation 3.99.

After thyristor Q_1 is turned off at $t' = 0$,

$$i_A = 0 \quad \text{A:} \qquad t' > 0 \quad \text{s} \tag{3.101}$$

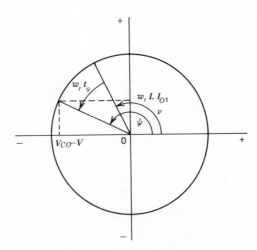

Fig. 3.41 Determination of angles for circuit of Fig. 3.40a.

And since diode D is conducting

$$v_O = -v_D = 0 \quad \text{V:} \qquad t' > 0 \quad \text{s} \qquad (3.102)$$

For the center mesh of the circuit of Fig. 3.40a, while thyristor Q_2 is still conducting

$$v_C = v_{AK} \quad \text{V:} \qquad 0 < t' < t_2 - t_1 \quad \text{s} \qquad (3.103)$$

Thus for the entire outer loop of the circuit

$$v_{AK} = V - L\frac{di_L}{dt'} \quad \text{V} \qquad (3.104)$$

The time for which v_{AK} is negative is shown as t_q on the curve of v_{AK} versus t in Fig. 3.40b. This time must exceed the turn-off time t_{off} for thyristor Q_1 if that thyristor is to remain off. Substitution in equation 3.104 for i_L from equation 3.97 and setting $t' = t_q$ yields

$$v_{AK} = V + \omega_r L I_{O1}\sin\omega_r t_q - (V - V_{CO})\cos\omega_r t_q = 0 \quad \text{V} \qquad (3.105)$$

From equation 3.99

$$\tan\psi = \frac{\omega_r L I_{O1}}{V_{CO} - V} \qquad (3.106)$$

from which it follows that

$$\sin\psi = \frac{\omega_r L I_{O1}}{V_x} \qquad (3.107)$$

$$\cos\psi = \frac{V_{CO} - V}{V_x} \qquad (3.108)$$

where

$$V_x = \left[(\omega_r L I_{O1})^2 + (V_{CO} - V)^2\right]^{1/2} \quad \text{V} \qquad (3.109)$$

so that rearrangement of equation 3.105 and substitution from equations 3.107 to 3.109 yields

$$V = -V_x[\cos\psi\cos\omega_r t_q + \sin\psi\sin\omega_r t_q]$$

$$= -V_x\cos(\psi - \omega_r t_q) \quad \text{V} \qquad (3.110)$$

From which

$$\omega_r t_q = \psi - \cos^{-1}\left(-\frac{V}{V_x}\right) = \psi - \nu \quad \text{rad} \tag{3.111}$$

Since $\cos\nu < 0$, angle ν lies in the second or third quadrant. However, since $\omega_r t_q > 0$, and ψ is a second-quadrant angle, it follows that ν must also be a second quadrant angle, as shown in Fig. 3.41. Thus from equations 3.99 and 3.111

$$t_q = t_2 - t_1 - \frac{\nu}{\omega_r} \quad \text{s} \tag{3.112}$$

where

$$90° < \nu < 180° \tag{3.113}$$

It remains to determine the final capacitor voltage V_{C2} when thyristor Q_2 turns off, since the capacitor will remain charged to this potential difference.

From equations 3.103 and 3.104,

$$v_C = V - L\frac{di_L}{dt'} \quad \text{V:} \qquad 0 < t' < t_2 - t_1 \quad \text{s} \tag{3.114}$$

Substitution in equation 3.114 for i_L from equation 3.97 yields

$$v_C = V - (V - V_{CO})\cos\omega_r t' + \omega_r L I_{O1}\sin\omega_r t' \quad \text{V} \tag{3.115}$$

and substitution from equation 3.107 and 3.108 in equation 3.115 yields

$$v_C = V + V_x[\cos\psi\cos\omega_r t' + \sin\psi\sin\omega_r t']$$

$$= V + V_x\cos(\psi - \omega_r t') \quad \text{V} \tag{3.116}$$

where V_x is defined in equation 3.109. The final capacitor voltage when Q_2 turns off at $t' = t_2 - t_1$ and $\omega_r t' = \psi$ is therefore

$$V_{C2} = V + V_x \quad \text{V} \tag{3.117}$$

The time variation of v_C is shown in Fig. 3.40b. Thus when both thyristors have turned off,

$$v_O = -v_D = 0 \quad \text{V:} \qquad t > t_2 \quad \text{s} \tag{3.118}$$

$$v_{AK} = V \quad \text{V:} \quad t > t_2 \quad \text{s} \tag{3.119}$$

$$v_2 = v_{AK} - v_C = -V_x \quad \text{V:} \qquad t > t_2 \quad \text{s} \tag{3.120}$$

It is also apparent from the curve of v_2 versus t in Fig. 3.40b that, before another pulse of i_A can be commutated, capacitor C must again be charged to a negative terminal voltage. The method of doing this is explained in Chapter 6.

Example 3.6 In the circuit of Fig. 3.40a, $V = 100$ V, $L = 100$ μH, $C = 25$ μF, and the capacitor voltage $V_{CO} = -100$ V. If the current i_A in thyristor Q_1 was 100 A when thyristor Q_2 was turned on, determine:

(a)   Peak current through Q_2
(b)   Conduction interval of Q_2
(c)   Turn-off time provided for Q_1
(d)   Final capacitor voltage
(e)   The quantities specified in (a) to (d) for the condition in which $i_A = 0$ when Q_2 is turned on.

Solution

(a)

$$I_{Lm} = \left[\left(\frac{V - V_{CO}}{\omega_r L} \right)^2 + I_{O1}^2 \right]^{1/2} \quad \text{A} \tag{3.98}$$

$$\omega_r L = \sqrt{L/C} = 2 \quad \Omega$$

$$I_{Lm} = \left[\left(\frac{200}{2} \right)^2 + 100^2 \right]^{1/2} = 141 \quad \text{A}$$

(b)

$$\omega_r(t_2 - t_1) = \tan^{-1} \frac{\omega_r L I_{O1}}{V_{CO} - V} = \psi \tag{3.99}$$

$$\omega_r = \frac{1}{\sqrt{LC}} = \frac{1}{(100 \times 10^{-6} \times 25 \times 10^{-6})^{1/2}} = \frac{10^6}{50}$$

$$\psi = \tan^{-1} \frac{10^6}{50} \frac{100 \times 10^{-6} \times 100}{-100 - 100} = \tan^{-1} -1 = \frac{3\pi}{4}$$

since ψ must be a second-quadrant angle.

$$t_2 - t_1 = \frac{\psi}{\omega_r} = \frac{50}{10^6} \times \frac{3\pi}{4} = 118 \quad \mu s$$

(c)

$$\omega_r t_q = \psi - \cos^{-1}\left(-\frac{V}{V_x}\right) = \psi - \nu \qquad (3.111)$$

where

$$V_x = \left[(\omega_r L I_{O1})^2 + (V_{CO} - V)^2\right]^{1/2} \quad V \qquad (3.109)$$

$$V_x = \left[\left(\frac{10^6}{50} \times 100 \times 10^{-6} \times 100\right)^2 + (-100 - 100)^2\right]^{1/2} = 282 \quad V$$

$$\nu = \cos^{-1}\left(-\frac{100}{282}\right) = 180.0 - 69.2 = 110.8°$$

since ν must be a second-quadrant angle.

$$\omega_r t_q = \psi - \nu = 135 - 110.8 = 24.2°$$

$$\therefore \quad t_q = \frac{24.2\pi}{180} \frac{50}{10^6} = 21.1 \quad \mu s$$

(d)

$$V_{C2} = V + V_x \qquad (3.117)$$

$$V_{C2} = 100 + 282 = 382 \quad V$$

(e)   From equation 3.98 $I_{Lm} = 100 \quad A$

From equation 3.99 $\tan\psi = 0$

$$\therefore \quad \psi = \pi \quad \text{rad}$$

since ψ must be a second-quadrant angle.

$$t_2 - t_1 = \frac{\psi}{\omega_r} = \frac{50}{10^6} \times \pi = 157 \quad \mu s$$

From equation 3.109

$$V_x = 200 \quad V$$

and from equation 3.111

$$v = \cos^{-1}\left(-\frac{100}{200}\right) = \pi - \frac{\pi}{3}$$

$$\therefore \quad \omega_r t_q = \psi - v = \frac{\pi}{3}$$

$$t_q = \frac{50}{10^6} \times \frac{\pi}{3} = 52.4 \quad \mu s$$

From equation 3.117

$$V_{C2} = V + V_x = 100 + 200 = 300 \quad V$$

PROBLEMS

3.1   In the circuit of Fig. P3.1, a pulse of gate current i_G lasting 100 μs is applied to thyristor Q at $\alpha = \pi/4$.

(a)   Derive an expression for current i.
(b)   Sketch to scale the time variations of v, v_{AK}, v_O, and i.
(c)   Determine the peak value of i and the instant at which it occurs.

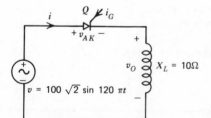

Fig. P3.1

3.2   The rms value of the source voltage in Fig. 3.5a is 220 V. Calculate the values of the average rectified current I_O and the rms output current I_R, and draw the waveforms of v_{AK} to scale for $\alpha = 60°$ and $\alpha = 135°$, if

(a) $R = 8 \quad \Omega, \qquad \omega L = 0$

(b) $R = 8 \quad \Omega, \qquad \omega L = 16 \quad \Omega$

3.3   In the circuit of Fig. P3.2, a pulse of gate current i_G lasting 100 μs is applied to thyristor Q at $\alpha = \pi/3$. Determine:

(a)   The conduction angle γ.
(b)   The extinction angle β.
(c)   The average and rms values of $i_O (I_O$ and $I_R)$.
(d)   The power delivered to the source V_C.
(e)   The power factor at the source v.

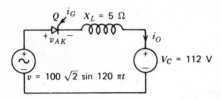

Fig. P3.2

3.4   Repeat problem 3.3 for $\alpha = \pi/4$.

3.5   Repeat problem 3.3 if the gating-current signal extends over the range $\pi/4 \leqslant \omega t \leqslant 5\pi/4$. (This type of gating by means of a prolonged application of i_G is called "continuous gating," as opposed to the "pulse gating" of problems 3.1 to 3.4).

3.6   In the circuit of Fig. P3.3, a pulse of gate current i_G is applied to thyristor Q at $\alpha = \pi/6$.

(a)   Calculate the average and rms values of current i.
(b)   Calculate the power delivered to source V_C and the power factor at the ac source terminals.
(c)   Sketch to the same scale the time variations of voltages v, v_L, and v_{AK}.

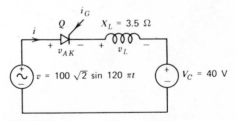

Fig. P3.3

3.7   In the circuit of Fig. P3.4, a pulse of gate current is applied to thyristor Q each cycle at $\alpha = \pi/2$.

(a)   Sketch to scale the time variation of the output voltage v_O.
(b)   Determine the average value of the output current I_O.

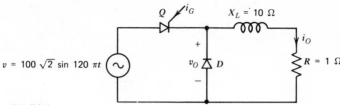

Fig. P3.4

3.8 For a certain thyristor-heat sink combination the thermal impedances are $\theta_{JC}=0.12$, $\theta_{CS}=0.07$, and $\theta_{SA}=0.25°C/W$ and the power loss in the thyristor is 250 W. Calculate the maximum ambient temperature permissible if the junction temperature is not to exceed 125°C. Is the heat sink employed satisfactory?

3.9 In the circuits of Fig. P3.5a to c, $L=100$ µH, $C=25$ µF. Sketch to scale the time variations of i, v_L, and v_C if the thyristor is turned on with the initial capacitor voltages as follows:

(a) In a, $v_C(0)=0$
(b) In b, $v_C(0)=-50$ V
(c) In c, $v_C(0)=-50$ V
(d) In c $v_C(0)=50$ V

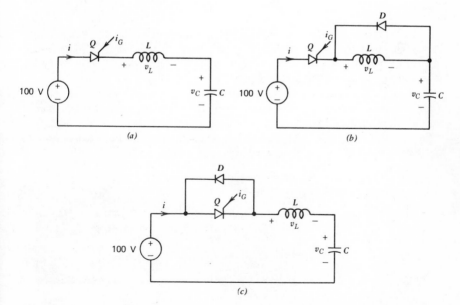

Fig. P3.5

3.10 In the circuits of Fig. P3.6a to d, $L = 100$ μH, $C = 25$ μF. Sketch to scale the time variations of i, i_1, v_L, and v_C if the thyristor is turned on with the initial capacitor voltage $v_C(0) = -100$ V in each case.

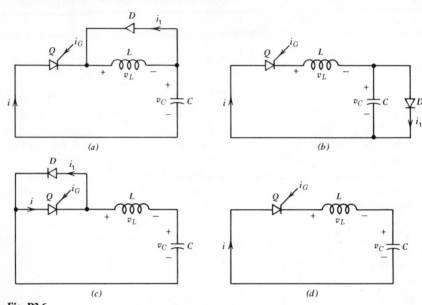

Fig. P3.6

3.11 For thyristors Q_1 and Q_2 in Figs. P3.7a and b dv/dt may not exceed 200 V/μs. The purpose of the circuits is to produce an alternating current in the capacitor C (i.e., it is a basic series inverter circuit) by turning on Q_1 at $t = 0$ and Q_2 at $t = 200$ μs. The initial capacitor voltage $V_{CO} = 0$.

(a) Analyze the first cycle of operation of circuit a and determine whether it will function satisfactorily.

(b) Analyze circuit b, calculate the time available for thyristor Q_1 to turn off, and show what will happen if t_{off} for Q_1 is greater than this time.

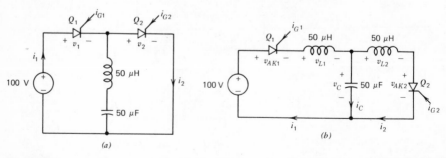

Fig. P3.7

3.12   In the circuits of Fig. P3.8*a* and *b*, the turn-off time of the thyristor is 20 μs and $V = 100$ V rms. State whether the circuits will rectify satisfactorily for the following sets of conditions:

(a)   In circuit *a*, $\omega = 120\pi$, $R = 2\ \Omega$, $\alpha = 0$
(b)   In circuit *a*, $\omega = 120\pi$, $R = 2\ \Omega$, $\alpha = \pi/2$
(c)   In circuit *b*, $\omega = 120\pi$, $R = 0$, $X_L = 10\ \Omega$, $\alpha = 0$
(d)   In circuit *b*, $\omega = 120\pi$, $R = 0$, $X_L = 10\ \Omega$, $\alpha = \pi/2$
(e)   In circuit *a*, $\omega = 120\pi, \times 10^2$, $R = 2\ \Omega$, $\alpha = 0$
(f)   In circuit *b*, $\omega = 120\pi, \times 10^2$, $R = 0.1\ \Omega$, $X_L = 10\ \Omega$, $\alpha = \pi/6$

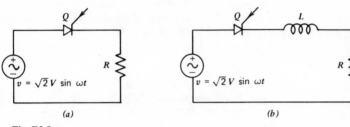

Fig. P3.8

3.13   In the circuit of Fig. P3.9, $L = 30\ \mu$H, $C = 120\ \mu$F. If thyristor Q is turned on at $t = 0$, sketch to scale the subsequent time variations of i, v_L, v_C, and i_D. Capacitor C is initially charged to $V_{CO} = -75$ V.

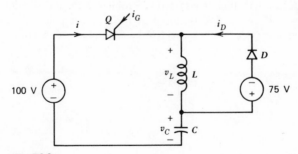

Fig. P3.9

3.14   In the circuit of Fig. P3.10, $L = 100\ \mu$H, $C = 1000\ \mu$F, and the initial capacitor voltage is $V_{CO} = -1000$ V. The purpose of the circuit is to return the energy stored in the capacitor to the source. Sketch to scale v_C, v_L, i_D, and v_{AK} clearly indicating the peak values of each variable and the times at which they occur.

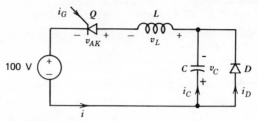

Fig. P3.10

3.15   The thyristor shown in Fig. P3.11 is type 261 M specified in Fig. 3.24a. The initial capacitor voltage is zero, and $C = 300$ μF. Calculate the minimum value of inductance L required to satisfy the di/dt rating of the thyristor when SW is closed at $t = 0$ and the thyristor is turned on at $t = 100$ μs.

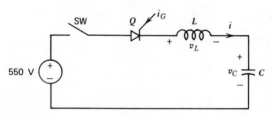

Fig. P3.11

3.16   The thyristor shown in Fig. P3.11 is type 261F specified in Fig. 3.24a. $C = 300$ μF, L has the value determined in Problem 3.15, and the initial capacitor voltage $V_{CO} = 0$. If switch SW is closed at $t = 0$, and a pulse of gate current is applied to thyristor Q at $t = 100$ μs, sketch the resulting time variations of i, v_L and v_C.

FOUR

AC VOLTAGE CONTROLLERS

AC voltage controllers, as the name implies, are employed to vary the rms value of the alternating voltage applied to a load circuit by introducing thyristors between it and a constant-voltage ac source. There are two methods of control:

1. On-off control
2. Phase control

In on-off control, the thyristors are employed as switches to connect the load circuit to the source for a few cycles of the source voltage and then to disconnect it for a comparable period. The thyristor thus acts as a high-speed contactor. In phase control, as has already been seen in Chapter 3, the thyristors are employed as switches to connect the load circuit to the source for a chosen portion of each cycle of the source voltage. The power circuit configurations for on-off control differ in no way from those for phase control. Moreover, the analysis of the performance of on-off controllers presents no difficulty that is not also met in that of phase-controlled systems. Attention is therefore concentrated on phase control in this chapter.

Applications of ac voltage controllers include the following:

1. Industrial heating.
2. Induction heating of metals.
3. Lighting controls.
4. Primary transformer control for electrochemical processes.
5. Transformer tap changing.
6. Speed control of induction-motor driven pumps and fans.

155

They may be employed in closed-loop control systems, where they function as high-power operational amplifiers in which the angle of retard α at which the thyristors begin to conduct is varied in response to an error signal. If it is essential that the current waveform of a controller shall have the least possible harmonic content, then a sinusoidal voltage controller may be used.

Since all of the systems listed in the preceding paragraph are ac excited, line commutation is possible. Consequently ac voltage controllers are among the least complicated of power conditioning systems to build. This simplicity does not extend to their analysis and design, however, since with other than purely resistive load circuits the extinction angle is not easily determined.

At this point it is advisable to define "angle of retard" to distinguish it from the "delay angle" defined in Section 3.4.4. It is defined as the interval in electrical angular measure by which the firing pulse is delayed by phase control in relation to natural operation that would occur with no controller circuit elements in the circuit and a purely resistive load.

4.1 TYPES OF AC VOLTAGE CONTROLLERS

The power circuits and output voltage or current waveforms of a full-wave and a half-wave voltage controller supplying a resistive load circuit are shown in Fig. 4.1a and b, respectively. The dc component introduced into the circuit by the asymmetrical waveform of the half-wave controller has the disadvantage already discussed in connection with half-wave rectifiers, in that the source must be "ideal." The current waveform of the full-wave controller has alternating symmetry and therefore has no direct component. For this reason the single-phase full-wave controller is a practical circuit for any kind of source or load network.

Circuits of three-phase full-wave and half-wave controllers for Δ-connected load circuits or Y-connected circuits in which the neutral point is inaccessible are shown in Fig. 4.2b and a. Either controller shown there could of course be employed with either load circuit. The half-wave controller in Fig. 4.2b economizes in the cost of devices and does not give rise to dc components in any part of the system. However, it introduces more harmonics into the line current than does the full-wave controller, and this is a serious disadvantage.

For Δ-connected load circuits in which each end of each phase is accessible, the arrangement shown in Fig. 4.3 may be employed, and has the advantage of reducing the current rating of the devices. For Y-connected load circuits in which the neutral point is accessible and can be opened, the controller circuit shown in Fig. 4.4 reduces the number of

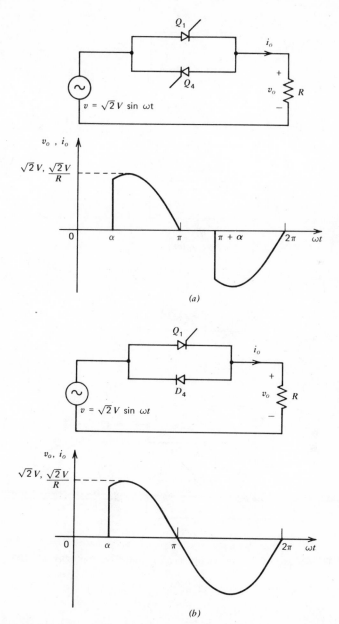

Fig. 4.1 Single-phase AC voltage controllers. (*a*) Full-wave controller. (*b*) Half-wave controller.

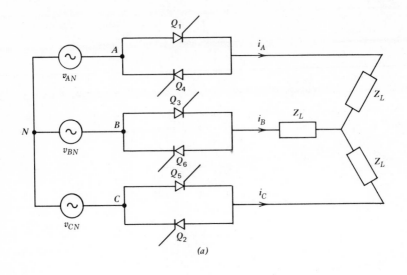

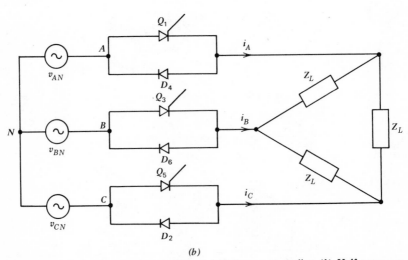

Fig. 4.2 Three-phase AC voltage controllers. (a) Full-wave controller. (b) Half-wave controller.

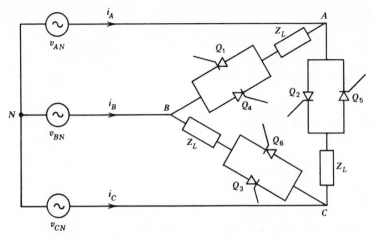

Fig. 4.3 Delta-connected controller.

thyristors required to three and considerably simplifies the control circuitry.

AC voltage controllers may of course be employed in the so-called "two-phase" or quarter-phase system and in any symmetrical n-phase system, where their operation can be analyzed in the same way as in a three-phase system.

It is possible also to economize in devices by constructing three-phase controllers similar to those in Fig. 4.2, but omitting one pair of devices and connecting one source terminal directly to one load-circuit terminal. When operated with phase control these circuits would be highly asymmetrical. They are, however, suitable for on-off control.

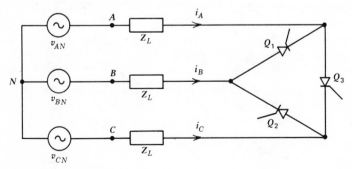

Fig. 4.4 Neutral-point controller.

   The various controller configurations that have been described differ
from one another in transfer characteristic relating load current to angle of
retard α as well as in the harmonic content introduced into the load and
line currents. The selection of a particular configuration depends on the
nature of the load circuit to be supplied and on the range of control
required. In the remainder of this chapter, the view is taken that
asymmetrical and half-wave configurations are suitable only for on-off
control, and therefore such controllers are not analyzed. The performance
of tap-changing transformers and sinusoidal voltage controllers are analyzed, and these may of course be employed in single-phase or three-phase
configurations. The general technique of analysis is common to all systems,
and much of it has already been introduced in the preceding chapters.

4.2 SINGLE-PHASE FULL-WAVE CONTROLLER

Figure 4.5a shows the circuit of a controller supplying a load network
possessing resistance and inductance. Only one of the two thyristors can be
conducting at any instant. If an interval is considered for which Q_1 is
conducting, then from equation 3.18

$$i_{A1} = \frac{\sqrt{2}\,V}{Z}\left[\sin(\omega t - \phi) - \sin(\alpha - \phi)\epsilon^{(R/L)(\alpha/\omega - t)}\right] \quad A \qquad (4.1)$$

where

$$Z = \left[R^2 + (\omega L)^2\right]^{1/2} \quad \Omega \qquad (4.2)$$

$$\phi = \tan^{-1}\frac{\omega L}{R} \qquad (4.3)$$

Each thyristor may now be visualized as acting during one half cycle like
the thyristor in the single-phase controlled rectifier described in Section
3.3.1, with the important constraint that the conduction angle γ cannot
exceed 180° if the gating control circuit is properly designed. On the basis
of the waveforms shown in Fig. 3.5, it will therefore be seen that the
waveforms for the circuit of Fig. 4.5a are as shown in Fig. 4.5b. From
these it may clearly be seen that as α is reduced until $\gamma = 180°$, the
waveforms of i_o and v_o approach the pure sinusoidal form for which $\alpha = \phi$.
This may be confirmed from the relationships between α, γ, ϕ and the
extinction angle β in equations 3.19 and 3.20, which state

$$\sin(\beta - \phi) = \sin(\alpha - \phi)\epsilon^{(R/L)[(\alpha - \beta)/\omega]} \qquad (4.4)$$

$$\gamma = \beta - \alpha \qquad (4.5)$$

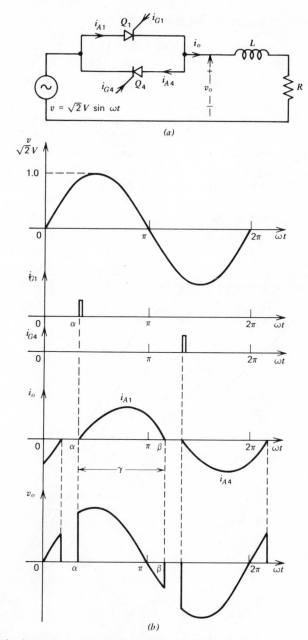

Fig. 4.5 Single-phase full-wave controller with RL load circuit.

If $\alpha = \phi$, then from equations 4.4 and 4.5

$$\sin(\beta - \phi) = \sin(\beta - \alpha) = 0 \qquad (4.6)$$

and

$$\beta - \alpha = \gamma = 180° \qquad (4.7)$$

Thus the curves in Fig. 3.6 showing the relationships between α and γ for various values of ϕ may be employed for the circuit of Fig. 4.5a, provided that it is borne in mind that γ cannot exceed 180°, which is the same thing as saying that α may not be less than ϕ. These modified curves are shown in Fig. 4.6. If indeed the gating signals were current pulses as indicated in Fig. 4.5, and α were made less than ϕ, only one thyristor would conduct, since by the time the first thyristor had turned off, the

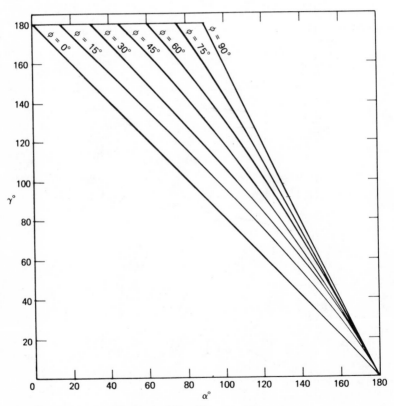

Fig. 4.6 γ versus α for the circuit of Fig. 4.5a.

gating signal of the second would already have been applied and removed. This makes desirable a prolongation of the gating signals in the manner further discussed in Section 4.2.2. When this is done, the effect is that for the range $0 < \alpha < \phi$ the output current remains constant, while both thyristors continue to conduct in turn commencing at the angles $\omega t = \phi$ and $\omega t = \pi + \phi$.

The average value of the thyristor currents may be determined as in Section 3.3.1, so that from equation 4.1 the normalized value of the average thyristor current is

$$I_N = \frac{1}{2\pi} \int_{\alpha}^{\alpha + \gamma} \left[\sin(\omega t - \phi) - \sin(\alpha - \phi)\epsilon^{(R/L)(\alpha/\omega - t)} \right] d(\omega t) \qquad (4.8)$$

and for any value of α, the value of I_N may be obtained from the curves of Fig. 3.7, always bearing in mind that α cannot be less than ϕ. The curves of Fig. 3.7 terminated at the point $\alpha = \phi$ are shown in Fig. 4.7.

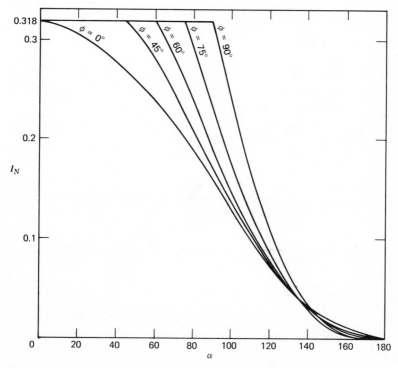

Fig. 4.7 I_N versus α for the circuit of Fig. 4.5*a*.

The normalized rms value of the thyristor current is from equation 4.1

$$I_{RN} = \left[\frac{1}{2\pi} \int_{\alpha}^{\alpha+\gamma} \left[\sin(\omega t - \phi) - \sin(\alpha - \phi)\epsilon^{(R/L)(\alpha/\omega - t)} \right]^2 d(\omega t) \right]^{1/2} \quad (4.9)$$

and for any permissible value of α, I_{RN} may be obtained from the curves of Fig. 3.8, which are shown in modified form in Fig. 4.8.

The normalized value of the rms output current is then

$$I_{RoN} = \left[I_{RN}{}^2 + I_{RN}{}^2 \right]^{1/2} = \sqrt{2}\, I_{RN} \quad (4.10)$$

so that the rms output current is

$$I_{Ro} = \frac{\sqrt{2}\,V}{Z} I_{RoN} \quad \text{A} \quad (4.11)$$

As may be seen from the wave form of v_o in Fig. 4.5b, the normalized value of the rms output voltage is

$$V_{RoN} = \left[\frac{1}{\pi} \int_{\alpha}^{\alpha+\gamma} \sin^2 \omega t\, d(\omega t) \right]^{1/2} = \left\{ \frac{1}{2\pi} \left[\gamma + \frac{1}{2}\sin 2\alpha - \frac{1}{2}\sin 2(\alpha+\gamma) \right] \right\}^{1/2}$$

$$(4.12)$$

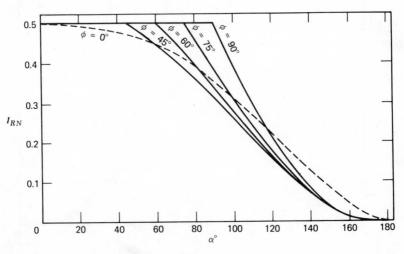

Fig. 4.8 I_{RN} versus α for the circuit of Fig. 4.5a.

so that the rms output voltage is

$$V_{Ro} = \sqrt{2} \, V V_{RoN} \quad \text{V} \tag{4.13}$$

and for any pair of values of α and γ from Fig. 4.6, this voltage may be calculated.

4.2.1 Harmonic Analysis The load or line-current waveform may be described by the Fourier series

$$i_o = \sum_{n=1}^{\infty} a_n \sin n\omega t + \sum_{n=1}^{\infty} b_n \cos n\omega t \quad \text{A} \tag{4.14}$$

where

$$a_n = \frac{1}{\pi} \int_0^{2\pi} i_o \sin n\omega t \, d(\omega t) = \frac{2}{\pi} \int_0^{\pi} i_o \sin n\omega t \, d(\omega t) \quad \text{A} \tag{4.15}$$

$$b_n = \frac{1}{\pi} \int_0^{2\pi} i_o \cos n\omega t \, d(\omega t) = \frac{2}{\pi} \int_0^{\pi} i_o \cos n\omega t \, d(\omega t) \quad \text{A} \tag{4.16}$$

and in Fig. 4.5a

$$i_o = i_{A1} \quad \text{A} \tag{4.17}$$

The rms value of the nth harmonic is then given by

$$I_{nR} = \frac{1}{\sqrt{2}} [a_n^2 + b_n^2]^{1/2} \quad \text{A} \tag{4.18}$$

From equations 4.14 to 4.17 the harmonic content of the line current for any value of α and ϕ may then be determined by obtaining γ from Fig. 4.6. If, for simplicity, a purely resistive load circuit is considered, so that $\phi = 0$, then from equation 4.1

$$i_o = \frac{\sqrt{2} \, V}{R} \sin \omega t \quad \text{A:} \qquad \alpha < \omega t < \pi \tag{4.19}$$

and substitution in equations 4.15 and 4.16 yields

$$a_n = \frac{2}{\pi} \int_\alpha^\pi \frac{\sqrt{2} \, V}{R} \sin \omega t \sin n\omega t \, d(\omega t) \quad \text{A} \tag{4.20}$$

$$b_n = \frac{2}{\pi} \int_\alpha^\pi \frac{\sqrt{2} \, V}{R} \sin \omega t \cos n\omega t \, d(\omega t) \quad \text{A} \tag{4.21}$$

By integration of equations 4.20 and 4.21,

$$a_n = \frac{\sqrt{2}\,V}{\pi R}\left[\frac{\sin(n+1)\alpha}{n+1} - \frac{\sin(n-1)\alpha}{n-1}\right] \quad \text{A:} \quad n \neq 1 \qquad (4.22)$$

$$b_n = \frac{\sqrt{2}\,V}{\pi R}\left[\frac{\cos(n+1)\alpha - \cos(n+1)\pi}{n+1} - \frac{\cos(n-1)\alpha - \cos(n-1)\pi}{n-1}\right] \quad \text{A:}$$

$$n \neq 1 \qquad (4.23)$$

For $n = 1$, equations 4.22 and 4.23 yield incommensurables; however, substitution of $n = 1$ in equations 4.20 and 4.21 and integration yields

$$a_1 = \frac{\sqrt{2}\,V}{\pi R}\left[\pi - \alpha + \frac{1}{2}\sin 2\alpha\right] \quad \text{A} \qquad (4.24)$$

$$b_1 = -\frac{\sqrt{2}\,V}{\pi R}\sin^2 \alpha \quad \text{A} \qquad (4.25)$$

Equations 4.22 to 4.25 have been employed to plot the curves of normalized harmonic rms values shown in Fig. 4.9, where the ordinate $H_{n\alpha}$ is defined as

$$H_{n\alpha} = \frac{\text{rms value of } n\text{th harmonic at angle } \alpha}{\text{rms value of line current at } \alpha = 0} \qquad (4.26)$$

Similar sets of curves may be produced for any value of $\phi \neq 0$.

The rms value of the nth harmonic of output voltage is given by

$$V_{nR} = I_{nR}\left[R^2 + (n\omega L)^2\right]^{1/2} \quad \text{V} \qquad (4.27)$$

In general current harmonic amplitudes are reduced when the load circuit inductance is increased.

Example 4.1 A single-phase full-wave controller is used to control the power from a 2300-V ac source into a resistive load that can vary from 1.15 to 2.30 Ω. The maximum output power desired is 2300 kW. Calculate the maximum value of thyristor voltage, the rms thyristor current I_R, and the average thyristor current I_Q, as well as the maximum rms value of third-harmonic current in the line for any operating condition of this system.

Solution

(a) For $R = 2.30\,\Omega$. When maximum power is to be delivered to the maximum load-circuit resistance, then $\alpha = 0$, and the output power is

$$P_o = R_o I_{Ro}^2$$

that is,

$$2300 \times 10^3 = 2.30 I_{Ro}^2$$

so that

$$I_{Ro} = 1000 \text{ A}$$

From equation 4.10 it may be seen that

$$I_{QR} = \frac{I_{Ro}}{\sqrt{2}} = 707 \quad \text{A}$$

For a resistive load circuit, $\beta = \pi$ and

$$i_o = \frac{\sqrt{2}\,V}{R} \sin \omega t \quad \text{A:} \qquad \alpha < \omega t < \pi$$

The average thyristor current is

$$I_Q = \frac{1}{2\pi} \int_\alpha^\pi \frac{\sqrt{2}\,V}{R} \sin \omega t\, d(\omega t) = \frac{\sqrt{2}\,V}{2\pi R} [\cos \alpha + 1] \tag{1}$$

The rms thyristor current is

$$I_{QR} = \left[\frac{1}{2\pi} \int_\alpha^\pi \left(\frac{\sqrt{2}\,V}{R} \sin \omega t \right)^2 d(\omega t) \right]^{1/2} = \frac{V}{\sqrt{2}\,R} \left[1 - \frac{\alpha}{\pi} + \frac{\sin 2\alpha}{2\pi} \right]^{1/2} \tag{2}$$

For

$$\alpha = 0, \qquad \frac{I_{QR}}{I_Q} = \frac{\pi}{2} \tag{3}$$

thus the maximum value of I_Q is

$$I_Q = \frac{2}{\pi} I_{QR} = \frac{2}{\pi} \times 707 = 450 \quad \text{A}$$

(b)   For $R = 1.15\,\Omega$.   When maximum power is delivered to the minimum load-circuit resistance, then $\alpha > 0$, and

$$2300 \times 10^3 = 1.15 I_{Ro}^{\ 2}$$

so that

$$I_{Ro} = 1414 \quad \text{A}$$

and the rms thyristor current is

$$I_{QR} = \frac{1414}{\sqrt{2}} = 1000 \quad \text{A}$$

The base value of the thyristor current is

$$I_B = \frac{\sqrt{2}\,V}{R} = \frac{\sqrt{2} \times 2300}{1.15} = 2830 \quad \text{A}$$

Thus

$$I_{RN} = \frac{1000}{2830} = 0.354$$

From Fig. 4.8, $\alpha = 90°$, and entering Fig. 4.7 with this value of α gives

$$I_N = 0.160$$

Thus

$$I_Q = I_N \times I_B = 0.160 \times 2830 = 450 \quad \text{A}$$

and this result may be confirmed by substitution in equation 1. Thus

$$\text{maximum value of } I_{QR} = 1000 \quad \text{A}$$

$$\text{maximum value of } I_Q = 450 \quad \text{A}$$

The peak forward and reverse voltages applied to the thyristors are $2300\sqrt{2} = 3250\,\text{V}$.

(c)   Third-Harmonic Current.   Inspection of the curves in Fig. 4.9 shows that the maximum value of third harmonic occurs when $\alpha = \pi/2$.

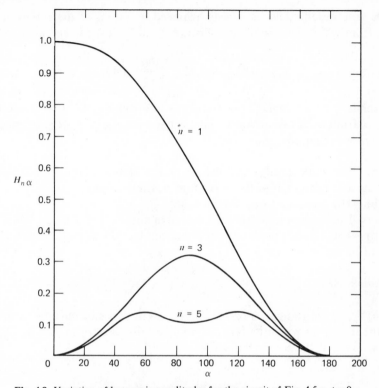

Fig. 4.9 Variation of harmonic amplitudes for the circuit of Fig. 4.5a: $\phi=0$.

Since this is the angle of retard employed when maximum power is delivered and $R = 1.15\,\Omega$, then from equation 4.22

$$a_3 = \frac{2300\sqrt{2}}{1.15\pi} \left[\frac{\sin 2\pi}{4} - \frac{\sin \pi}{2} \right] = 0 \quad \text{A}$$

and from equation 4.23

$$b_3 = \frac{2300\sqrt{2}}{1.15\pi} \left[\frac{\cos 2\pi - \cos 4\pi}{4} - \frac{\cos \pi - \cos 2\pi}{2} \right] = 900 \quad \text{A}$$

Thus the rms value of the third harmonic is from equation 4.18

$$I_{3R} = \frac{1}{\sqrt{2}} \left[a_3{}^2 + b_3{}^2 \right]^{1/2} = \frac{900}{\sqrt{2}} = 637 \quad \text{A}$$

This same result could have been achieved by means of the curves of Fig. 4.9, from which for $\alpha = \pi|2$ and $n = 3$, $H_{n\alpha} = 0.32$. Thus from equation 4.26

$$I_{3R} = 0.32 \frac{V}{R} = 0.32 \times \frac{2300}{1.15} = 640 \quad A$$

Example 4.2 A single-phase full-wave controller is used to control the power from a 2300-V ac source into a load circuit of 2.30-Ω resistance and 2.30-Ω inductive reactance. Determine:

(a) The control range (i.e., the range of variation of α necessary to vary the current from zero to the maximum possible value).
(b) The maximum rms line current.
(c) The maximum power and power factor.
(d) The rms thyristor current, the conduction angle and the power factor at the source for $\alpha = \pi/2$.

Solution

(a) Control Range.   For zero power, each thyristor must be turned on at the instant at which its forward voltage falls to zero, so that no current flows. Thus if the source voltage is

$$v = 2300\sqrt{2} \, \sin \omega t$$

then for

$$P = 0 \qquad \alpha = \alpha_{max} = \pi$$

As has been explained in Section 4.2, the lowest value to which α may be reduced is given by the angle of the load impedance ϕ. Thus for $P = P_{max}$

$$\alpha = \alpha_{min} = \phi = \tan^{-1} \frac{\omega L}{R} = \frac{\pi}{4}$$

Thus the control range is $\pi/4 \leqslant \alpha \leqslant \pi$.

(b) Maximum Current.   When $\alpha = \alpha_{min}$, the load current and load voltage waveforms are both pure sinusoids with a phase displacement of ϕ, and the source and load interact as if the regulator were not present. Thus

$$I_{Ro} = \frac{V}{\left[R^2 + (\omega L)^2 \right]^{1/2}} = \frac{2300}{[2.3^2 + 2.3^2]^{1/2}} = 707 \quad A$$

(c) Maximum Power. This occurs at maximum output current, and

$$P_{max} = RI_{Ro}^2 = 2.30 \times 707^2 = 1150 \times 10^3 \quad W$$

$$\text{power factor} = \frac{\text{active power}}{\text{apparent power}} = \frac{1150 \times 10^3}{2300 \times 707} = 0.707$$

For this particular condition of operation, the power factor is $\cos\phi$. This is not true for the nonsinusoidal condition of operation when $\alpha > \alpha_{min}$.

(d) For $\alpha = \pi/2$. From Fig. 4.8, for $\alpha = \pi/2$, $\phi = \pi/4$,

$$I_{RN} = 0.31$$

The base value of the thyristor current is

$$I_B = \frac{\sqrt{2} \; V}{Z} = \frac{\sqrt{2} \times 2300}{[2.3^2 + 2.3^2]^{1/2}} = 1000 \quad A$$

Thus

$$I_{QR} = I_{RN} \times I_B = 0.31 \times 1000 = 310 \quad A$$

From Fig. 4.6, for $\alpha = \pi/2$, $\phi = \pi/4$, the conduction angle is

$$\gamma = 130°$$

The rms output current is

$$I_{Ro} = \sqrt{2} \; I_{QR} = \sqrt{2} \times 310 = 425 \quad A$$

The output power is

$$P_o = RI_{Ro}^2 = 2.3 \times 425^2 = 415 \times 10^3 \quad W$$

$$\text{power factor} = \frac{P_o}{VI_{Ro}} = \frac{415 \times 10^3}{2300 \times 425} = 0.425$$

4.2.2 Gating Signals The gating signals for the two thyristors in the circuit of Fig. 4.5a must be isolated from one another, since if they are not, the two cathodes will be connected together and both thyristors will be shorted out of circuit.

When $L = 0$ and the load circuit is purely resistive, then $\alpha_{min} = \phi = 0$, and each thyristor ceases to conduct at the end of a half cycle of the supply voltage. Under these circumstances, pulse gating may be employed as illustrated in Fig. 4.10. The gate current required to turn on a thyristor is

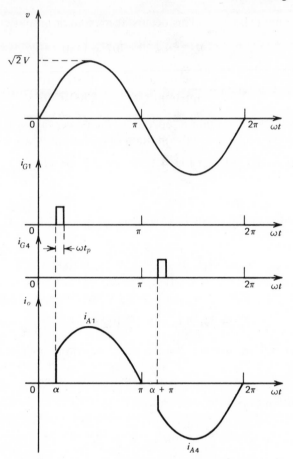

Fig. 4.10 Pulse control for the circuit of Fig. 4.5a: $\phi = 0$.

typically of the order of 100 to 400 mA, and the pulse duration t_p must be at least $5\mu s$.

Pulse gating is not suitable for RL load circuits. The reason for this is shown in Fig. 4.11, where at $\omega t = \alpha + \pi$ thyristor Q_1 is still conducting; that is, the effect of the load-circuit inductance is such that at this instant $v_o = v$, and the voltage across both thyristors is zero. By the time Q_1 has ceased conducting, the pulse of i_{G4} has ceased, and consequently Q_4 does not turn on. Thus the controller operates with an asymmetrical waveform due to conduction of Q_1 only, and this produces an undesirable dc component of load and source current. This difficulty could be removed by using "continuous gating," that is, by making the gating pulse last for a

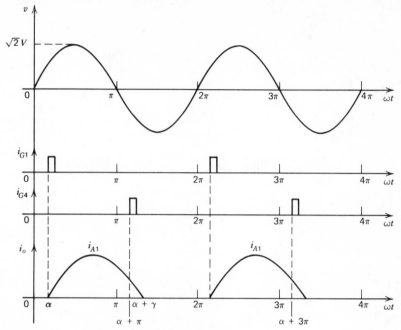

Fig. 4.11 Unsuitability of pulse control when $\phi \neq 0$.

period of $(\pi - \alpha)/\omega$ s, so that as soon as i_{A1} fell to zero, Q_4 would then turn on. However, due to the need to isolate the gating signals of the two thyristors, it is desirable that these signals should be supplied to the two thyristors via isolating transformers. Such transformers are small when only a short pulse must be transmitted, but become large when a long pulse is required, so that continuous gating is undesirable on these grounds.

The technique which ensures turn-on of Q_4 and at the same time requires only a small isolating transformer is "high-frequency carrier gating," in which a series of short pulses lasting throughout the intervals $\alpha < \omega t < \pi$ for thyristor Q_1 and $\alpha + \pi < \omega t < 2\pi$ for thyristor Q_4 are applied. These pulses normally have a frequency of the order of 30 kHz. The three types of gating signal are illustrated in Fig. 4.12, being (a) pulse gating; (b) continuous gating, and (c) high-frequency carrier gating.

Figure 4.13 shows the effect of high-frequency carrier gating when the controller is supplying an RL load circuit and $0 \leqslant \alpha \leqslant \phi$. If the controller is switched on at $\omega t = 0$, Q_1 will turn on when $\omega t = \alpha$; Q_4 will turn on as soon as i_{A1} falls to zero. For a few cycles after switching on $i_o = i_{A1} - i_{A4}$ will have an asymmetrical waveform, but this transient condition is damped

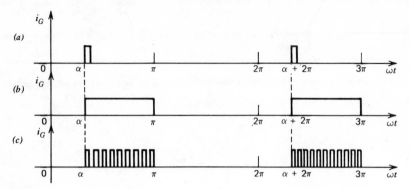

Fig. 4.12 Types of gating signal.

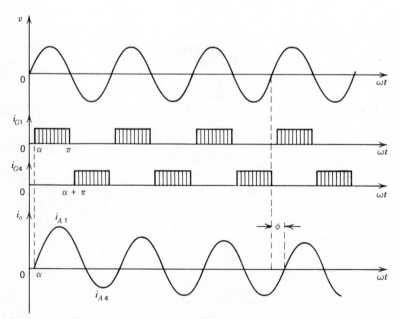

Fig. 4.13 High-frequency carrier gating for $\phi \neq 0$; $\alpha \leqslant \phi$.

174

out by R, and eventually a symmetrical and sinusoidal waveform of i_o results. In the range $\phi < \alpha < \pi$, the current i_o is discontinuous as illustrated in Fig. 4.5b, and there is no asymmetrical switching transient comparable to that in Fig. 4.13.

4.3 THREE-PHASE, FULL-WAVE, WYE-CONNECTED CONTROLLER

The circuit of Fig. 4.2a is repeated in more detail in Fig. 4.14. The controller that is shown is a three-wire system, so that if load current is to flow, then thyristors in at least two lines must be conducting. Let the line-to-line source voltages be

$$v_{AB} = \sqrt{2}\ V \sin \omega t \quad \text{V}$$

$$v_{BC} = \sqrt{2}\ V \sin (\omega t - 2\pi|3) \quad \text{V} \qquad (4.28)$$

$$v_{CA} = \sqrt{2}\ V \sin (\omega t - 4\pi|3) \quad \text{V}$$

Then each line-to-line source voltage drives current through two branches of the load in series.

It is convenient to refer the angles of retard of all thyristors to the same datum, and for this the value of $\alpha = 0$ for thyristor Q_1 is employed. In establishing this datum it is as well to recall the definition of the angle of

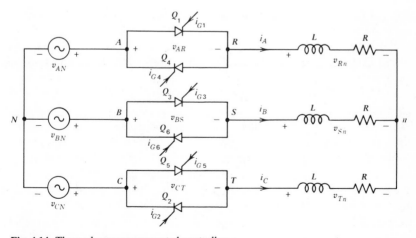

Fig. 4.14 Three-phase wye-connected controller.

retard. It is the interval in electrical angular measure by which the firing pulse is delayed by phase control in relation to natural operation that would occur with no controller circuit elements in circuit and a purely resistive load.

Figure 4.15 illustrates the condition of natural operation described in this definition. The currents are in phase with the line-to-neutral voltages, and it might appear that it would be convenient to adopt the zero value of v_{AN} as specifying the origin of the ωt scale. However, the discussion of the operation of the controller under any conditions other than those illustrated in Fig. 4.15 must be conducted in terms of the line-to-line voltages, and for that reason the origin specified by equations 4.28 is employed. Since

$$\overline{\mathbf{V}}_{AN} = \frac{1}{\sqrt{3}} \overline{\mathbf{V}}_{AB} \underline{/-30°} \quad \text{V} \tag{4.29}$$

then

$$i_A = \frac{\sqrt{2} \, V}{\sqrt{3} \, R} \sin(\omega t - \pi|6) \quad \text{A} \tag{4.30}$$

Zero angle of retard for thyristor Q_1 is given by the point in Fig. 4.15 at which current i_A commences its positive half cycle. So that at $\omega t = \pi|6$, $\alpha = 0$, and for any other value of α

$$\alpha = \omega t - \frac{\pi}{6} \tag{4.31}$$

The gating signals of the thyristors in the three branches must have the same sequence and phase displacements as do the source voltages. Thus if the angle of retard of thyristor Q_1 is α, then that of Q_3 must be $\alpha + 2\pi|3$, and that of Q_5 must be $\alpha + 4\pi|3$. The delay angle of Q_4 in line A must be $\alpha + \pi$, that of Q_6 must be $\alpha + 2\pi|3 + \pi$, and that of Q_2 must be $\alpha + 4\pi|3 + \pi$. The resulting sequence of gating signals is shown in Fig. 4.16, where a Q_1 angle of retard $\alpha = \pi|2$ has been shown.

While the necessary sequence and phase displacements of the gating signals are known, however, it is not immediately obvious what the range of α or the duration of the gating signals must be. If the gating signals are to be so far retarded that zero current flows, but any reduction in α would result in load current, then the thyristors must be turned on just at the instant at which the forward voltage applied to them disappears.

In Fig. 4.16, at $\omega t = 0^+$, v_{CA} is positive. The thyristors through which positive v_{CA} tends to drive current are Q_5 and Q_4 in series. A gating signal

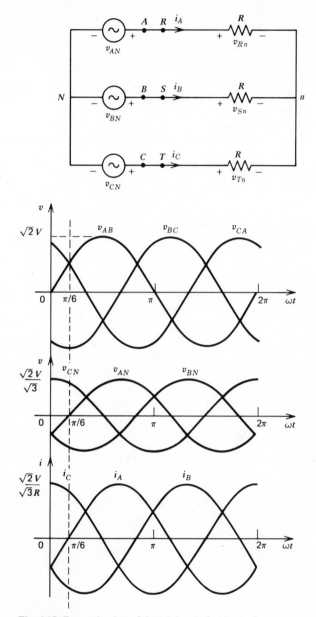

Fig. 4.15 Determination of datum for angle of retard α.

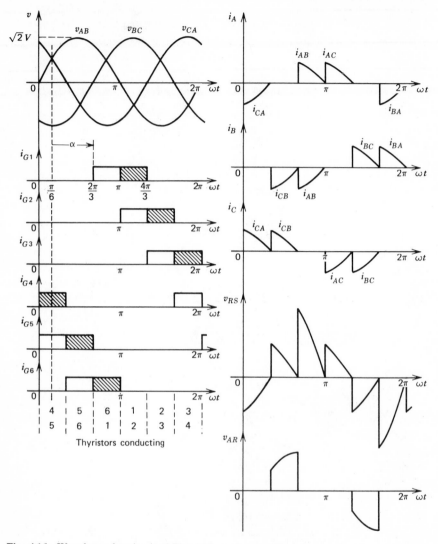

Fig. 4.16 Waveforms for circuit of Fig. 4.14: $\phi = 0$, $\alpha = \pi/2$.

must therefore be applied to, or already exist on, these thyristors at $\omega t = \pi/3$, where v_{CA} ceases to be positive. This fixes the end point of the pulse of i_{G4} and the starting point of that of i_{G5}. The starting points and end points of all the other gating signals for zero load current may be related to these two, and the resulting gating signals for the maximum angle of retard that would give zero output are shown in Fig. 4.16 by the shaded parts of the rectangular pulses of i_{G1} to i_{G6}. Any increase of output calls for an extension to the left of these zero-output gating signals. If the shaded parts of the pulses were not present, then over the range $\pi/2 \leqslant \alpha \leqslant 5\pi/6$ only one thyristor would be turned on, and no current would flow.

The shaded part of the pulse of i_{G1} shows that for a resistive load circuit the maximum angle of retard of thyristor Q_1 is

$$\alpha_{max} = \pi - \frac{\pi}{6} = \frac{5\pi}{6} \qquad (4.32)$$

so that the range of α required is $0 \leqslant \alpha \leqslant 5\pi|6$.

If the controller is operating with a purely resistive load circuit, as the angle of retard of i_{G1} is reduced from $\alpha = 5\pi|6$ the thyristors begin to conduct two at a time, and this mode of operation continues until α is reduced to $\pi|2$, the condition shown in Fig. 4.16. The waveforms of the line currents are also shown in Fig. 4.16, and these are made up of pulses of currents in two lines, as indicated by the subscripts on each waveform. The line-to-neutral voltages v_{Rn}, v_{Sn}, v_{Tn}, have the same waveforms as the line currents. The line-to-line load voltages may be determined from equations such as

$$v_{RS} = v_{Rn} - v_{Sn} = Ri_A - Ri_B \qquad (4.33)$$

and the waveform of voltage v_{RS} is shown in Fig. 4.16.

When α is reduced below the value $\pi|2$, three thyristors will conduct simultaneously for intervals of ωt whose length depends on α. Thus there will be parts of the cycle when three thyristors conduct, and this will be called Mode I operation, and parts where only two conduct, Mode II operation. The line-to-line source-voltage waveforms and those of the gating signals and line currents for $\alpha = \pi|6$ are shown in Fig. 4.17. If the load circuit is purely resistive, then during Mode I operation the system will function as if no thyristors were present. The line current waveforms will be identical with those of the line-to-neutral voltages, and the current amplitude and phase angle in each line may be determined as in Fig. 4.15.

During Mode II operation, one line-to-line source voltage produces current in two load-circuit branches in series. The line-to-line voltage that is effective in each interval of Mode II operation may be determined from the table of "Thyristors Conducting" in Fig. 4.17 and the circuit of Fig. 4.14. For example, during the interval $\pi|6 < \omega t < \pi|3$, thyristors Q_5 and Q_6 are conducting; therefore, voltage v_{BC} is effective. From Fig. 4.14 and equation 4.28, the current from B to C must then be

$$i_{BC} = i_B = -i_C = \frac{v_{BC}}{2R} = \frac{\sqrt{2}\,V}{2R} \sin(\omega t - 2\pi|3) \quad A \qquad (4.34)$$

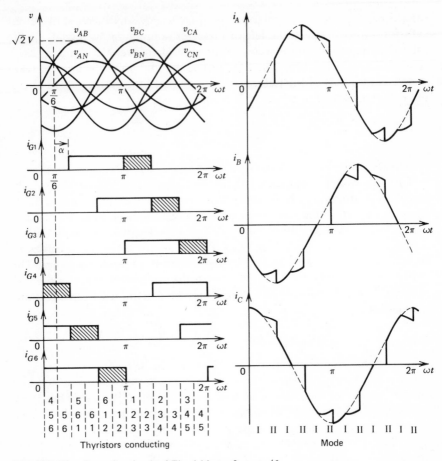

Fig. 4.17 Waveforms for circuit of Fig. 4.14: $\phi = 0$, $\alpha = \pi/6$.

Also

$$i_A = 0 \qquad (4.35)$$

The currents during the remaining parts of Mode II operation may be similarly determined and are shown in Fig. 4.17. The waveforms of the line-to-line voltages at the load may be determined from relationships such as that of equation 4.33. At $\alpha = 0$, Mode I operation continues throughout the whole cycle, and the current waveforms for this condition are indicated in broken line in Fig. 4.17.

Normalized values of average and rms thyristor currents may be calculated from the waveforms of Figs. 4.16 and 4.17, employing the methods explained in Section 3.3. Curves of I_N versus α and I_{RN} versus α for the thyristors are shown marked $\phi = 0$ in Figs. 4.18 and 4.19 where the base current employed is $\sqrt{2}\ V|\sqrt{3}\ R$. These curves may be employed in designing controllers for resistive load circuits.

When the load circuit possesses inductance, the analysis becomes very complicated, since conduction does not cease at the instant at which a line-to-line or line-to-neutral voltage becomes zero. Typical current waveforms are therefore shown in the oscillograms of Fig. 4.20, and experimentally determined curves of I_N versus α and I_{RN} versus α for various values of ϕ are shown in Figs. 4.18 and 4.19 where the base current employed is $\sqrt{2}\ V|\sqrt{3}\ Z$. The minimum value of α required for a load circuit that possesses inductance as well as resistance is equal to ϕ, and when α is less than this value, the system operates as if the controller were not present.

It is also necessary when choosing a thyristor for a particular application to know what is the maximum voltage that will be applied to it. While it is clear that in Mode I operation, when thyristors in all three lines are conducting, this voltage will always be zero, it is by no means obvious what voltage will be applied to the thyristors in the nonconducting line during Mode II operation. The nature of the load circuit does not affect this voltage, so that it may be determined for a resistive load circuit in Fig. 4.14 and from the corresponding waveforms of Fig. 4.16.

For the entire outer loop of Fig. 4.14

$$v_{CA} + v_{AR} + v_{Rn} - v_{Tn} - v_{CT} = 0 \quad \text{V} \tag{4.36}$$

While thyristor Q_4 is conducting,

$$v_{AR} = 0 \quad \text{V:} \qquad 0 < \omega t < \frac{\pi}{3} \tag{4.37}$$

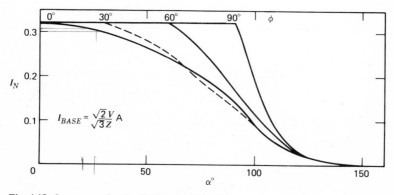

Fig. 4.18 I_N versus α for the thyristors of Fig. 4.14.

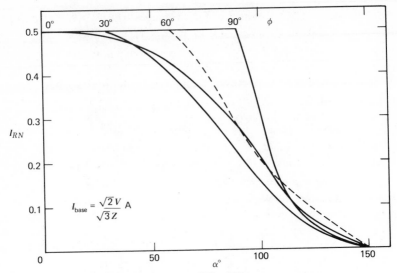

Fig. 4.19 I_{RN} versus α for the thyristors of Fig. 4.14.

At $\omega t = \pi/3$, thyristor Q_6 turns on, and Q_4 turns off. Thus

$$v_{AR} \neq 0 \quad \text{V:} \qquad \frac{\pi}{3} < \omega t < \frac{2\pi}{3} \tag{4.38}$$

During this second interval, while Q_5 and Q_6 are conducting, and while $i_A = 0$

$$v_{BS} = 0$$

$$v_{CT} = 0 \qquad \frac{\pi}{3} < \omega t < \frac{2\pi}{3} \tag{4.39}$$

$$v_{Rn} = 0$$

Thus for the lower mesh of Fig. 4.14

$$v_{BC} + v_{Tn} - v_{Sn} = 0 \tag{4.40}$$

and since

$$v_{Tn} = -v_{Sn} \quad \text{V} \tag{4.41}$$

it follows that

$$v_{Tn} = -\frac{v_{BC}}{2} \quad \text{V} \tag{4.42}$$

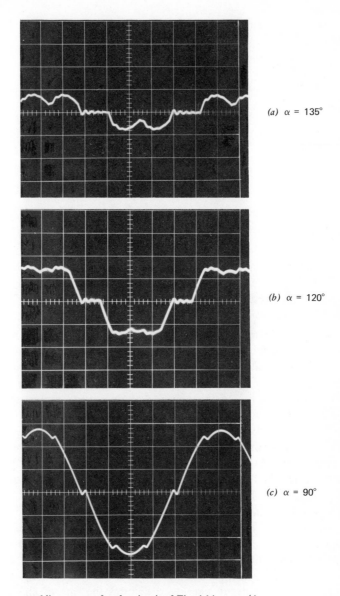

(a) α = 135°

(b) α = 120°

(c) α = 90°

Fig. 4.20 Oscillograms of line current for the circuit of Fig. 4.14: $\phi = \pi/4$.

183

Substitution from equations 4.39 and 4.42 in equation 4.36 then yields

$$v_{CA} + v_{AR} + \frac{v_{BC}}{2} = 0 \quad \text{V} \tag{4.43}$$

Therefore

$$v_{AR} = -\frac{v_{BC}}{2} - v_{CA}$$

$$= -\sqrt{2}\ V\left[\frac{1}{2}\sin\left(\omega t - \frac{2\pi}{3}\right) + \sin\left(\omega t - \frac{4\pi}{3}\right)\right]$$

$$= -\sqrt{2}\ V\frac{\sqrt{3}}{2}\cos\left(\omega t + \frac{\pi}{3}\right) \quad \text{V:} \qquad \frac{\pi}{3} < \omega t < \frac{2\pi}{3} \tag{4.44}$$

and this expression for v_{AR} applies also during the interval $4\pi/3 < \omega t < 5\pi/3$. The corresponding waveform of v_{AR} is shown in Fig. 4.16, where it is seen that the maximum value occurs at $\omega t = 2\pi/3$ and $\omega t = 5\pi/3$ and is $\pm\sqrt{2}\ V \times (\sqrt{3}/2)$.

As has been remarked in Section 4.2.1 in connection with the single-phase, full-wave controller, a purely resistive load results in the greatest harmonic content of the line currents, and the harmonics for this case may be obtained by Fourier analysis of waveforms such as are shown in Figs. 4.16 and 4.17. Due to the absence of analytical expressions for the line currents when the load circuit possesses inductance, harmonic analysis for such load circuits can only be carried out by numerical methods. The fact that current harmonics are much reduced by the presence of inductance may however be verified from the oscillograms of Fig. 4.20, which show waveforms which are considerable less "spiky" than those of current in Figs. 4.15 and 4.16. In the waveforms of Fig. 4.20a and b it will be observed that an oscillatory current appears in the line after each thyristor has turned off. This is due to the fact that the thyristor does not turn off at zero current, as is assumed in the ideal model, but at a small negative current. The energy stored in the load-circuit inductance at the instant of turn off is then dissipated in the RLC circuit formed by that inductance and the snubber circuit employed to limit the rate of change of thyristor voltage, as described in Section 3.4.5.

4.3.1 Induction Motor Drive   The three-phase full-wave voltage controller is often employed as a means of obtaining a variable speed drive with Design D three-phase squirrel-cage induction motors. Such motors may have speed-torque characteristics of the form illustrated in Fig. 4.21, where output torque T_O is shown in per unit of the rated shaft torque, and motor speed ω_m is shown in per unit of the synchronous speed of the motor ω_{syn}, defined as

$$\omega_{syn} = \frac{2}{p} \omega_s \quad \text{rad/s} \tag{4.45}$$

In equation 4.45, ω_s is the angular frequency of the stator supply, and p is the number of poles on the stator winding. An alternative to the speed scale is the scale of slip s, defined as

$$s = \frac{\omega_{syn} - \omega_m}{\omega_{syn}} \tag{4.46}$$

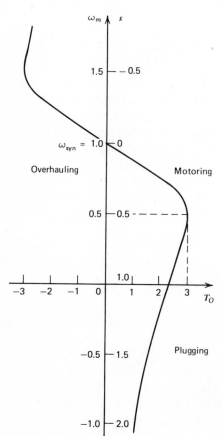

Fig. 4.21 Speed torque characteristic of a design D squirrel-cage induction motor.

Design D motors are also available in which maximum torque appears at a lower speed than that shown in Fig. 4.21—even at standstill. The Design D motor consequently has a high full-load operating slip in the range $0.1 < s < 0.15$. Rated speeds for such motors operating at 60 Hz are typically as shown in Table 4.1.

Table 4.1 Rated Speeds of Design D Motors (rpm)

No. of Poles	Synchronous Speed	Rated Speed
2	3600	3150
4	1800	1575
6	1200	1050
8	900	790
	etc.	

Not all three quadrants of the characteristic shown in Fig. 4.21 are necessarily employed in a particular drive system. In particular, if the load has no tendency to overhaul the motor, that is, to drive it in a positive direction at more than synchronous speed, then the second quadrant of Fig. 4.21 may be ignored. The part of the characteristic in the fourth quadrant represents a condition in which the motor has a direction of rotation opposite to that which would be produced by its own developed torque. This condition of operation, called "plugging," may be usefully employed. The first quadrant of Fig. 4.21 represents normal motoring operation.

When the terminal voltage applied to an induction motor is reduced, the torque developed by the motor is reduced at all speeds. The output torque is approximately proportional to the square of the applied voltage. A family of speed-torque curves for a Design D induction motor operating at a number of different terminal voltages is shown in Fig. 4.22. If the line marked T_L, representing a hypothetical mechanical load that requires rated motor output torque at rated motor speed, is imposed on the motor characteristics, then the possibility of speed control is immediately apparent. This motor-load system will run at a steady state under the conditions represented by an intersection of the load characteristic with any motor characteristic, provided only that the intersection takes place at a point such as p_1, where the slip is less than that at which maximum torque is produced, that is, a point on the so-called "stable" part of the motor characteristic. An intersection at a point such as p_2, where the slip is greater than that for maximum torque, does not represent a possible steady-state condition of operation, since above that intersection the motor has excess torque available to accelerate the load to the point p_1.

For speed control over the range from standstill up to rated motor speed, there are two alternative closed-loop systems. The first of these may be employed with loads of low inertia that tend to decelerate rapidly when the driving torque is reduced. Such a system uses only the first quadrant of the motor characteristic, and its basic principle is indicated in the schematic diagram of Fig. 4.23. The controller shown in Fig. 4.23 could be one of those illustrated in Figs. 4.2 to 4.4, where Z_L would represent the per-phase motor impedance referred to the stator. Since however standard induction motor stators are usually connected internally, so that only three terminals are brought out, the arrangements of Fig. 4.2 are the most

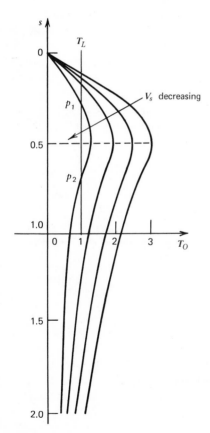

Fig. 4.22 Induction motor speed control by voltage variation.

probable. The feedback signal shown in Fig. 4.23 is speed, but other variables may be employed, depending on the nature of the driven load. In the event that operation is required at a point on the unstable part of the motor characteristic, then the motor may be held at that speed by terminal voltage variation.

The second alternative closed-loop system is employed with high-inertia loads which decelerate very slowly when driving torque is removed and which therefore must be braked when a speed reduction is called for. Such a system uses the first and fourth quadrants of the motor characteristic. Moreover, for this application, the motor must be of a design that provides maximum torque at approximately standstill, since otherwise fourth-quadrant operation would result in excessive motor currents. A family of characteristics for such a motor with a hypothetical load characteristic superimposed is shown in Fig. 4.24. The first quadrant of this diagram requires no further explanation. The parts of the motor characteristics in the fourth quadrant are brought into use by reversing the phase sequence of the voltages at the motor terminals. In effect, this rotates the motor characteristics through 180°, transferring them to the second and third quadrants.

Consider that the system of Fig. 4.24 is operating at motor voltages V_1 with positive sequence, the slip and output torque having the values at point p_1. If a speed reduction is now called for, and the phase sequence of the motor excitation is reversed without change in the magnitude of the voltages, then the motor operating characteristic becomes that shown in the second and third quadrants. Under these conditions, a braking torque

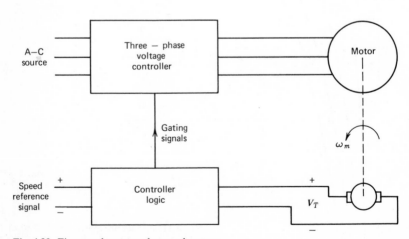

Fig. 4.23 First-quadrant speed-control system.

is applied to the load of which the magnitude is given by the horizontal intercept between the s axis and p_2.

Operation of the motor in the fourth quadrant, that is, plugging, is not without its dangers. Under these conditions both the energy supplied to the motor by the electrical system and the energy supplied via the shaft by the decelerating mechanical load must be dissipated as heat within the Design D motor. Even with the kind of characteristic shown in Fig. 4.24, motor currents will be large, and so will be the RI^2 losses in the motor. Where plugging is expected to be particularly severe, a wound-rotor induction motor may be employed with sufficient external resistance permanently connected in the rotor circuit to result in maximum torque at standstill. Much of the energy supplied to the motor during plugging is then dissipated in the external rotor-circuit resistor and does not heat up and endanger the motor.

The schematic diagram of Fig. 4.23 may be considered to represent the two-quadrant drive, provided it is understood that the three-phase voltage

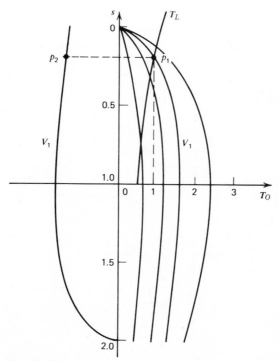

Fig. 4.24 Two quadrant speed-control system.

controller must be capable of phase-sequence reversal of its output voltages. The method by which this may be achieved is shown in Fig. 4.25, and this diagram should be compared with Fig. 4.14. For first-quadrant operation, thyristor pairs A, B, C are employed and provide positive-sequence voltages at terminals R, S, T. For fourth-quadrant operation, thyristor pairs A^*, B^*, C are employed and provide negative-sequence voltages at terminals R, S, T. The controller logic must be so designed that short-circuit of the source by turning on pairs A^* and B or A and B^* simultaneously is prohibited.

It may readily be seen that if a reversing, four-quadrant drive is required, the system of Figs. 4.23 to 4.25 provides that also.

4.3.2 Induction Motor Model The model employed in calculating the performance of an induction motor is the per-phase equivalent circuit of the motor referred to the stator, and this is shown in Fig. 4.26. The parameters of this equivalent circuit are obtained from the no-load and locked-rotor tests described in standard texts on electric machines. The

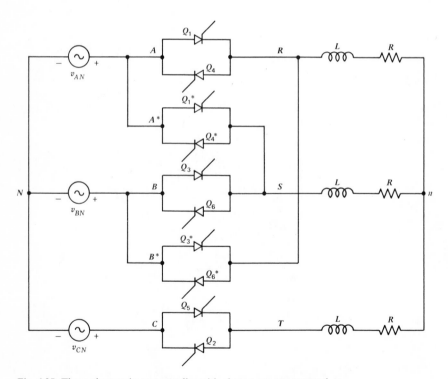

Fig. 4.25 Three-phase voltage controller with phase-sequence reversal.

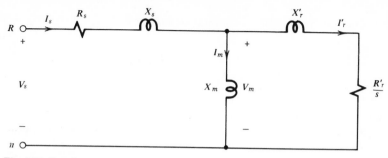

Fig. 4.26 Per-phase equivalent circuit of an induction motor.

rotational losses of the motor are also obtained from these tests, being taken as equal to the power input to the motor on no-load P_{NL}. These losses, friction, windage, and rotor core loss, are often assumed to be constant at all speeds, but this is a considerable approximation, particularly when motors are employed with a variable-voltage supply. If the motor manufacturer provides a curve of no-load input power over a wide range of terminal voltages, more accurate methods of including rotational losses in performance calculations are possible.

When a three-phase induction motor is supplied from a voltage controller such as that shown in Fig. 4.14, then the complete motor will be represented by three per-phase equivalent circuits in which the three terminals marked n are connected together to form the star or neutral point of the load circuit. At any chosen speed, each per-phase circuit may be reduced to an RL impedance, such as is shown in Fig. 4.14.

The power transmitted across the air gap of the motor to the rotor and there converted to mechanical form, absorbed in rotational losses, or dissipated as heat in the rotor windings, is that dissipated in the three fictitious resistances R_r'/s of the three phases. This is called the "air-gap power" and is

$$P_{ag} = 3\frac{R_r'}{s}(I_r')^2 \quad \text{W} \tag{4.47}$$

This power may be divided into two parts,

$$P_{ag} = 3R_r'(I_r')^2 + 3(1-s)\frac{R_r'}{s}(I_r')^2 \quad \text{W} \tag{4.48}$$

Of the two terms on the right-hand side of this equation, the first is clearly the power dissipated as heat in the rotor windings. The second represents the mechanical power output at the motor coupling plus the rotational losses. The mechanical power output is therefore

$$P_O = 3(1-s)\frac{R_r'}{s}(I_r')^2 - P_{NL} \quad \text{W} \qquad (4.49)$$

The maximum terminal voltage applied to the motor will be the rated terminal voltage, and this will give maximum forward speed. The maximum current drawn by the motor does not appear at maximum speed, but it is important to be able to determine at least approximately what that maximum will be. For this purpose it is convenient to have an analytical expression of load torque T_L as a function of speed. A load commonly requiring a first-quadrant drive is a fan or centrifugal pump, for which to a good approximation

$$T_L = k_L \omega_m^2 \quad \text{n-m} \qquad (4.50)$$

where k_L is a constant for a particular load. If motor rotational losses are neglected, then from equation 4.49 the motor output torque is

$$T_O = \frac{P_O}{\omega_m} = \frac{3}{\omega_m}(1-s)\frac{R_r'}{s}(I_r')^2 \quad \text{n-m} \qquad (4.51)$$

If now it is assumed that in Fig. 4.26,

$$X_m \gg \left|\frac{R_r'}{s} + jX_r'\right| \quad \Omega \qquad (4.52)$$

then

$$I_s \cong I_r' \quad \text{A} \qquad (4.53)$$

From equation 4.46

$$\omega_m = (1-s)\omega_{\text{syn}} \quad \text{rad/s} \qquad (4.54)$$

so that substitution from equations 4.53 and 4.54 in equation 4.51 yields

$$T_O = \frac{3}{\omega_{\text{syn}}}\frac{R_r'}{s}I_s^2 \quad \text{n-m} \qquad (4.55)$$

Under steady-state conditions or moderate acceleration of such low-inertia loads $T_O \cong T_L$, so that from equations 4.50, 4.54, and 4.55,

$$k_L(1-s)^2\omega_{\text{syn}}^2 = \frac{3}{\omega_{\text{syn}}}\frac{R_r'}{s}I_s^2 \quad \text{n-m} \qquad (4.56)$$

from which

$$I_s = \left[\frac{k_L}{3} \frac{s}{R_r'} (1-s)^2 \omega_{syn}^3 \right]^{1/2}$$

$$= K_1 (1-s) s^{1/2} \quad \text{A} \tag{4.57}$$

where K_1 is a constant. Differentiation of this expression for I_s with respect to s and equating the derivative to zero shows that the maximum value of I_s will occur at approximately

$$s = \tfrac{1}{3} \tag{4.58}$$

An alternative to the fan or pump type of load is one in which the load torque is approximately constant at all speeds. In such a case, equation 4.56 becomes

$$T_L = \frac{3}{\omega_{syn}} \frac{R_r'}{s} I_s^2 \quad \text{n-m} \tag{4.59}$$

and

$$I_s = K_2 \sqrt{s} \quad \text{A} \tag{4.60}$$

From equation 4.60, I_s is seen to be a maximum when s is a maximum. For a first-quadrant drive, I_s will be a maximum at standstill. For a drive employing plugging, I_s will reach its maximum if the phase sequence is reversed when the motor is running at rated speed. The slip under these conditions will be

$$s = 2 - s_{rated} \tag{4.61}$$

where from equations 4.46

$$s_{rated} = \frac{\omega_{syn} - \omega_{rated}}{\omega_{syn}} \tag{4.62}$$

The specification of the controller for one of these types of variable speed drive is best illustrated by means of numerical examples.

Example 4.4 A centrifugal pump has a speed-torque characteristic that may be described by the relationship

$$T_L = 1.40 \times 10^{-3} \omega_m^2 \quad \text{n-m}$$

The maximum permissible speed is 1550 rpm.

Determine the horsepower rating of a suitable 60-Hz, Design D, squirrel-cage induction motor.

Solution   The required mechanical power input to the load is

$$P_L = T_L \omega_m$$

$$= 1.40 \times 10^{-3} \left(1550 \times \frac{2\pi}{60} \right)^3 = 5990 \quad \text{W}$$

The required motor horsepower is

$$\text{HP} = \frac{5990}{746} = 8.03$$

Since this is not a standard rating, a 10-HP motor will be used.

Example 4.5   A suitable 220-V, Design D motor for the system of Example 4.4 has the following per-phase equivalent-circuit parameters referred to the stator of the machine:

$$R_s = 0.251 \quad \Omega \qquad R_r' = 0.650 \quad \Omega$$

$$X_s = 0.361 \quad \Omega \qquad X_r' = 0.361 \quad \Omega$$

$$X_m = 17.3 \quad \Omega \qquad P_{NL} = 553 \quad \text{W}$$

If the motor is to be supplied from a three-phase full-wave voltage controller in a first-quadrant drive, and the pump speed is to be varied from 1550 down to 775 rpm, determine the output rating of the controller and the range of angle of retard α which must be provided by the control circuit.

Solution   The maximum speed of 1550 rpm calls for a four-pole induction motor with a synchronous speed of 1800 rpm. The slip at 1550 rpm is thus

$$s_1 = \frac{1800 - 1550}{1800} = 0.139$$

The fictitious rotor-circuit resistance at 1550 rpm is

$$\frac{R_r'}{s_1} = \frac{0.650}{0.139} = 4.68 \quad \Omega$$

The motor equivalent circuit at this speed is shown in Fig. E4.5a. The impedance between terminals R and n may be calculated, and is

$$\overline{\mathbf{Z}}_1 = 4.42 + j1.81 = 4.77 \; \underline{/22.3°} \quad \Omega$$

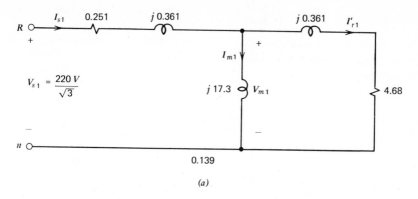

(a)

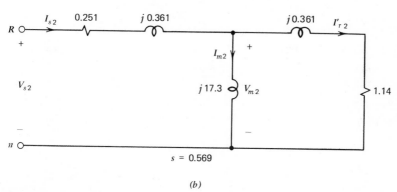

(b)

Fig. E4.5 (a) and (b)

so that

$$\phi_1 = 22.3°$$

and

$$I_{s1} = \frac{220}{\sqrt{3}} \times \frac{1}{4.77} = 26.7 \quad A$$

The rotor current at this speed will be

$$I_r' = \left| \frac{j17.3}{4.68 + j17.7} \right| \times 26.7 = 25.2 \quad A$$

Thus from equation 4.49, the output power is

$$P_{O1} = 3(1 - 0.139) \times 4.68 \times 25.2^2 - 553 = 7120 \quad W$$

This is greater than the power required by the pump at maximum speed and confirms that the motor is suitable for this service. It also shows that under steady-state conditions the maximum output voltage will not be required from the converter, even at maximum speed.

From Fig. 4.19, the base current at 1550 rpm is

$$I_{B1} = \frac{\sqrt{2}\, V}{\sqrt{3}\, Z_1} = \frac{\sqrt{2} \times 220}{\sqrt{3} \times 4.77} = 37.7 \quad A$$

Thus

$$I_{RN1} = \frac{1}{\sqrt{2}} \frac{I_{s1}}{I_{B1}} = \frac{1}{\sqrt{2}} \frac{26.7}{37.7} = 0.5$$

Interpolation in Fig. 4.19 between the curves for $\phi = 0°$ and $\phi = 30°$ on the line $I_{RN} = 0.5$ yields a required delay angle at maximum speed of

$$\alpha_1 = 22.3°$$

This conclusion could have been reached directly by observing that at the maximum output voltage delivered by the controller

$$\alpha_1 = \phi_1$$

At 775 rpm the slip is

$$s_2 = \frac{1800 - 775}{1800} = 0.569$$

Thus

$$\frac{R_r'}{s_2} = \frac{0.650}{0.569} = 1.14 \quad \Omega$$

and from Fig. E4.5b

$$\overline{Z}_2 = 1.34 + j0.787 = 1.56 \underline{/30.4°} \quad \Omega$$

so that

$$\phi_2 = 30.4°$$

At 775 rpm, the power delivered to the pump must be

$$P_{L2} = P_{O2} = 1.40 \times 10^{-3} \left(775 \times \frac{2\pi}{60} \right)^3 = 748 \quad W$$

From equation 4.49, if it is assumed that rotational losses of the motor are constant

$$3(1-s_2)\frac{R'_r}{s_2}(I'_{r2})^2 = P_{O2} + P_{NL}$$

so that

$$I'_{r2} = \left[\frac{748 + 553}{3(1 - 0.569) \times 1.14}\right]^{1/2} = 29.7 \quad \text{A}$$

The line current may then be obtained from the relationship

$$I'_{r2} = \left|\frac{j17.3}{1.14 + j17.7}\right| I_{s2}$$

so that

$$I_{s2} = \frac{17.7}{17.3} \times 29.7 = 30.4 \quad \text{A}$$

The base current at 775 rpm is

$$I_{B2} = \frac{\sqrt{2} \times 220}{\sqrt{3} \times 1.56} = 115 \quad \text{A}$$

$$I_{RN2} = \frac{1}{\sqrt{2}}\frac{I_{s2}}{I_{B2}} = \frac{1}{\sqrt{2}}\frac{30.4}{115} = 0.187$$

From Fig. 4.19, for $I_{RN} = 0.187$, $\phi = 30.4$, α may be read off and is

$$\alpha_2 = 93°$$

The range of α required is thus

$$22.3° < \alpha < 93°$$

In fact the curves of Fig. 4.19 give rather higher values of α at the low-speed end of the range than would be required in practice. This is because current I_{s2} calculated in this example is a sinusoidal current, while the minimum-speed output current from the controller will contain harmonics and a fundamental component of rms value equal to I_{s2}. It would thus have a higher value of I_{RN} than has been employed in these calculations.

The motor terminal voltage at 775 rpm is also of interest, and its line-to-line value will be

$$\sqrt{3}\ V_{s2} = \sqrt{3}\ Z_2 I_{s2} = \sqrt{3}\ \times 1.56 \times 30.4 = 82.1 \quad V$$

It has been shown that an approximate value of the maximum motor current may be obtained from the conditions at $s = 1/3$ for this type of load, that is, at

$$\omega_m = (1 - s)\omega_{syn} = \frac{2}{3} \times \frac{2\pi}{60} \times 1800 = 40\pi \quad rad/s$$

At this speed, the power delivered to the pump must be

$$P_{L3} = P_{O3} = 1.40 \times 10^{-3}(40\pi)^3 = 2780 \quad W$$

From equation 4.49,

$$3 \times \frac{2}{3} \times 3 \times 0.650(I'_{r3})^2 = 2780 + 553 \quad W$$

so that

$$I'_{r3} = 29.2 \quad A$$

The line current may then be obtained from the relationship

$$I'_{r3} = \left| \frac{j17.3}{3 \times 0.650 + j17.7} \right| I_{s3} \quad A$$

so that

$$I_{s3} = \frac{17.8}{17.3} \times 29.2 = 30.1 \quad A$$

and this is virtually equal to the current at minimum speed. In fact the calculated values of both of these currents will tend to be high due to the assumption of constant rotational losses. More accurate data on these losses would result in lower calculated currents at both speeds, the fall in calculated value being greater at the lower speed.

4.4 THREE-PHASE FULL-WAVE DELTA-CONNECTED CONTROLLER

Any delta connected load circuit may be supplied by a three-phase controller of the type discussed in Section 4.3. However if the terminals of the three branches of such a load circuit are accessible and may be opened, then the circuit arrangement shown in Fig. 4.27 may be adopted, with the result that thyristors of lower current ratings than would be needed in the lines may be employed.

Let the line-to-line source voltages be

$$v_{AB} = \sqrt{2} \, V \sin \omega t \quad V$$

$$v_{BC} = \sqrt{2} \, V \sin(\omega t - 2\pi/3) \quad V \qquad (4.63)$$

$$v_{CA} = \sqrt{2} \, V \sin(\omega t - 4\pi/3) \quad V$$

The waveforms of these voltages are shown in Fig. 4.28, and once again the gating signals of the thyristors must have the same sequence and phase displacements as do the source voltages. However in this circuit the angle of retard for thyristor Q_1 is zero at $\omega t = 0$, since each branch of the delta

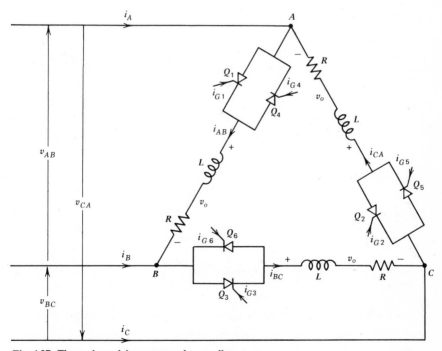

Fig. 4.27 Three-phase delta-connected controller.

operates as a single-phase controller excited by a line-to-line source voltage. The gating signals for a delay angle of $\alpha = 2\pi/3$ are shown in Fig. 4.28.

The branch current waveforms for this angle of retard and a purely resistive load circuit are shown in Fig. 4.28. To obtain full control of the output voltage v_o with a resistive load, the angle of retard must be varied over the range $0 \leqslant \alpha \leqslant \pi$. Each line-to-line voltage of the source and the branch of the delta to which it is applied may be analyzed as already described in Section 4.2. The line currents are

$$i_A = i_{AB} - i_{CA} \quad \text{A}$$

$$i_B = i_{BC} - i_{AB} \quad \text{A} \qquad\qquad (4.64)$$

$$i_C = i_{CA} - i_{BC} \quad \text{A}$$

and their waveforms for a resistive load circuit are also shown in Fig. 4.28. As before, when α is reduced to ϕ the circuit operates as if no controller were present and the maximum output is obtained.

The harmonics of the branch or line currents for load circuits with or without inductance may be determined by Fourier analysis, as described in Section 4.2.1. However owing to the delta connection, the triplen harmonic components (i.e., those of order $n = 3m$, where m is an integer) of branch currents flow round the delta without appearing in the line, since these zero-sequence harmonics are in phase in all three branches. The rms branch current is thus

$$I_{AB} = \left[I_{1R}^2 + I_{3R}^2 + I_{5R}^2 + I_{7R}^2 + I_{9R}^2 + \cdots \right]^{1/2} \quad \text{A} \qquad (4.65)$$

while the rms line current is

$$I_A = \sqrt{3} \left[I_{1R}^2 + I_{5R}^2 + I_{7R}^2 + I_{11R}^2 + \cdots \right]^{1/2} \quad \text{A} \qquad (4.66)$$

so that

$$I_A < \sqrt{3}\, I_{AB} \quad \text{A} \qquad\qquad (4.67)$$

Example 4.6 For a three-phase wye connected controller with the circuit shown in Fig. 4.14, $V_{AB} = 230\,\text{V}$, $R = 1\,\Omega$, and $\omega L = 1\,\Omega$. Calculate the following:

(a) The maximum rms thyristor current, $I_{R\,\text{max}}$.
(b) The maximum instantaneous thyristor voltage, $v_{AK\,\text{max}}$.

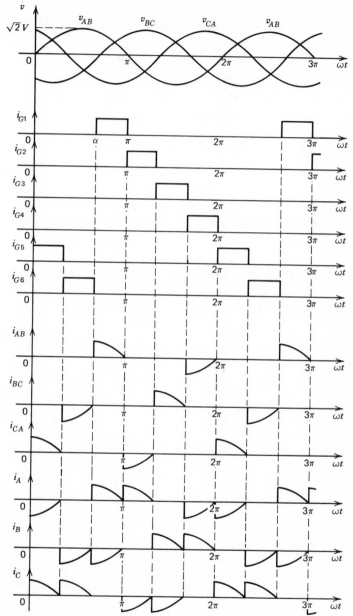

Fig. 4.28 Waveforms for circuit of Fig. 4.27: $\phi = 0$, $\alpha = 2\pi/3$.

201

(c) The thyristor rating of the controller, R_Q.

(d) The thyristor derating factor, D_R.

(e) The control range for the angle of retard α.

Where

$$R_Q = N v_{AK\,max} I_{QR\,max} \qquad (4.68)$$

and N is the number of thyristors in the controller.

$$D_R = \frac{R_Q}{P_{max}} \qquad (4.69)$$

and P_{max} is the maximum three-phase power absorbed in the load circuit.

Solution

(a) The maximum line current and thyristor current flows when

$$\alpha = \phi = \frac{\pi}{4} = 45°$$

The system then operates as if no controller were present, and

$$I_{A\,max} = \frac{V}{\sqrt{3}\ Z} = \frac{230}{\sqrt{3} \times \sqrt{2}} = 94 \quad A$$

Since this current flows in each thyristor for alternate half cycles only, then

$$I_{QR\,max} = \left[\frac{1}{2\pi} \int_0^\pi \left(\frac{\sqrt{2}\ V}{\sqrt{3}\ Z} \sin \omega t \right)^2 d(\omega t) \right]^{1/2}$$

$$= \frac{1}{\sqrt{2}} \frac{V}{\sqrt{3}\ Z} = 66.5 \quad A$$

(b) From equation 4.44 and the related discussion, it will be seen that the maximum instantaneous voltage that may be applied to the thyristor at any value of α is

$$v_{AK\,max} = \sqrt{2}\ V \frac{\sqrt{3}}{2} = 282 \quad V$$

(c) $R_Q = 6 \times 282 \times 66.5 = 107 \times 10^3 \quad VA$

(d) $P_{max} = 3 \times 94^2 \times 1 = 26.5 \times 10^3$ W

$$D_R = \frac{107 \times 10^3}{26.5 \times 10^3} = 4.04$$

(e) From the curves of Figs. 4.18 and 4.19 it may be seen that the effective range of variation of α is

$$45° < \alpha < 150°$$

Example 4.7 For a three-phase delta-connected controller with the circuit shown in Fig. 4.27, $V_{AB} = 230$ V, $R = 3\Omega$, and $\omega L = 3\Omega$. Calculate the following:

(a) The maximum rms thyristor current, I_{Rmax}.
(b) The maximum instantaneous thyristor voltage, v_{AKmax}.
(c) The thyristor rating of the controller, R_Q, as defined in Example 4.6.
(d) The thyristor derating factor, D_R, as defined in Example 4.3.
(e) The control range for the angle of retard α.

Solution

(a) The maximum line current and thyristor current flows when

$$\alpha = \phi = \frac{\pi}{4} = 45°$$

The system then operates as if no controller were present, and

$$I_{ABmax} = \frac{V}{Z} = \frac{230}{3\sqrt{2}} = 53.6 \quad A$$

$$I_{QRmax} = \frac{1}{\sqrt{2}} I_{ABmax} = \frac{53.6}{\sqrt{2}} = 37.8 \quad A$$

(b) Since each branch of the load circuit and the associated line-to-line source voltage acts like a single-phase, full-wave controller, then the maximum instantaneous voltage that will be applied to any one thyristor is

$$v_{AKmax} = \sqrt{2}\, V = 230\sqrt{2} = 325 \quad V$$

(c) $R_Q = 6 \times 37.8 \times 325 = 73.8 \times 10^3 \quad VA$

(d) $P_{max} = 3 \times (53.6)^2 \times 3 = 26.5 \times 10^3$ W

$$D_R = \frac{73.8 \times 10^3}{26.5 \times 10^3} = 2.78$$

(e)   The effective range of variation of α is

$$45° < \alpha < 180°$$

The values of the derating factors D_R determined in parts d of the foregoing examples show that much better utilization of thyristor capacity is obtained with the delta-connected circuit than with the wye-connected circuit.

4.5 SINGLE-PHASE TRANSFORMER TAP CHANGER

The circuit of a single-phase transformer tap changer is shown in Fig. 4.29, where for simplicity only two transformer turns ratios are employed, and the secondary winding is tapped at its center point. Thus for each half of the secondary winding, the induced voltage is

$$v = \sqrt{2}\, V \sin \omega t \quad \text{V} \tag{4.70}$$

Thus if thyristors Q_1 and Q_2 are turned off for the whole cycle of the source voltage v_s, then thyristors Q_3 and Q_4 may operate as a single-phase voltage controller circuit, giving a range of rms output voltage of $0 \le V_{oR} \le V$ V. On the other hand, if Q_3 and Q_4 are turned off for the whole cycle of the source voltage v_s, while Q_1 and Q_2 are turned on throughout

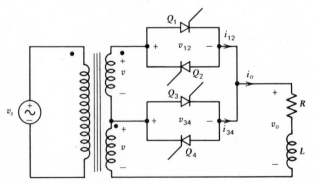

Fig. 4.29 Single-phase transformer tap changer.

alternate half cycles of v_s, then the load voltage is $v_o = 2v$, corresponding to an rms output voltage of $2V$ V.

Provided that certain precautions are observed, it is possible also to vary the rms output voltage over the range $V \leqslant V_{oR} \leqslant 2V$ V. The simplest case to consider is that in which the load circuit is purely resistive.

(a) *Load-circuit inductance zero.* If thyristors Q_1 and Q_2 are turned off for the whole cycle of v_s, while Q_3 and Q_4 are turned on throughout alternate half cycles, then

$$i_o = \frac{\sqrt{2}\,V}{R}\sin \omega t \quad \text{V} \tag{4.71}$$

Now let Q_1 be turned on at $\omega t = \alpha$, where $\alpha < \pi$. Then

$$v_{12} = 0 \quad \text{V:} \qquad \alpha < \omega t < \pi \tag{4.72}$$

and for the circuit mesh including Q_1 and Q_3,

$$v_{12} - v_{34} - v = 0 \quad \text{V} \tag{4.73}$$

so that

$$v_{34} = -v \quad \text{V} \tag{4.74}$$

and Q_3 is commutated.

At $\omega t = \pi$, Q_1 turns off, and Q_4 is turned on. At $\omega t = \pi + \alpha$, Q_2 is turned on, and

$$v_{12} = 0 \quad \text{V:} \qquad \pi + \alpha < \omega t < 2\pi \tag{4.75}$$

and for the circuit mesh including Q_2 and Q_4, equations 4.73 and 4.74 apply. Since $v < 0$, then $v_{34} > 0$, and Q_4 is commutated. At $\omega t = 2\pi$, Q_2 turns off, Q_3 is turned on, and the cycle is repeated. The resulting variations of load voltage and current are shown in Fig. 4.30, from which it may be seen that the rms value of the output voltage is within the range $V < V_{oR} < 2V$ V.

(b) *Load circuit with inductance.* If again thyristors Q_1 and Q_2 are turned off for the whole cycle of v_s, while Q_3 and Q_4 are turned on throughout alternate half cycles, then

$$i_o = \frac{\sqrt{2}\,V}{Z}\sin(\omega t - \phi) \quad \text{V} \tag{4.76}$$

where

$$Z = \left[R^2 + (\omega L)^2\right]^{1/2} \quad \Omega \tag{4.77}$$

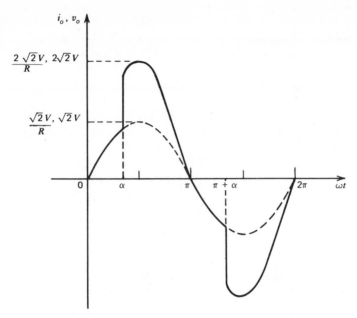

Fig. 4.30 Waveforms of v_o and i_o for circuit of Fig. 4.29: $L = 0$, $\alpha = \pi/3$.

and

$$\phi = \tan^{-1}\frac{\omega L}{R} \tag{4.78}$$

The time variations of load voltage and current under these conditions are shown in Fig. 4.31.

Now let Q_1 be turned on at $\omega t = \alpha$, where $\alpha < \phi$. Then since Q_4 is still conducting, $v_{34} = 0$, and the upper half of the transformer secondary winding is short circuited through Q_1 and Q_4. Thus the control circuit must be so designed as to ensure that Q_1 may not be turned on until $i_o \geqslant 0$, and similarly Q_2 may not be turned on until $i_o \leqslant 0$. In other words, an upward transition, increasing the magnitude of the output voltage v_o can only take place when i_o and v_o are either both positive or both negative. Subject to this constraint, the system may be analyzed in the same way as was the single-phase full-wave voltage controller.

Waveforms of v_o and i_o for an RL load circuit are shown in Fig. 4.32. The chief advantage of the transformer tap changer as compared with a single-phase full-wave ac voltage controller lies in the great reduction of harmonics in the load and line currents. This is particularly marked when

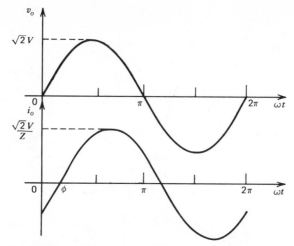

Fig. 4.31 Waveforms of v_o and i_o for circuit of Fig. 4.29: $L \neq 0$, $\alpha = \pi$, Q_1 and Q_2 off.

the tap changer does not have merely two taps, as shown in Fig. 4.29, but several, as shown in Fig. 4.33.

If in the circuit of Fig. 4.33 the thyristors for tap 3 are turned on throughout each half cycle of the source voltage, while those for tap 2 are turned on at $\omega t = \alpha$ and $\pi + \alpha$ to give a somewhat higher value of V_{oR} than would be obtained from tap 3 alone, then only the induced voltage v_{23} between taps 2 and 3 is being switched to contribute harmonics to the load voltage v_o. If $v_{30} \gg v_{23}$, then the harmonics will form a very small part of v_o, and the current harmonics will consequently be low even in the worst possible case when the load circuit is purely resistive and $\alpha = \pi/2$. In other words, the voltage v_{30} goes entirely to reinforce the fundamental component of v_o. If the ratio $v_{23} : v_{30}$ is known, then the harmonic content of the output voltage may be determined, employing the curves shown in Fig. 4.9. From this and the resistance of the load circuit, the output current harmonics may then be calculated for a resistive load circuit.

4.6 SINGLE-PHASE SINUSOIDAL VOLTAGE CONTROLLER

If continuous voltage control over a wide range with low harmonic content is required, then the number of thyristors employed in a transformer tap changer may become excessive. A more economical alternative converter is shown in Fig. 4.34. The secondary windings $1, 2, 3 \ldots n$ of the transformer have voltages that increase in geometric progression with a common ratio of 2. Thus if v is the voltage of secondary number 1, then the voltage of the

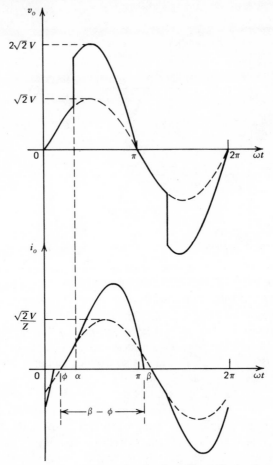

Fig. 4.32 Voltage and current waveforms for circuit of Fig. 4.29: $L \neq 0$, α for Q_1 and Q_2 is $\pi/3$.

nth secondary is $2^{n-1}v$. By means of the pairs of parallel thyristors, any series combination of secondaries 1 to n may be achieved, so that voltages from v to $(2^n - 1)v$ in discrete steps equal to v may be obtained.

In addition to secondary windings 1 to n, which conduct for the entire cycle of the load current or not at all, a further phase-controlled secondary winding A may be employed as a vernier to permit continuous variation between the discrete values of output voltage. Necessarily, this winding contributes harmonics to the line and load currents. However, as has already been explained for the case of the transformer tap changer, the

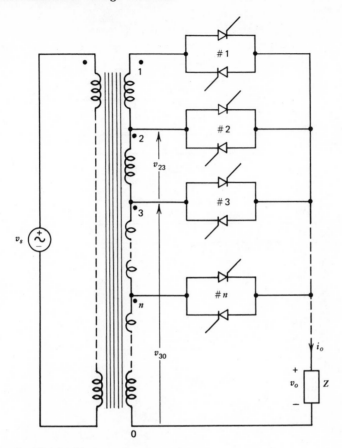

Fig. 4.33 Multitap transformer tap changer.

harmonic content will be very much lower than would be the case if a normal single-phase, full-wave voltage controller were employed.

The waveforms of voltage and current for purely resistive and for RL load circuits differ in no way from those shown in Figs. 4.30 and 4.32, which would correspond to an output voltage range requiring only windings A and 1 of the controller of Fig. 4.34 to carry current for any or all of the cycle.

The operation of the vernier winding in conjunction with any series grouping of the remaining secondary windings may be understood from the circuit shown in Fig. 4.35 where in fact it is assumed that all other secondaries are conducting. If the controller is supplying a purely resistive

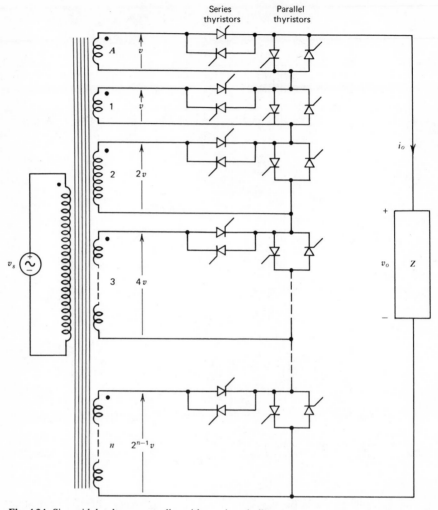

Fig. 4.34 Sinusoidal voltage controller with vernier winding.

load, then during the positive half cycle of v_s, thyristor Q_3 is conducting, and

$$v_o = (2^n - 1)v \quad \text{V:} \qquad 0 < \omega t < \alpha \tag{4.79}$$

At $\omega t = \alpha$, Q_1 is turned on, and Q_3 is commutated, so that

$$v_o = 2^n v \quad \text{V:} \qquad \alpha < \omega t < 2\pi \tag{4.80}$$

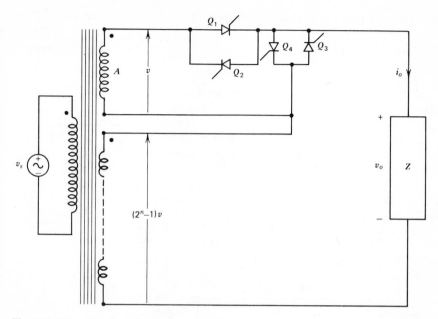

Fig. 4.35 Operation of vernier winding.

At the end of the positive half cycle of v_s, i_o falls to zero, Q_1 turns off, and Q_4 is turned on. At $\omega t = \pi + \alpha$, Q_2 is turned on, and Q_4 is commutated. At the end of the cycle, Q_2 turns off, and the cycle is repeated.

With a load circuit that possesses inductance, short-circuit possibilities arise, and operating constraints apply similar to those described in Section 4.5(b) for the tap changer. If for $\omega t > 0$, thyristor Q_4 is still conducting, and thyristor Q_1 is turned on, then the vernier secondary winding is short-circuited through these two thyristors. Thus the control circuit must be so designed as to ensure that thyristors Q_1 and Q_2 may not be turned on until i_o becomes zero. In other words, an upward transition, increasing the magnitude of the output voltage v_o can only take place when i_o and v_o are either both positive or both negative. It naturally follows that all of the remaining series thyristors in secondary circuits 2 to n of Fig. 4.34 must also be turned on when $i_o = 0$.

PROBLEMS

4.1 The circuit of Fig. 4.1a is employed to control by the on-off method the power delivered to a resistive load. The repeated cycle of control is such that the thyristors are on for $1/6$ s and off for $1/12$ s. Calculate the ratio of the power delivered to the resistive load to that which would be delivered with the thyristors permanently turned on.

4.2 In the controller circuit of Fig. 4.5a, $V = 230$V, $\omega = 120\pi$ (i.e., 60-Hz supply), $L = 0$, $R = 2.3\,\Omega$. Determine

(a) The angle α at which the greatest forward or reverse voltage is applied to either of the thyristors and the magnitudes of those voltages.

(b) The greatest rms and average thyristor current for any angle α.

(c) The minimum time provided for turn off of the thyristors for any angle α.

(d) The required value of α to give an rms output voltage of 115V.

(e) The amplitude of the third harmonic current when $\alpha = \pi/2$.

4.3 In the controller circuit of Fig. 4.5a, $V = 230$V, $\omega = 120\pi$, $X_L = 2.3\,\Omega$, $R = 0.01\,\Omega$. If the thyristors are supplied with continuous gating signals and the circuit has reached a steady-state condition of operation, sketch to scale the thyristor voltages, output voltage, and load current for each of the three cases $\alpha = 0$, $\alpha = \pi/2$, $\alpha = 3\pi/4$.

Determine also

(a) The angle α at which the greatest forward or reverse voltage is applied to either of the thyristors and the magnitudes of those voltages.

(b) The maximum possible value of rms thyristor current.

(c) The maximum possible value of average thyristor current.

(d) The control range for this load circuit.

4.4 Repeat problem 3 with $X_L = 2.3\,\Omega$ and $R = 2.3\,\Omega$.

4.5 For the systems described in problems 4.2, 4.3, and 4.4, calculate the maximum value of di/dt occurring in the thyristors.

4.6 For the single-phase ac voltage controller of Fig. 4.5a, explain the difficulty that would arise if the gating currents i_{G1} and i_{G4} were supplied from a common terminal.

4.7 For the three-phase wye-connected controller in Fig. 4.14, sketch the waveforms of currents i_A, i_B, i_C and voltages v_{RS} and v_{AR} if $L = 0$ and angle of retard $\alpha = \pi/3$.

4.8 In the controller of Fig. 4.14, the line-to-line source voltage is 110V, $L = 0$, and $R = 2.2\,\Omega$. Determine

(a) The maximum forward and reverse thyristor voltages, rms thyristor currents and average thyristor currents for any value of angle of retard α.

(b) The control range.

(c) The value of α to give a half-cycle average line-to-neutral load voltage of 55V.

4.9 Repeat problem 4.8 if $R = 2.2\,\Omega$ and $X_L = 4.7\,\Omega$.

4.10 Calculate the average current, rms current, and maximum voltage ratings of the thyristors in the controller of Fig. 4.14 if it were employed to supply a 50-kW three-phase 220-V induction motor. The motor full load power factor and efficiency are 0.866 and 0.9, respectively.

4.11 Employ suitable diagrams or expressions to explain why the line currents in the three-phase delta-connected controller of Fig. 4.27 contain no third harmonic component at any value of α.

4.12 For the three-phase delta-connected controller in Fig. 4.27, sketch the waveforms of currents i_A, i_B, and i_C if $L=0$ and $\alpha=\pi|6$.

4.13 In the controller of Fig. 4.27, the line-to-line source voltage is 110V, $R=2.2\Omega$ and $X_L=4.7\Omega$. Determine

(a) The maximum forward and reverse thyristor voltages, rms thyristor currents and average thyristor currents for any value of α.
(b) The control range.
(c) The value of α to give an rms voltage of 55 V across each branch of the load. Neglect harmonics with $n>1$.

4.14 Calculate the average current, rms current, and maximum voltage ratings of the thyristors in the controller of Fig. 4.27 if it were employed to supply a 50-kW three-phase 220-V induction motor. The motor full-load power factor and efficiency are 0.866 and 0.9, respectively.

4.15 Show that in a neutral-point controller not more than two thyristors can conduct at any instant, and sketch the gating signals required if Z is purely resistive and the line current i_A is a maximum. (This is the condition which defines the zero value of α).

4.16 Assuming that each gating signal extends over an angular interval of π rad, sketch the gating signals for a neutral-point controller when $\alpha=\pi|2$ rad, and show that the line currents contain even harmonic components.

4.17 A 40-HP 220-V three-phase delta-connected induction motor supplied from the controller circuit shown in Fig. 4.14 is used to maintain constant pressure on the delivery side of a centrifugal pump. A 220-V line-to-line source is available and the current limit of the controller is set to 150% of the rated current of the motor. If the rated power factor and efficiency of the motor are 0.85 and 0.75 respectively, determine:

(a) The rms current rating of the thyristors.
(b) The peak forward or reverse voltage applied to the thyristors.
(c) The required control range of angle of retard α.

4.18 The ends of the three-phase stator windings of the motor of problem 4.17 are brought out to six terminals, so that it may be employed with the controller circuit shown in Fig. 4.27. Repeat problem 4.17 for this arrangement.

FIVE

CONTROLLED RECTIFIERS

Controlled rectifiers form the large majority of converters employing power semiconductors. They are used to vary the average value of the direct voltage applied to a load circuit by introducing thyristors between that load circuit and a constant voltage ac source. For this purpose the thyristors are phase controlled.

Applications of controlled rectifiers include the following:

1. dc motor speed control systems, widely used in steel mills, paper mills, and such.
2. Electrochemical and electrometallurgical processes.
3. Magnet power supplies.
4. Converters at the input end of dc transmission lines.
5. Portable hand tool drives.

Like ac voltage controllers, controlled rectifiers may be employed in closed-loop control systems, where they function as high-power operational amplifiers in which the angle α at which the thyristors are turned on is varied in response to an error signal. In general a single-phase ac source is adequate for rectifier ratings of 1 or 2 kW, but for higher powers a three-phase ac source is normally used. Once again, the problem of current harmonics introduced into the supply system and load circuit arises, and their magnitude must be determined.

In this chapter, the principal types of controlled rectifier circuits are first described and their mode of operation briefly explained. A detailed analysis of the commonest circuit configurations is then carried out, and the procedures for determining the quantities essential in designing a rectifier

214

are indicated. The principles employed in analyzing and designing these selected circuits are applicable to any other controlled rectifier circuits. They are also applicable to uncontrolled rectifiers embodying diodes only.

5.1 TYPES OF CONTROLLED RECTIFIERS

A large variety of controlled rectifier circuits can be built, and each one may be classified in two ways. It may be classified according to the number of phases of the alternating voltage source supplying it, or it may be classified according to the number of pulses of current that pass through the load circuit during one cycle of the source voltage. It is helpful to determine both classifications for each circuit discussed.

Each of the possible controlled rectifier circuits differs from the others in transfer characteristic relating output voltage and angle α, in input harmonics, output ripple, and the required control range (i.e., range of variation of α) which determines the control circuit design. The selection of a particular configuration depends on the application requirements.

The single-phase half-wave circuit has already been discussed in detail in Section 3.3. As was explained there, a half-wave controlled rectifier provides only one pulse of load current during each cycle of the voltage source. An alternative classification for this circuit is therefore a "one-pulse" rectifier. The operation of such a rectifier with a resistive load is illustrated in Fig. 5.1. As already mentioned in Section 3.3, the direct

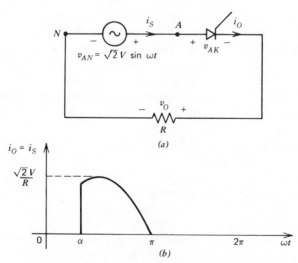

Fig. 5.1 Single-phase half-wave controlled rectifier.

component of source current introduced into the circuit by the assymetrical waveform of this controller has the disadvantage that the ac source must be ideal, and this renders the circuit impractical for most purposes.

The direct component of source current is eliminated by means of the bridge circuit shown in Fig. 5.2a. In that circuit one pair of thyristors conduct during each half cycle, giving full-wave rectification and a source current with alternating symmetry, as shown in Fig. 5.2b. This source current has no direct component, and consequently a nonideal source such as a transformer may be employed. There are now two pulses of load current per cycle of the voltage source, and this circuit may be classified either as a "single-phase full-wave" controlled rectifier, or as a "two-pulse" controlled rectifier.

The direct component of the source current may also be eliminated by employing a transformer with a centre-tapped secondary winding, as shown in Fig. 5.3. In effect the transformer has the function of transforming the source from a single-phase to a "two-phase" source, so that strictly speaking this is a "two-phase half-wave" controlled rectifier circuit. The time variation of load current obtained with this circuit is the same as that for the circuit of Fig. 5.2, as also is the waveform of the source current. Thus this is a "two-pulse" rectifier. However, since the ultimate voltage source is in fact only single-phase, this converter also is commonly called a "single-phase full wave" controlled rectifier.

The decision as to which of the circuits shown in Figs. 5.2 and 5.3 should be employed in a particular application depends on a number of factors, of which the chief are the cost of the various circuit elements, the source voltage available, and the load voltage required.

A reduction in the load-voltage fluctuation or ripple may be obtained if a three-phase source is available. This may be employed with the circuit of Fig. 5.4a, where the three voltage sources represent a balanced three-phase voltage source. The time variation of load voltage and current produced in a resistive load by this circuit is shown in Fig. 5.4b where, if the angle α were reduced to the minimum value, the voltage variation shown partly in broken line would apply. Figure 5.4b also shows that this is a three-pulse controlled rectifier. However, this circuit suffers from the disadvantage that it is a half-wave rectifier, since, with a resistive load, current only flows in any one of the three sources when the source voltage is positive. This means that the source currents have a dc component, and the three-phase source, like the single-phase source of Fig. 5.1, must be ideal.

Adoption of a bridge configuration of the three-phase circuit results in full-wave rectification and elimination of the dc component from the source current. Figure 5.5a shows such an arrangement in which the three voltage sources may be considered as the line-to-neutral secondary volt-

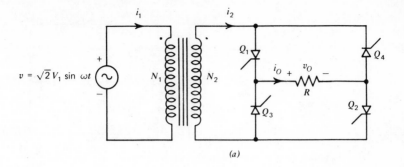

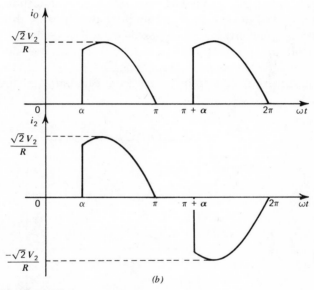

Fig. 5.2 Single-phase full-wave controlled bridge rectifier.

ages of a three-phase transformer. The curves showing the time variations of the load voltage and current and one of the line currents in Fig. 5.5b indicate that this is a six-pulse controlled rectifier. Again if the angle α is reduced to the minimum value the voltage variation shown partly in broken line would apply. By comparison with the voltage curve of Fig. 5.4b it may be seen that the ripple is much reduced. As the angle α is increased, the output voltage ripple also increases, but only at very large values of α does the load current become discontinuous.

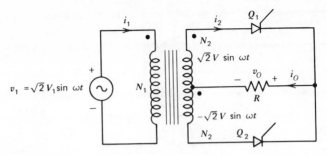

Fig. 5.3 Single-phase full-wave controlled rectifier.

The circuit of Fig. 5.6 also gives full-wave rectification and economizes in the cost of devices by replacing three thyristors by three diodes. This is called a semiconverter, as opposed to the full converter of Fig. 5.5. The main disadvantage of the semiconverter is that it is not capable of the inverter action explained in Section 5.2.3.

Since there is little to be gained from examining circuits that are rarely

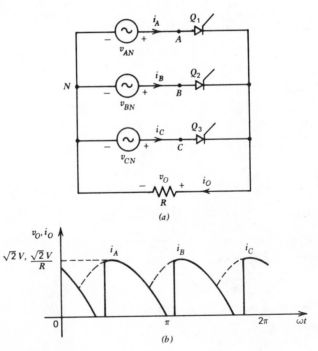

Fig. 5.4 Three-phase half-wave controlled rectifier.

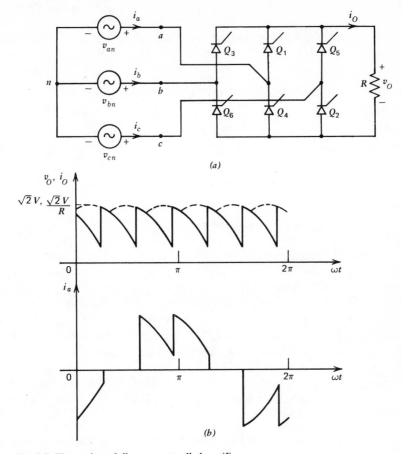

(a)

(b)

Fig. 5.5 Three-phase full-wave controlled rectifier.

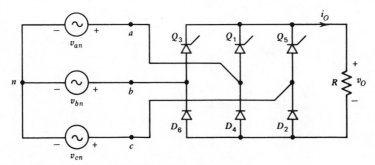

Fig. 5.6 Three-phase "semiconverter" rectifier.

used in practice, the remainder of this chapter is devoted to analyzing single-phase and three-phase full-wave converters.

5.2 SINGLE-PHASE FULL-WAVE CONTROLLED RECTIFIERS

If the transformers and thyristors of the circuits in Figs. 5.2a and 5.3 are regarded as ideal, then each of these circuits may be represented by the equivalent circuit of Fig. 5.7, where

$$v_{AN} = \sqrt{2} \ V \sin \omega t \quad \text{V} \tag{5.1}$$

$$v_{BN} = \sqrt{2} \ V \sin (\omega t + \pi) = -\sqrt{2} \ V \sin \omega t \quad \text{V} \tag{5.2}$$

This is the equivalent circuit of a "two-pulse" rectifier and should be compared with the one-pulse circuit of Fig. 5.1a and the three-pulse circuit of Fig. 5.4a. It will also be seen that thyristor Q_1 of Fig. 5.7 is equivalent to thyristor Q_1 of Fig. 5.3, but it is also equivalent to thyristors Q_1 and Q_2 in series of Fig. 5.2a. From this it may be concluded that the maximum reverse voltage applied to a thyristor in the circuit of Fig. 5.2a is only half of that applied to a thyristor in the circuit of Fig. 5.3, and this is one of the advantages of the bridge circuit. However, this advantage must be weighed against the inseparable disadvantage that the introduction into the circuit of two thyristors in series doubles the heat losses due to the internal resistance of the devices.

5.2.1 RL Load Circuit with Electromotive Force In the analysis that follows, the transformer is assumed to be ideal, but the effect of its leakage inductance is later considered qualitatively. The circuit to be analyzed is therefore that shown in Fig. 5.8 where, with switch SW open, no current flows in any branch.

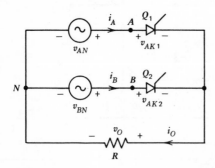

Fig. 5.7 Equivalent circuit for Figs. 5.2a and 5.3.

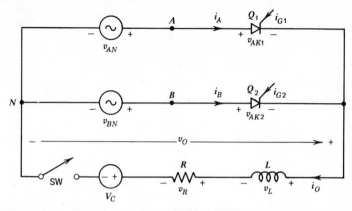

Fig. 5.8 Single-phase full-wave controlled rectifier with EMF in load circuit.

At this point it becomes advisable to repeat the definition of angle α at which thyristor Q_1 is turned on. It is convenient to choose $\alpha = 0$ as the operating condition in which the rectifier delivers maximum output current. For the converters whose common equivalent circuit is shown in Fig. 5.8, this means that $\alpha = 0$ at $\omega t = 0$, and these two quantities have a common origin in Fig. 5.9. The definition is thus: α is the interval in electrical angular measure by which the starting point of conduction is delayed by phase control in relation to the operation of the same circuit in which the thyristors are replaced by diodes. This angle is called the "delay angle", to distinguish it from the "angle of retard" of Chapter 4.

The circuit of Fig. 5.8 has two distinct modes of operation. For any given set of load-circuit parameters R, L, and V_C, it may broadly be said that when α is large, the load current is discontinuous, being made up of a series of pulses, each lasting for less than π radians. Under these circumstances, the two thyristor branches act alternately with the load circuit branch as completely independent single-phase half-wave controlled rectifiers. The range of α over which this mode of operation takes place depends on the values of the load-circuit parameters, and the determination of this range is discussed in the following parts of this section.

A single-phase half-wave controlled rectifier operating with discontinuous load current has been analyzed in detail in Section 3.3.5. The relationships determined in that section, as well as families of curves similar to those shown in Figs. 3.13 to 3.15 but prepared for a series of values of ϕ may be employed to describe the operation of the circuit of Fig. 5.8 in the discontinuous mode. However, since with this circuit there are two pulses of load current per cycle of source voltage, the normalized value of the average output current I_N will be twice that for the half-wave

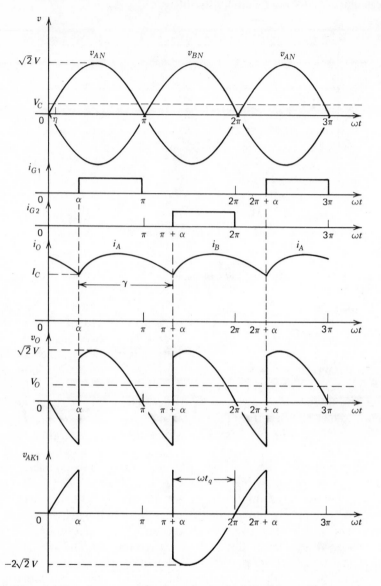

Fig. 5.9 Time variation of currents and voltages in circuit of Fig. 5.8—rectifier operation with continuous current.

circuit, so that the abscissae of Fig. 3.14 must be doubled. Also the normalized value of the rms output current I_{RN} will be $\sqrt{2}$ times that for the half-wave circuit, so that the abscissae of Fig. 3.15 must be multiplied by $\sqrt{2}$ in order to apply to the circuit of Fig. 5.8.

Figure 3.13 shows that the single-phase half-wave controlled rectifier can operate with a conduction angle γ that exceeds π rad. It is therefore necessary to determine how the circuit of Fig. 5.8 will operate when ϕ, m, and α are such as to give $\gamma > \pi$ in Fig. 3.13. Under these conditions, Q_1 will still be conducting when Q_2 is turned on at $\omega t = \pi + \alpha$. As may be seen from Fig. 5.9, at this instant $v_{AN} < 0$, and $v_{BN} > 0$; consequently Q_1 is commutated, and the load current i_O is transferred to Q_2. This transfer of current is also indicated in Fig. 5.9. In this mode of operation, the load current i_O is continuous, and the curves of Figs. 3.13 to 3.15 can no longer be applied to the circuit of Fig. 5.8.

In Fig. 5.10 is shown a diagram that defines the possible and permissible conditions of operation of the converter. A point $[\alpha, m]$ in this diagram for which $m \geqslant 1$ represents a condition in which no thyristors will turn on, since at no instant in the cycle is a forward voltage applied to them. A point $[\alpha, m]$ for which $m \leqslant -1$ represents a condition in which no thyristors will turn off, since at no instant is a reverse voltage applied to them. Under such conditions short circuit of the ac source through the thyristors results. An operating boundary therefore exists at $m = -1$, or in practice slightly above this line, since the thyristors need a finite time in which to turn off.

In the two quadrants of Fig. 5.10 is a family of curves drawn for a series of values of ϕ. For any chosen values of ϕ and α, these curves define the value of m at which the transition from discontinuous-current operation to continuous-current operation takes place as m becomes increasingly negative. These curves are computed from the expression for the normalized output current under discontinuous-current operation that, from equations 3.36 and 3.37, is

$$\frac{Z}{\sqrt{2} \, V} i_O = \sin(\omega t - \phi) - \left\{ \frac{m}{\cos \phi} - \left[\frac{m}{\cos \phi} - \sin(\alpha - \phi) \right] \epsilon^{(\alpha - \omega t)/\tan \phi} \right\} :$$

$$\alpha < \omega t < \alpha + \gamma \tag{5.3}$$

The method of computation is shown in the flow chart of Fig. 5.11.

A point $[\alpha, m]$ in the area marked 'A' in the small diagram of Fig. 5.10 represents a condition of discontinuous-current operation in which the thyristors turn on at the instant $\omega t = \sin^{-1} m = \eta$. As may be seen from Fig. 5.9, at any instant $\omega t < \eta$, a reverse voltage is applied to Q_1, since $V_C > v_{AN}$.

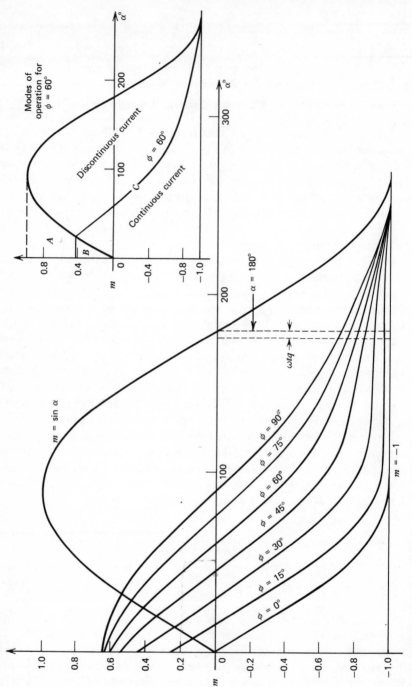

Fig. 5.10 Operating diagram for the circuit of Fig. 5.8.

A point $[\alpha, m]$ in the area marked 'B' in the small diagram of Fig. 5.10 represents a condition of continuous-current operation in which the thyristors turn on at $\omega t = \alpha \leqslant \sin^{-1} m$, since turn-on of one thyristor will commutate the other.

A point $[\alpha, m]$ in the area marked 'C' in the small diagram of Fig. 5.10, that is in the total area enclosed within the boundaries defined by $m = \sin \alpha$ and $m = -1$, represents a condition in which the thyristors will turn on at $\omega t = \alpha$, no matter whether the current is continuous or discontinuous.

To the left of the boundary formed by the line $\alpha = 180°$, the converter will operate stably over the range $0 \geqslant m > -1$ either in the discontinuous-current or the continuous-current mode. To the right of that boundary, the converter will operate stably only in the discontinuous-current mode. The reason for this may be seen from Figs. 5.8 and 5.9. If thyristor Q_1 is turned on at $\alpha < 180°$, then $v_{AN} > 0$, and $v_{BN} < 0$, so that for the mesh formed by the two thyristor branches

$$v_{AN} - v_{BN} + v_{AK2} - v_{AK1} = 0 \quad \text{V} \tag{5.4}$$

and since when Q_1 is turned on, $v_{AK1} = 0$, then

$$v_{AK2} = v_{BN} - v_{AN} < 0 \quad \text{V:} \qquad 0 < \alpha < 180° \tag{5.5}$$

and consequently Q_2 is commutated. If thyristor Q_1 is turned on at $\alpha > 180°$, then $v_{AK2} > 0$, and Q_2 continues to conduct. The ac source is then short circuited through the thyristors. Thus for any system of load-circuit angle ϕ, operation at a point $[\alpha, m]$ for $\alpha > 180°$ that lies below the curve for that value of ϕ is not permissible.

The first quadrant of Fig. 5.10 represents rectifier operation, since source V_C is absorbing energy. Figure 5.9 may be taken as a typical condition of continuous-current operation. In the fourth quadrant, the source V_C is delivering energy, and there are two possible conditions of operation. One is that the load circuit as a whole is delivering energy to the ac sources. In other words, the system is regenerating and functioning as an inverter from dc to fixed-frequency ac. The second is that the load circuit as a whole is absorbing energy. In other words, both ac and dc sources are supplying energy to the resistance of the load circuit. This is a condition intermediate between rectifier and inverter operation.

Under continuous-current operating conditions it is not difficult to decide what is happening in the fourth quadrant. An inspection of the waveform of v_O in Fig. 5.9 shows that, if the delay angle is increased until $\alpha > \pi/2$, then V_O becomes negative, that is, for continuous current operation, when $\alpha > \pi/2$ the system acts as an inverter. When $0 \leqslant \alpha \leqslant \pi/2$ and V_O

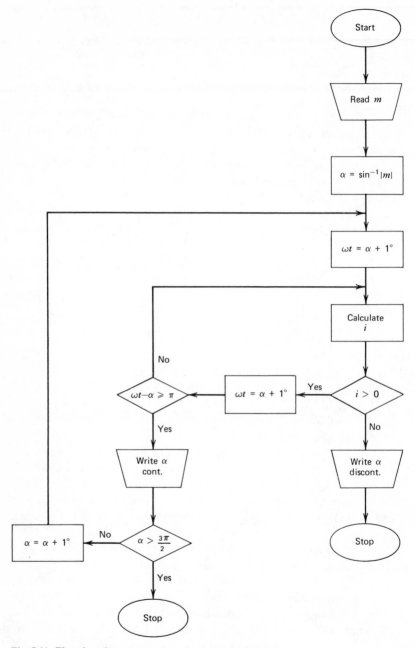

Fig. 5.11 Flowchart for computation of curves in Fig. 5.10.

is positive, the load-circuit resistance absorbs power from both dc and ac sources.

A condition of discontinuous-current operation in the fourth quadrant is illustrated in Fig. 5.12. The waveform of v_O in that figure is made up of three parts, which may be specified as follows:

$$v_O = V_C \text{ V:} \qquad i_O = 0 \text{ A} \tag{5.6}$$

$$v_O = v_{AN} \text{ V:} \qquad i_O = i_A \neq 0 \text{ A} \tag{5.7}$$

$$v_O = v_{BN} \text{ V:} \qquad i_O = i_B \neq 0 \text{ A} \tag{5.8}$$

Here the criterion which determines whether or not the system is inverting is supplied by the average power output of the converter. This is

$$P_O = \frac{1}{\pi} \int_\alpha^{\alpha + \gamma} v_O i_O d(\omega t) \quad \text{W} \tag{5.9}$$

If $P_O < 0$, then the system is inverting. If $P_O > 0$, then the system is in the intermediate condition of operation. To determine P_O it is necessary to obtain γ from equation 3.38.

As may be seen from Fig. 5.10, the range of α required to give complete control depends on the value of m and the mode of operation. For rectifier operation with $m = 0$, the required range is $0 \leqslant \alpha \leqslant 180°$. However if $m < 0$, and it is desired to reduce i_O smoothly to zero while the system is operating as an inverter, then values of $\alpha > 180°$ will be required. Such ranges are not shown in Figs. 5.9 and 5.12.

From the point of view of power circuit design and the rating of circuit components, the continuous current mode of operation is of the greatest importance. This mode is therefore discussed in detail.

As shown in Fig. 5.9, the load voltage v_O for continuous current operation is simply defined and may be analyzed into the series

$$v_O = V_O + \sum_{n=1}^{\infty} c_n \cos(n\omega t - \theta_n) \quad \text{V} \tag{5.10}$$

where the first term is the average output voltage

$$V_O = \frac{1}{\pi} \int_\alpha^{\pi + \alpha} \sqrt{2} \, V \sin \omega t \, d(\omega t) = \frac{2\sqrt{2}}{\pi} V \cos \alpha \quad \text{V} \tag{5.11}$$

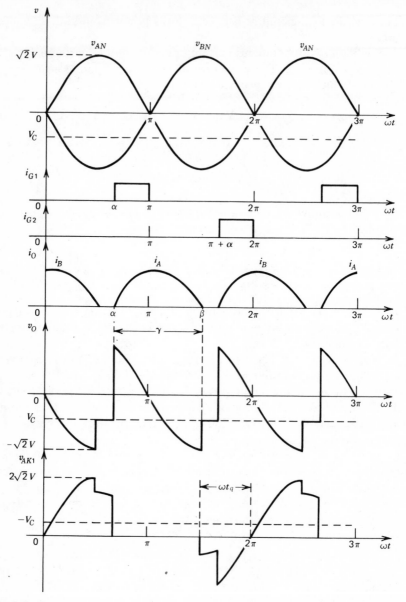

Fig. 5.12 Time variation of currents and voltages in circuit of Fig. 5.8—inverter operation with discontinuous current.

while

$$c_n = [a_n^2 + b_n^2]^{1/2} \quad V \tag{5.12}$$

$$\theta_n = \tan^{-1} \frac{a_n}{b_n} \tag{5.13}$$

and in equation 5.12

$$a_n = \frac{2}{\pi} \int_\alpha^{\alpha+\pi} v_O \sin n\omega t \, d(\omega t)$$

$$= \frac{2\sqrt{2}\,V}{\pi} \left[\frac{\sin(n+1)\alpha}{n+1} - \frac{\sin(n-1)\alpha}{n-1} \right] \quad V \tag{5.14}$$

$$b_n = \frac{2}{\pi} \int_\alpha^{\alpha+\pi} v_O \cos n\omega t \, d(\omega t)$$

$$= \frac{2\sqrt{2}\,V}{\pi} \left[\frac{\cos(n+1)\alpha}{n+1} - \frac{\cos(n-1)\alpha}{n-1} \right] \quad V \tag{5.15}$$

An inspection of the waveform of v_O in Fig. 5.9 shows that its fundamental frequency is twice that of the ac source. This means that all harmonics of the output voltage must be of the order $n = 2m$, where m is an integer. No odd harmonics in terms of the ac source frequency are therefore possible. In Fig. 5.13 are shown curves of the normalized harmonic amplitude $c_n/\sqrt{2}\,V$ versus α with n as a parameter. In Fig. 5.14 are shown corresponding curves of θ_n versus α.

The rms value of the load-circuit voltage v_O is

$$V_R = \left[\frac{1}{\pi} \int_\alpha^{\pi+\alpha} [\sqrt{2}\,V \sin \omega t]^2 \, d(\omega t) \right]^{1/2} = V \quad V \tag{5.16}$$

and the ripple voltage is

$$V_{RI} = [V_R^2 - V_O^2]^{1/2} = V \left[1 - \frac{8\cos^2\alpha}{\pi^2} \right]^{1/2} \quad V \tag{5.17}$$

The voltage ripple factor is then

$$K_v = \frac{V_{RI}}{V_O} \tag{5.18}$$

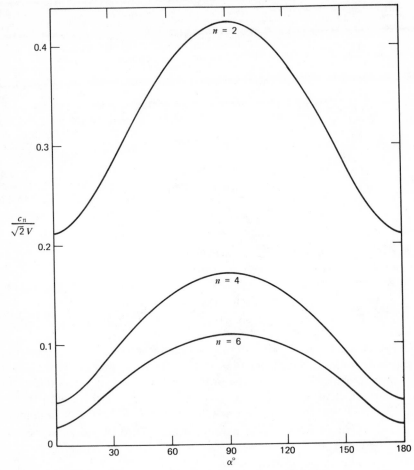

Fig. 5.13 $c_n/\sqrt{2}\ V$ versus α with n as parameter for circuit of Fig. 5.8.

A series describing the load current i_O may be obtained from equation 5.10 and the parameters of the load circuit, giving

$$i_O = I_O + \sum_{n=1}^{\infty} d_n \cos(n\omega t - \theta_n - \phi_n) \quad \text{A} \tag{5.19}$$

where

$$I_O = \frac{V_O - V_C}{R} \quad \text{A} \tag{5.20}$$

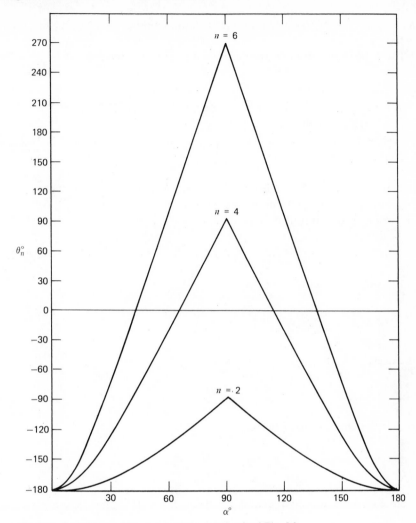

Fig. 5.14 θ_n versus α with n as parameter for circuit of Fig. 5.8.

$$d_n = \frac{c_n}{\left[R^2 + (n\omega L)^2 \right]^{1/2}} \quad \text{A} \tag{5.21}$$

$$\phi_n = \tan^{-1} \frac{n\omega L}{R} \tag{5.22}$$

These expressions, in conjunction with the curves of Figs. 5.13 and 5.14, may be employed to determine the instantaneous value of i_O at any point in the cycle.

The rms value of each harmonic component of the current is given by

$$I_{nR} = \frac{d_n}{\sqrt{2}} \quad \text{A} \tag{5.23}$$

and the ripple current is then

$$I_{RI} = \left[\Sigma I_{nR}{}^2 \right]^{1/2} \quad \text{A} \tag{5.24}$$

This quantity may be determined to any desired degree of accuracy by calculating the required number of terms of the series in equation 5.19. The rms value of the output current is

$$I_R = \left[I_O{}^2 + I_{RI}{}^2 \right]^{1/2} \quad \text{A} \tag{5.25}$$

and the current ripple factor is

$$K_i = \frac{I_{RI}}{I_O} \tag{5.26}$$

The average thyristor current is necessarily

$$I_Q = \frac{I_O}{2} \quad \text{A} \tag{5.27}$$

so that the rms thyristor current is

$$I_{QR} = \frac{I_R}{\sqrt{2}} \quad \text{A} \tag{5.28}$$

From Fig. 5.9 the time available for turn-off of the thyristors is seen to be

$$t_q = \frac{\pi - \alpha}{\omega} \quad \text{s} \tag{5.29}$$

and this must exceed t_{off} for the thyristors employed.

When thyristor Q_2 in Fig. 5.8 is conducting, v_{AK2} is zero, so that for the mesh formed by the two thyristor branches

$$v_{AK1} = v_{AN} - v_{BN} \quad \text{V} \tag{5.30}$$

The time variation of v_{AK1} is shown in Fig. 5.9, and from this it may be seen that the maximum value of forward or reverse voltage applied to the thyristors of Fig. 5.8 is

$$v_{AK\max} = \pm 2\sqrt{2}\ V \quad \text{V} \tag{5.31}$$

The waveform of v_{AK1} in Fig. 5.9 is that which would appear across the thyristor of Fig. 5.3; however as already remarked, these have a voltage applied to them which is twice that applied to the thyristors of Fig. 5.2a.

In the circuit of Fig. 5.2a the bridge circuit might be excited simply by means of an alternating voltage source. It is normally desirable that a converter shall be excited via a transformer for isolation, even when the supply voltage is of the required value, and a one-to-one ratio transformer must be fitted. The transformer of the circuit of Fig. 5.3 is of course an indispensable component.

When designing a converter, it is important to determine the rating of the transformer required, and for this purpose the actual circuit configuration to be employed must be taken into account.

For the circuit of Fig. 5.2a, the rms current in the secondary winding of the transformer is the same as the rms load current, so that

$$I_2 = I_R \quad \text{A} \tag{5.32}$$

and the rating of the secondary winding is

$$S_2 = V I_R \quad \text{VA} \tag{5.33}$$

If as usual the turns ratio is defined as $n = N_1 | N_2$, where N_1 and N_2 are the numbers of turns in the primary and secondary windings respectively, then the rating of the primary winding is

$$S_1 = nV\frac{I_R}{n} = V I_R = S_2 \quad \text{VA} \tag{5.34}$$

For the circuit of Fig. 5.3, the rms current in the secondary windings of the transformer is the same as that in the thyristors, that is,

$$I_2 = I_{QR} = \frac{I_R}{\sqrt{2}} \quad \text{A} \tag{5.35}$$

but since for the same source voltage and rms load circuit voltage there must be twice as many turns on the secondary side of this transformer as would be required on the transformer to supply the bridge circuit of Fig. 5.2a, it follows that the rating of the whole secondary winding is

$$S_2 = 2 V I_{QR} = \sqrt{2}\ V_R I_R \quad \text{VA} \tag{5.36}$$

The n must now be the turns ratio between the primary winding and one half of the secondary winding. And since current flows in the primary winding during both half cycles, but in each half of the secondary winding during alternate half cycles, it follows that

$$I_1 = \frac{\sqrt{2}\,I_2}{n} = \frac{I_R}{n} \quad A \tag{5.37}$$

The rating of the primary winding is therefore

$$S_1 = n\frac{VI_R}{n} = VI_R \quad VA \tag{5.38}$$

While the ratings of the primary windings of the transformers are the same for the two circuits, it may be seen from equations 5.33 and 5.36 that the circuit of Fig. 5.3 requires a larger transformer than does the bridge circuit of Fig. 5.2a, since the rating of the secondary winding is greater.

Example 5.1   For the single-phase bridge controlled rectifier shown in Fig. E5.1 $L = 20$ mH, $R = 4.35\,\Omega$, $V_C = 0$, and the delay angle $\alpha = 75°$. Calculate the following:

(a)   The average output current.
(b)   The rms output current.
(c)   The average and rms thyristor currents.
(d)   The power factor at the ac source.

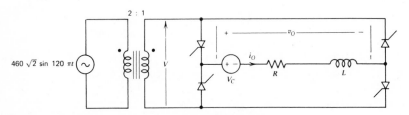

Fig. E5.1 Single-phase full-wave controlled bridge rectifier.

Solution   Since the transformer ratio is $2:1$, $V = 230\,V$

$$\phi = \tan^{-1}\frac{\omega L}{R} = \tan^{-1}\frac{120\pi \times 20 \times 10^{-3}}{4.35} = \tan^{-1}\frac{7.54}{4.35} = 60°$$

For $\alpha = 75°$ and $\phi = 60°$, Fig. 5.10 shows that the current is discontinuous, thus the curves of Figs. 3.7 and 3.8 may be employed. The base current is

$$\frac{\sqrt{2}\,V}{Z} = \frac{\sqrt{2} \times 230}{[4.35^2 + 7.54^2]^{1/2}} = 37.7 \quad A$$

(a) From Fig. 3.7, for $\alpha = 75°$, $\phi = 60°$, $I_N = 0.25$. Since this is a two-pulse rectifier, I_N must be multiplied by two. Thus

$$I_O = 2 \times 0.25 \times 37.7 = 18.9 \quad A$$

(b) From Fig. 3.8, for $\alpha = 75°$, $\phi = 60°$, $I_{RN} = 0.42$. Since this is a two-pulse rectifier, I_{RN} must be multiplied by $\sqrt{2}$. Thus

$$I_R = \sqrt{2} \times 0.42 \times 37.7 = 22.4 \quad A$$

(c) Average thyristor current is

$$I_Q = \frac{I_O}{2} = \frac{18.9}{2} = 9.45 \quad A$$

The rms thyristor current is

$$I_{QR} = \frac{I_R}{\sqrt{2}} = \frac{22.4}{\sqrt{2}} = 15.8 \quad A$$

(d) Active power delivered to the load circuit is

$$P_L = RI_R^2 = 4.35 \times 22.4^2 = 2180 \quad W$$

Input apparent power is

$$S = VI_R = 230 \times 22.4 = 5150 \quad VA$$

$$\text{input power factor} = \frac{2180}{5150} = 0.423$$

Example 5.2 For the converter and load circuit of Example 5.1, $L = 40$ mH, $R = 4\Omega$, $V_C = 80$ V and $\alpha = 30°$. Calculate the following:

(a) The average output current.
(b) The rms output current.

(c) The power delivered to source V_C.

(d) The power factor at the ac source.

Solution   As before, $V = 230\,\text{V}$.

$$\phi = \tan^{-1} \frac{\omega L}{R} = \tan^{-1} \frac{15.1}{4} = 75°$$

$$m = \frac{V_C}{\sqrt{2}\,V} = \frac{80}{\sqrt{2} \times 230} = 0.25$$

For $m = 0.25$ and $\phi = 75°$, Fig. 5.10 shows that a delay angle of $\alpha = 30°$ is possible and that the load current will be continuous.

(a) From equation 5.11

$$V_O = \frac{2\sqrt{2}}{\pi}\,V \cos\alpha = \frac{2\sqrt{2}}{\pi} \times 230 \cos 30° = 179 \quad \text{V}$$

From equation 5.20

$$I_O = \frac{V_O - V_C}{R} = \frac{179 - 80}{4} = 24.8 \quad \text{A}$$

(b) The rms output current may be determined from equation 5.25 but it must first be decided how many terms in the series of equation 5.19 should be evaluated to give a sufficiently accurate value of I_{RI}. Consider the second harmonic: from Fig. 5.13 for $\alpha = 30°$

$$\frac{c_n}{\sqrt{2}\,V} = 0.28$$

$$\therefore c_n = 0.28\sqrt{2} \times 230 = 91.0$$

From equation 5.21

$$d_n = \frac{c_n}{\left[R^2 + (n\omega L)^2\right]^{1/2}} = \frac{91.0}{\left[4^2 + (2 \times 15.1)^2\right]^{1/2}} = 2.83 \quad \text{A}$$

From equation 5.23 the rms value of the second harmonic is then

$$I_{2R} = \frac{2.83}{\sqrt{2}} = 2.0 \quad \text{A}$$

Since the second harmonic rms value is already small in comparison with I_O, higher harmonics may be neglected, and from equation 5.25

$$I_R = [24.8^2 + 2.0^2]^{1/2} = 24.8 \quad \text{A}$$

(c)   The power delivered to source V_C is

$$P_C = I_O V_C = 80 \times 24.8 = 1980 \quad \text{W}$$

(d)   Power delivered to the load-circuit resistance is

$$P_R = RI_R^2 = 4 \times 24.8^2 = 2460 \quad \text{W}$$

Input apparent power is

$$S = VI_R = 230 \times 24.8 = 5700 \quad \text{VA}$$

$$\text{input power factor} = \frac{P_C + P_R}{S} = \frac{1980 + 2460}{5700} = 0.78$$

Example 5.3   For the converter and load circuit of Example 5.1, $L = 40$ mH, $R = 4\,\Omega$, $V_C = -80$ V, and $\alpha = 30°$. Calculate the following:

(a)   The average output current.
(b)   The rms output current.
(c)   The average power delivered by source V_C.
(d)   The power factor at the ac source.

From Example 5.2, $\phi = 75°$, $m = -0.25$, $V_O = 179$ V, $I_{2R} = 2.0$ A.
    For $m = -0.25$ and $\phi = 75°$, Fig. 5.10 shows that the load current will be continuous.

(a)   From equation 5.20

$$I_O = \frac{V_O - V_C}{R} = \frac{179 + 80}{4} = 64.8 \quad \text{A}$$

(b)   As in Example 5.2, the effect of the harmonic currents on the rms magnitude of the load current is negligible, and

$$I_R \cong I_O = 64.8 \quad \text{A}$$

(c)   Power delivered by source V_C is

$$P_C = I_O V_C = 80 \times 64.8 = 5180 \quad \text{W}$$

(d)   Power delivered to the load-circuit resistance is

$$P_R = RI_R{}^2 = 4 \times 64.8^2 = 16.8 \times 10^3 \quad \text{W}$$

Input active power from the ac source is

$$P = P_R - P_C = (16.8 - 5.2)10^3 = 11,600 \quad \text{W}$$

Input apparent power is

$$S = VI_R = 230 \times 64.8 = 14.9 \times 10^3 \quad \text{VA}$$

Input power factor is

$$\frac{P}{S} = \frac{11.6}{14.9} = 0.778$$

5.2.2 Effect of Transformer Leakage Inductance   The circuit of Fig. 5.8 is an ideal model based on the assumption that both the source and the transformer have zero internal impedance. While for all practical purposes this may be true for the source, it is not true for the transformer, nor yet is it desirable that it should be. It is as well at this stage therefore to determine what the effect of a nonideal transformer will be.

Figure 5.15 shows the equivalent circuit for a full-wave single-phase rectifier with an inductance L_a included in each of the thyristor branches. This inductance represents the effect of the leakage inductances of the transformer referred to the secondary winding. The remaining transformer equivalent-circuit parameters are omitted, since in comparison with the leakage inductance they have a negligible effect on the operation of the circuit.

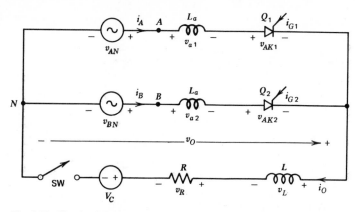

Fig. 5.15 Circuit of Fig. 5.8 with transformer leakage inductance added.

The effect of L_a is that the instantaneous commutation hitherto assumed cannot in fact take place, since when one thyristor is turned on, the two thyristor currents cannot change instantaneously, one falling to zero, the other rising to the value of i_O at $\omega t = n\pi + \alpha$.

Figure 5.16 shows the time variation of the voltages and currents in the circuit of Fig. 5.15 a long time after switch SW has been closed. It may there be seen that, although each thyristor commences to conduct at the instant at which it is turned on, its current must increase from zero during a finite overlap interval of μ/ω s, while during the same interval of time the current in the other thyristor decreases to zero. The load current during this interval of overlap is therefore the sum of the two thyristor currents.

Comparison of the curves for v_O in Figs. 5.9 and 5.16 also shows that a consequence of the presence of leakage inductance is a reduction in the average output voltage V_O below the value predicted in equation 5.11. With normal values of transformer leakage inductance this reduction does not exceed 5% of the theoretical value of V_O for the type of converter discussed in this section.

The presence of leakage inductance has the disadvantage that under worst case conditions a step of voltage equal to $\sqrt{2}\ V$ is applied to the nonconducting thyristors. The thyristors may be protected against this as described in Sections 3.4.4 and 3.4.5. On the other hand, some leakage inductance is desirable to provide di|dt protection of the thyristors as well as some input line filtering. Indeed if at any time it is possible to build a rectifier without an isolating transformer, it is desirable to introduce some inductance into the source circuit to achieve just this filtering effect.

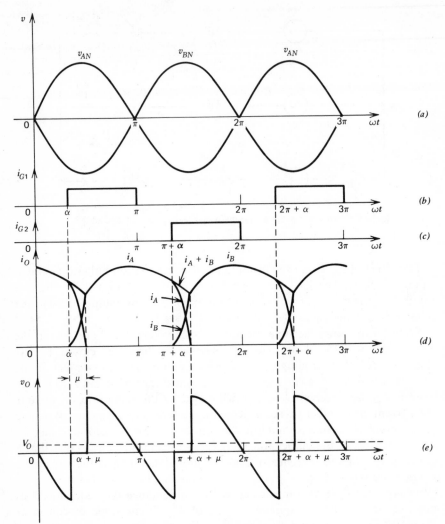

Fig. 5.16 Time variation of v_O and i_O in circuit of Fig. 5.15.

5.3 THREE-PHASE CONTROLLED RECTIFIERS

It has already been pointed out in Section 5.1 that the three-phase half-wave controlled rectifier of Fig. 5.4 is impractical for most purposes, since the source currents would have dc components, and the three-phase voltage source would have to be ideal. Since this is so, there is little point in examining that circuit, other than to note that it is a three-pulse converter, and that its equivalent circuit fits into a sequence which begins with those of Figs. 5.1a and 5.7.

The circuit of Fig. 5.5a is a three-phase full-wave converter in which ideal sources are not required, and this is therefore the arrangement most commonly met in practice. As in the case of the single-phase full-wave controlled rectifier, the ac source would normally be a transformer fitted for isolation, if for no other purpose. Although it was stated in Section 5.1 that this was a six-pulse converter, and this statement was illustrated by Fig. 5.5b, it was not shown why this should be the case. Before analyzing the circuit therefore it is necessary first to make this clear. For this purpose it is best to consider the simple case of a purely resistive load circuit. The converter is shown with a general load circuit in Fig. 5.17a, and for the immediately following discussion it is assumed that $L = 0$, and $V_C = 0$.

Consider in Fig. 5.17a that all thyristors are turned on throughout the entire cycle, or alternatively that the thyristors are replaced by diodes giving an uncontrolled rectifier. The factor determining which thyristors will conduct at any instant is the combination of the three source voltages v_{an}, v_{bn}, v_{cn}, which at that instant give the largest value of voltage $v_{pq} = v_O$. The "combinations of the source voltages" are of course simply the line-to-line voltages v_{ab}, v_{bc}, and v_{ca}, and voltage v_{pq} will reach a maximum value when any one of these three line-to-line voltages is at its positive or negative maximum.

The line-to-line voltages are shown in Fig. 5.17b. When v_{ab} is at its positive maximum at $\omega t = \pi/2$, thyristors Q_1 and Q_6 are conducting, giving

$$v_{pq} = v_O = v_{ab} \quad \text{V:} \qquad \omega t = \pi/2 \tag{5.39}$$

When v_{ab} is at its negative maximum—or v_{ba} is at its positive maximum— thyristors Q_3 and Q_4 are conducting, giving

$$v_{pq} = v_O = -v_{ab} = v_{ba} \quad \text{V:} \qquad \omega t = 3\pi/2 \tag{5.40}$$

This same argument may be pursued for each of the other two line-to-line voltages.

If the waveforms of v_{ba}, v_{cb}, and v_{ac} are added to Fig. 5.17b, as shown in broken line, then the appropriate pair of thyristors may be considered to be conducting when any one of the six voltage waveforms has a greater positive value than any of the other five, and this will be the value of v_O. From this it may be seen that six pulses of current per cycle will flow through the load circuit. It follows that the equivalent circuit for a three-phase full-wave rectifier is that shown in Fig. 5.18.

As in the case of the single-phase full-wave bridge rectifier, each thyristor shown in the equivalent circuit represents two thyristors in series in the

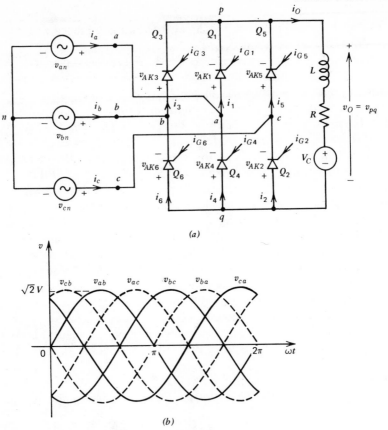

(a)

(b)

Fig. 5.17 Three-phase full-wave controlled rectifier.

actual circuit. These two thyristors are not turned on and commutated simultaneously however, as they were in the single-phase converter of Fig. 5.2a. Table 5.1 shows the correspondence between real and equivalent voltages and thyristors in the two circuits.

The waveforms of the equivalent source voltages are shown in Fig. 5.18b, and the phasor diagrams of the real three-phase source of Fig. 5.17a and the equivalent six-phase source of Fig. 5.18a are shown superposed in Fig. 5.19. Thus if

$$v_{an} = \frac{\sqrt{2}\,V}{\sqrt{3}} \sin(\omega t - 30°) \quad \text{V} \tag{5.41}$$

then

$$v_{AN} = v_{ab} = \sqrt{2}\,V \sin \omega t \quad \text{V} \tag{5.42}$$

Table 5.1

Interval	Actual Circuit		Equivalent Circuit	
	Voltage Applied to Load	Thyristors Conducting	Voltage Applied to Load	Thyristor Conducting
$\pi/3 < \omega t < 2\pi/3$	v_{ab}	Q_6, Q_1	v_{AN}	Q_1
$2\pi/3 < \omega t < \pi$	v_{ac}	Q_1, Q_2	v_{BN}	Q_2
$\pi < \omega t < 4\pi/3$	v_{bc}	Q_2, Q_3	v_{CN}	Q_3
$4\pi/3 < \omega t < 5\pi/3$	v_{ba}	Q_3, Q_4	v_{DN}	Q_4
$5\pi/3 < \omega t < 2\pi$	v_{ca}	Q_4, Q_5	v_{EN}	Q_5
$2\pi < \omega t < 7\pi/3$	v_{cb}	Q_5, Q_6	v_{FN}	Q_6

and

$$v_{BN} = \sqrt{2}\, V \sin(\omega t - \pi/3) \quad \text{V}$$

$$v_{CN} = \sqrt{2}\, V \sin(\omega t - 2\pi/3) \quad \text{V}$$

$$v_{DN} = \sqrt{2}\, V \sin(\omega t - \pi) \quad \text{V} \tag{5.43}$$

$$v_{EN} = \sqrt{2}\, V \sin(\omega t - 4\pi/3) \quad \text{V}$$

$$v_{FN} = \sqrt{2}\, V \sin(\omega t - 5\pi/3) \quad \text{V}$$

The definition of the angle α at which thyristor Q_1 of Fig. 5.18a is turned on is exactly the same as for the single-phase full-wave controlled rectifier, and this was given in Section 5.2.1. In this case, if the thyristors of Fig. 5.18a were replaced by diodes, then that replacing Q_1 would begin to conduct at $\omega t = \pi/3$, the instant at which v_{AN} acquires a greater positive value than any of the other five source voltages whose waveforms are shown in Fig. 5.18b. Thus for this converter, $\alpha = 0$ at $\omega t = \pi/3$ as shown in Fig. 5.20. With this reference point determined, the procedure for the analysis of the six-pulse converter differs in no essential way from that of the two-pulse converter of Fig. 5.8.

The operation of the converter with the general load circuit shown in Fig. 5.17 may now be discussed. As in the case of the single-phase

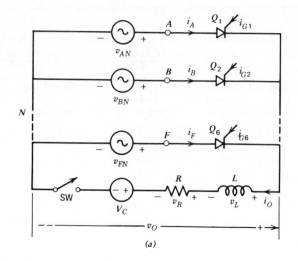

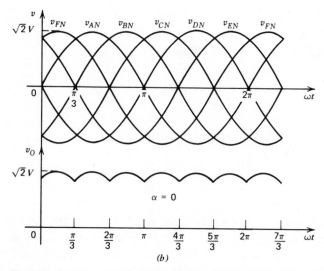

Fig. 5.18 Equivalent circuit for the converter of Fig. 5.17.

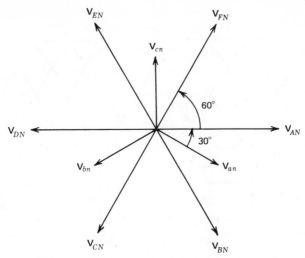

Fig. 5.19 Phasor diagram of real and equivalent voltage sources.

full-wave controlled rectifier, the circuit of Fig. 5.17 will operate with discontinuous load current if α is sufficiently great and the inductance of the load circuit is low. Results similar to those obtained in Section 3.3.5 for the single-phase half-wave rectifier but prepared for a series of values of ϕ may be employed to describe the operation of the equivalent circuit of Fig. 5.18 in the discontinuous mode. Since with this circuit there are six pulses of load current per cycle of source voltage, the normalized value of the average output current I_N will be six times that of the single-phase half-wave circuit, so that the abscissae of Fig. 3.14 must be multiplied by six. Furthermore, the values of α in Fig. 3.14 must be reduced by 60° owing to the datum from which α is now measured. Thus the curves marked $\alpha = 0$ and $\alpha = 30°$ in Fig. 3.14 do not apply to the three-phase full-wave rectifier. Since the line-to-line voltage of the three-phase ac source is applied to the load circuit of Fig. 5.17a, it also follows that

$$I_O = \frac{\sqrt{2}\,V}{Z} I_N \quad \text{A} \tag{5.44}$$

The normalized value of the rms output current I_{RN} will be $\sqrt{6}$ times that for the single-phase half-wave circuit. Thus the abscissae of Fig. 3.15 must be multiplied by $\sqrt{6}$ to apply to the three-phase full-wave converter, and the angle α must be reduced by 60°. Once again, the curves marked

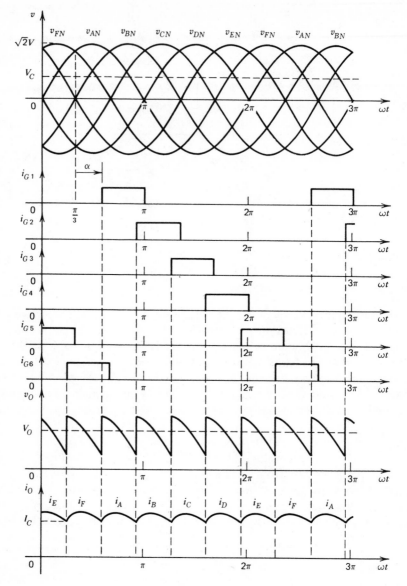

Fig. 5.20 Time variations of voltages and currents in the equivalent circuit of Fig. 5.18*a*—rectifier operation.

$\alpha = 0$ and $\alpha = 30°$ do not apply, and

$$I_R = \frac{\sqrt{2}\,V}{Z} I_{RN} \quad \text{A} \tag{5.45}$$

Since in the three-phase full-wave converter the range of α over which discontinuous operation takes place is much smaller than that for the single-phase full-wave rectifier, this mode of operation is even less important from the point of view of the power-circuit designer than in the single-phase case.

Figure 5.20 shows the time variations of certain of the circuit variables for the equivalent circuit of Fig. 5.18 when the converter is operating as a controlled rectifier in the continuous-current mode. The gating signals are shown in their phase relationship to the line-to-neutral voltage v_{AN} of Fig. 5.18.

In Fig. 5.21 is shown a diagram corresponding to that of Fig. 5.10 for the single-phase full-wave controlled rectifier. Much of the description of Fig. 5.10 applies to Fig. 5.21 also. Owing to the datum chosen for $\alpha = 0$, the upper boundary of the field of operation is now defined by $m = \sin(\alpha + \pi/3)$. The lower boundary is defined as before by $m = -1$. This field is again divided into two subfields by the line $\alpha = 180°$, and the reason for this is explained in the following when the time available for turn off is discussed in relation to Fig. 5.25.

The family of curves for the series of values of ϕ again define the condition at which the transition takes place from discontinuous to continuous-current operation. They are obtained from equation 5.3 by means of the computation illustrated in Fig. 5.11, with allowance made for the change in the datum of α and the possible maximum value of γ, which is here $60°$.

The first quadrant of Fig. 5.21 represents rectifier operation, since source V_C is absorbing energy. Fig. 5.20 may be taken as a typical condition of continuous-current operation. In the fourth quadrant, the source V_C is delivering energy, and there are two possible conditions of operation. One is that the load circuit as a whole is delivering energy to the ac sources. In other words the system is regenerating and functioning as an inverter from dc to fixed-frequency ac. The second is that the load circuit as a whole is absorbing energy. In other words both ac and dc sources are supplying energy to the resistance of the load circuit. This is a condition intermediate between rectifier and inverter operation.

Under continuous-current operating conditions it is not difficult to decide what is happening in the fourth quadrant. An inspection of the waveform of v_O in Fig. 5.20 shows that, if the delay angle is increased until

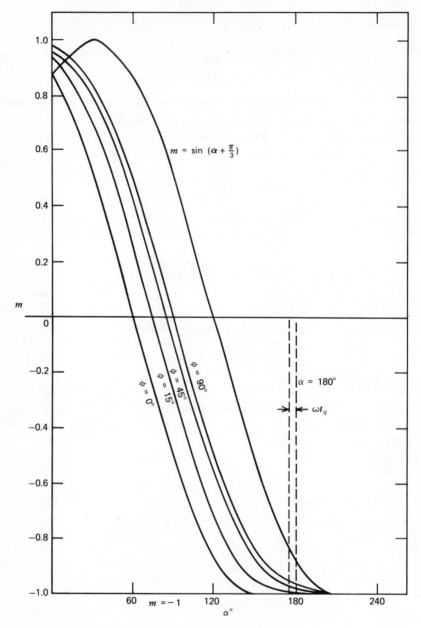

Fig. 5.21 Operating diagram for the circuit of Fig. 5.18.

248

$\alpha > \pi/2$, then V_O becomes negative, that is, for continuous current opera-
tion, when $\alpha > \pi/2$ the system acts as an inverter. When $0 \leqslant \alpha \leqslant \pi/2$ and V_O
is positive, the load-circuit resistance absorbs power from both dc and ac
sources.

A condition of discontinuous-current operation in the fourth quadrant is
illustrated in Fig. 5.22. The waveform of v_O in that figure is made up of
two parts, which may be specified as follows:

$$v_O = V_C \text{ V:} \quad i_O = 0 \quad \text{A} \tag{5.46}$$

$$v_O = v_{AN} \text{ V:} \quad i_O = i_A \neq 0 \quad \text{A} \tag{5.47}$$

Here the criterion which determines whether or not the system is inverting
is supplied by the average power output of the converter. This is

$$P_O = \frac{3}{\pi} \int_\alpha^{\alpha+\gamma} v_O i_O d(\omega t) \quad \text{W} \tag{5.48}$$

If $P_O < 0$, then the system is inverting. If $P_O > 0$, then the system is in the
intermediate condition of operation.

As may be seen from Fig. 5.21, the range of α required to give complete
control depends on the value of m and the mode of operation. For rectifier
operation with $m = 0$, the required range is $0 \leqslant \alpha \leqslant 120°$. For operation in
the fourth quadrant of Fig. 5.21, larger values of α will be required to
achieve complete turn off of the converter.

Now that the significance of the various areas of Fig. 5.21 has been
established, the remainder of this section is devoted to continuous-current
operation.

Once again, v_O may be represented by the series

$$v_O = V_O + \sum_{n=1}^{\infty} c_n \cos(n\omega t - \theta_n) \quad \text{V} \tag{5.49}$$

where in this case, from Fig. 5.20,

$$V_O = \frac{3}{\pi} \int_{\alpha+(\pi/3)}^{\alpha+(2\pi/3)} \sqrt{2} \, V \sin \omega t \, d(\omega t) = \frac{3\sqrt{2} \, V}{\pi} \cos \alpha \quad \text{V} \tag{5.50}$$

It may be noted that as in the case of the single-phase converter, for which
the corresponding expression is given in equation 5.11, V_O is a function of
cosine α.

An inspection of the waveform of v_O in Fig. 5.20 shows that its
fundamental frequency is six times that of the ac source. This means that

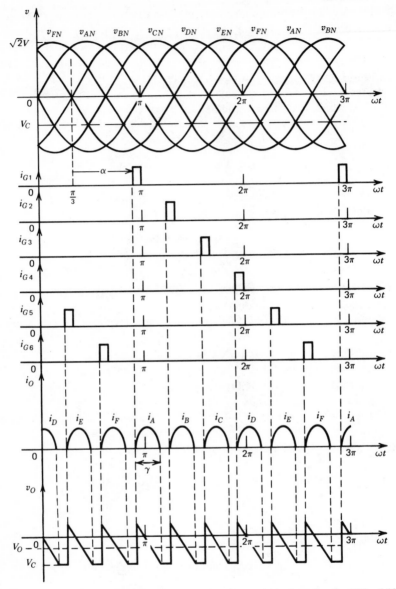

Fig. 5.22 Time variations of voltages and currents in the equivalent circuit of Fig. 5.18a—inverter operation.

all harmonics of the output voltage must be of the order $n = 6m$, where m is an integer. This illustrates very clearly the advantage of a polyphase source.

In equation 5.49

$$c_n = \left[a_n^2 + b_n^2 \right]^{1/2} \quad \text{V} \tag{5.51}$$

and

$$\theta_n = \tan^{-1} \frac{a_n}{b_n} \quad \text{rad} \tag{5.52}$$

As a consequence of the fact that the fundamental frequency of v_O is 6ω, it is possible to define the coefficients a_n and b_n in the more readily integrable forms that follow, namely:

$$a_n = \frac{6}{\pi} \int_{\alpha + (\pi/3)}^{\alpha + (2\pi/3)} v_O \sin n\omega t \, d(\omega t) \quad \text{V:} \qquad n = 6, 12, 18 \ldots \tag{5.53}$$

$$b_n = \frac{6}{\pi} \int_{\alpha + (\pi/3)}^{\alpha + (2\pi/3)} v_O \cos n\omega t \, d(\omega t) \quad \text{V:} \qquad n = 6, 12, 18 \ldots \tag{5.54}$$

In Fig. 5.23 are shown curves of the normalized harmonic amplitude $c_n | \sqrt{2} \, V$ versus α with n as a parameter. In Fig. 5.24 are shown corresponding curves of θ_n versus α.

The effect of transformer leakage reactance on the value of V_O is again to reduce it below the theoretical value given by equation 5.50. For a given per-unit leakage reactance per phase, the effect is more pronounced for a polyphase than for a single-phase converter, due to the increased number of commutations per cycle.

The rms value of the load-circuit voltage v_O is

$$V_R = \left[\frac{3}{\pi} \int_{\alpha + (\pi/3)}^{\alpha + (2\pi/3)} v_{AN}^2 \, d(\omega t) \right]^{1/2}$$

$$= \sqrt{2} \, V \left[0.5 + \frac{3\sqrt{3}}{4\pi} \cos 2\alpha \right]^{1/2} \quad \text{V} \tag{5.55}$$

and the ripple voltage is, as before

$$V_{RI} = \left[V_R^2 - V_O^2 \right]^{1/2} \quad \text{V} \tag{5.56}$$

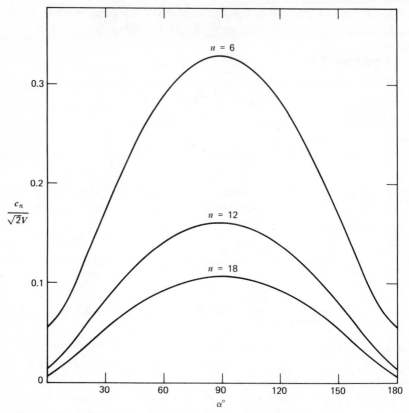

Fig. 5.23 $c_n/\sqrt{2}\,V$ versus α with n as parameter for circuits of Figs. 5.17 and 5.18.

The voltage ripple factor is then

$$K_v = \frac{V_{RI}}{V_O} \tag{5.57}$$

From the series of equation 5.49 a series describing i_O may be obtained and employed to determine the various output current values by means of equations 5.20 to 5.26.

While the output quantities are most readily determined from the equivalent circuit of Fig. 5.18a, it is most convenient to work with the actual circuit of Fig. 5.17a when determining thyristor and source voltages and currents. The time variations of the line-to-line source voltages and the load current are therefore shown in Fig. 5.25 and should be compared with the corresponding curves in Fig. 5.20. The thyristor currents that make up

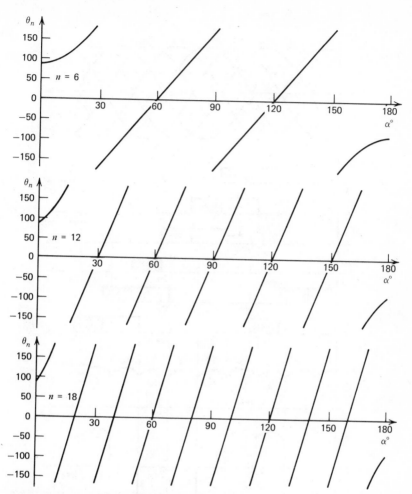

Fig. 5.24 θ versus α for circuits of Figs. 5.17 and 5.18.

the load current i_O are indicated on each cusp of the i_O curve. As may be seen from Table 5.1, current will flow in line a whenever the line-to-line voltage appearing at the load-circuit terminals is v_{ab}, v_{ac}, v_{ba}, or v_{ca}; that is, during two thirds of the cycle. The resulting line a current is shown in Fig. 5.25, and the other two line currents will have similar waveforms with phase shifts of $\pm 120°$.

At this point it becomes apparent that in one important respect the circuit of Fig. 5.18a is not truly "equivalent" to that of Fig. 5.17a. As may be seen from Table 5.1, while each thyristor of the equivalent circuit

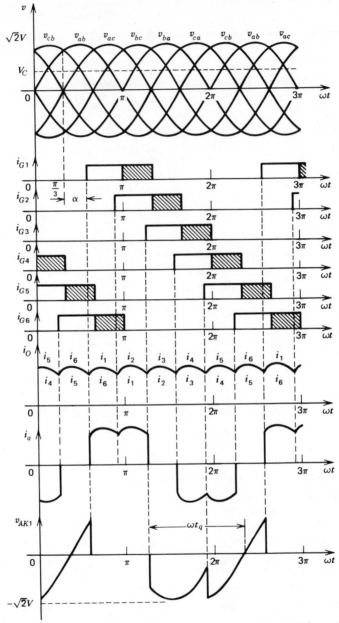

Fig. 5.25 Time variations of voltages and currents in the circuit of Fig. 5.17a.

conducts only one of the six pulses of current occurring in each cycle, each thyristor of the actual circuit of Fig. 5.17a conducts two of the six pulses, and this requires a modification to the gating signals shown in Fig. 5.20.

If the gating signals are to be so far delayed that zero current flows, but any reduction in α would result in load current, then the thyristors must be turned on just at the instant at which the forward voltage applied to them disappears. In Fig. 5.25 at $\omega t = \pi$, that is $\alpha = 2\pi/3$, v_{ab} is going negative. If it is assumed that V_C is zero, then a reduction of α below the value $2\pi/3$ must result in turn on of thyristors Q_6 and Q_1; thus a gating signal must be present on these two thyristors at $\omega t = \pi$. This fixes the end point of the required pulse of i_{G6} and the starting point of that of i_{G1}. The starting points and end points of all the other gating signals for zero load current may be related to these two, and the resulting gating signals for maximum delay angle, giving zero output current, are shown in Fig. 5.25 by the shaded parts of the rectangular pulses of i_{G1} to i_{G6}. Any increase of output calls for an extension to the left of these zero-output gating signals. If the shaded parts of the pulses were not present, then over the range $\pi/3 \leqslant \alpha < 2\pi/3$ only one thyristor would be gated, and no current would flow.

The forward or reverse voltage applied to a thyristor at any instant will depend on the source voltages and on which thyristors are conducting at that instant. Voltage v_{AK1} in Fig. 5.17a may be determined by considering the intervals in which i_1, i_3, and i_5 are flowing. Clearly when i_1 is flowing

$$v_{AK1} = 0 \quad \text{V:} \qquad i_1 \neq 0 \quad \text{A} \tag{5.58}$$

when i_3 is flowing, $v_{AK3} = 0$, and for the loop embodying Q_1, Q_3, v_{an}, and v_{bn},

$$v_{AK1} = v_{an} - v_{bn} \quad \text{V} \tag{5.59}$$

Expressions for the line-to-neutral source voltages may be determined from an inspection of the phasor diagram of Fig. 5.19, and substitution of those expressions in equation 5.59 yields

$$v_{AK1} = \frac{\sqrt{2}\,V}{\sqrt{3}} \sin(\omega t - 30°) - \frac{\sqrt{2}\,V}{\sqrt{3}} \sin(\omega t - 150°)$$

$$= \sqrt{2}\,V \sin \omega t \quad \text{V:} \qquad i_3 \neq 0 \quad \text{A} \tag{5.60}$$

When i_5 is flowing, $v_{AK5} = 0$, and for the loop embodying Q_1, Q_5, v_{an}, and v_{cn},

$$v_{AK1} = v_{an} - v_{cn} \quad \text{V} \tag{5.61}$$

so that from Fig. 5.19

$$v_{AK1} = \frac{\sqrt{2}\,V}{\sqrt{3}}\sin(\omega t - 30°) - \frac{\sqrt{2}\,V}{\sqrt{3}}\sin(\omega t + 90°)$$

$$= \sqrt{2}\,V\sin(\omega t - 60°) \quad V: \quad i_s \neq 0 \quad A \tag{5.62}$$

The three expressions for v_{AK1} given in equations 5.58, 5.60, and 5.62 have been employed to produce the waveform of v_{AK1} in Fig. 5.25. From this waveform it may be seen that, as α is varied, the maximum forward or reverse voltage that may be applied to a thyristor is

$$v_{AK\max} = \pm\sqrt{2}\,V \quad V \tag{5.63}$$

The time available for turn off of the thyristors during continuous-current operation may be obtained from the interval ωt_q shown on the waveform of v_{AK1} in Fig. 5.25. When this interval is determined as a function of α, it is found that a discontinuity appears at $\alpha = \pi/3$. Figure 5.25 shows a delay angle slightly less than $\pi/3$ rad. It will be seen from the waveform of v_{AK1} that, as α is increased above $\pi/3$, a positive value of v_{AK1} appears at $\omega t = 2\pi$, and this causes an abrupt reduction in ωt_q. It may further be seen that when α is increased to π radians, ωt_q becomes zero, and no time is available for turn off. For this value of α, therefore, short-circuit of the ac source in the continuous current mode would occur, and this defines the boundary $\alpha = 180°$ shown in Fig. 5.21. The relationship between ωt_q and α is

$$\omega t_q = \frac{4\pi}{3} - \alpha: \quad 0 \leqslant \alpha < 60°$$

$$\tag{5.64}$$

$$\omega t_q = \pi - \alpha: \quad 60° < \alpha < 180°$$

Each thyristor conducts two of the six current pulses occurring in one cycle, so that the average thyristor current is

$$I_Q = \frac{I_O}{3} \quad A \tag{5.65}$$

and the rms thyristor current is

$$I_{QR} = \frac{I_R}{\sqrt{3}} \quad A \tag{5.66}$$

Current flows in each source phase or line during four of the six pulses occurring in one cycle, so that the rms line current, which may be assumed to be the output current of a three-phase transformer or transformer bank, is

$$I_2 = \sqrt{2}\, I_{QR} = \frac{\sqrt{2}}{\sqrt{3}} I_R \quad \text{A} \tag{5.67}$$

Because the circuit is a three-wire system, the line currents cannot contain triplen (zero-sequence) harmonic components. This means that the waveforms of the transformer winding currents will not be affected by the choice of a wye or delta connection of the windings. In determining the required three-phase transformer rating, therefore, the wye-wye connection may be assumed, and the rating is

$$S_2 = \sqrt{3}\, VI_2 = \sqrt{2}\, VI_R = S_1 \quad \text{VA} \tag{5.68}$$

Example 5.4 For the three-phase full-wave (six-pulse) controlled rectifier of Fig. 5.17a, $R = 4\,\Omega$, $L = 0$, $V_C = 0$, $V_{ab} = 230\,\text{V}$, and delay angle $\alpha = 75°$.

(i) Calculate the following:

   (a) Average output current.
   (b) RMS output current.
   (c) Average and rms thyristor currents.
   (d) Power factor at the ac source.

(ii) Sketch to scale the time variations of v_{ab}, i_{G1}, i_O, and i_a.

Solution

(i)(a) Since $L = 0$, $\phi = 0$, and $Z = 4\,\Omega$. For $\alpha = 75°$ and $\phi = 0$, Fig. 5.21 shows that current i_O is discontinuous. Thus Fig. 3.7 may be used to calculate I_N. In entering Fig. 3.7 the value of α to be employed is $75° + 60° = 135°$, yielding $I_N = 0.04$. And since this is a six-pulse converter

$$I_N = 6 \times 0.04 = 0.24$$

$$I_{\text{BASE}} = \frac{230\sqrt{2}}{4} \quad \text{A}$$

Thus

$$I_O = \frac{230\sqrt{2}}{4} \times 0.24 = 20 \quad A$$

(b)   Figure 3.8 may be used to calculate I_{RN}. Entering Fig. 3.8 with $\alpha = 135°$ yields $I_{RN} = 0.16$. For the six-pulse converter

$$I_N = \sqrt{6} \times 0.16 = 0.39$$

Thus

$$I_R = \frac{230\sqrt{2}}{4} \times 0.39 = 32 \quad A$$

(c)   The average thyristor current is $I_Q = I_O/3 = 6.5\,A$. The rms thyristor current is $I_{QR} = I_R/\sqrt{3} = 18\,A$.

(d)   Output power $P_O = RI_R^2 = 4 \times 32^2 = 4096\,W$

$$I_a = I_R\sqrt{\frac{2}{3}} = 26 \quad A$$

$$S = \sqrt{3}\ VI_a = 10.3 \times 10^3$$

$$\text{power factor} = \frac{P_O}{S} = 0.40$$

(ii)   The required curves are shown in Fig. E5.4.

Example 5.5   For the rectifier of Fig. 5.17a, $R = 0.8\,\Omega$, $L = 5$ mH, $V_C = 200\,V$, $V_{ab} = 230\,V$, $\omega = 120\pi$ rad|s, and delay angle $\alpha = 30°$.

(i)   Calculate the following:

   (a)   Average output current.
   (b)   RMS output current.
   (c)   Average and rms thyristor currents.
   (d)   Power factor at the ac source.

(ii)   Sketch curves showing the time variations of v_{ab}, i_{G1}, i_O, and i_a on

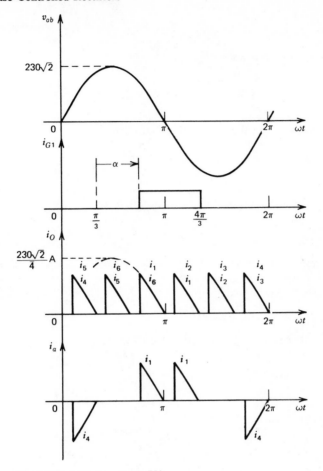

Fig. E5.4 $\phi=0$, $V_C=0$, $\alpha=75°$.

the assumption that the load circuit is sufficiently inductive to maintain the output current virtually constant.

Solution

(i)(a)

$$\phi=\tan^{-1}\frac{\omega L}{R} = \tan^{-1}\frac{120\pi \times 5 \times 10^{-3}}{0.8} = 67°$$

$$m=\frac{V_C}{\sqrt{2}\,V} = \frac{200}{\sqrt{2}\times 230} = 0.615$$

For $\alpha = 30°$, $\phi = 67°$, $m = 0.615$, Fig. 5.21 shows that current i_0 is continuous and that conduction will commence at $\omega t = 90°$. From equation 5.50, the average output voltage is

$$V_O = \frac{3\sqrt{2}}{\pi} V \cos \alpha = \frac{3\sqrt{2}}{\pi} \times 230 \cos 30° = 269 \quad V$$

From equation 5.20, the average output current is

$$I_O = \frac{V_O - V_C}{R} = \frac{269 - 200}{0.8} = 86.2 \quad A$$

(b)   The rms output current is obtained from

$$I_R = [I_O^2 + I_{RI}^2]^{1/2} \quad A \tag{5.25}$$

where I_{RI} may be determined to any degree of accuracy by including the desired number of harmonics in

$$I_{RI} = [\Sigma I_{nR}^2]^{1/2} \quad A \tag{5.24}$$

For the sixth harmonic at $\alpha = 30°$, from Fig. 5.23

$$\frac{c_6}{\sqrt{2} \, V} = 0.17$$

From this result and equation 5.21

$$d_6 = \frac{0.17\sqrt{2} \times 230}{\left[0.8^2 + (6 \times 1.88)^2\right]^{1/2}} = 4.89$$

and

$$I_{6R} = \frac{d_6}{\sqrt{2}} = 3.46 \quad A$$

As the rms value of the sixth harmonic is already small compared with I_O, higher harmonics may be neglected, and from equation 5.25

$$I_R = [86.2^2 + 3.46^2]^{1/2} = 86.3 \quad A$$

(c)   Average thyristor current is

$$I_Q = \frac{I_O}{3} = \frac{86.2}{3} = 28.7 \quad A$$

RMS thyristor current is

$$I_{QR} = \frac{I_R}{\sqrt{3}} = \frac{86.3}{\sqrt{3}} = 49.8 \quad A$$

(d) Output power delivered to the load circuit is

$$P_O = RI_R^2 + V_C I_O = 0.8 \times 86.3^2 + 200 \times 86.2$$

$$= 23.2 \times 10^3 \quad W$$

From equation 5.68 the three-phase apparent power is

$$S = \sqrt{2} \, VI_R = \sqrt{2} \times 230 \times 86.3 = 28.2 \times 10^3 \quad VA$$

and the power factor

$$PF = \frac{P_O}{S_2} = \frac{23.2}{28.2} = 0.823$$

(ii) The required curves are shown in Fig. E5.5.

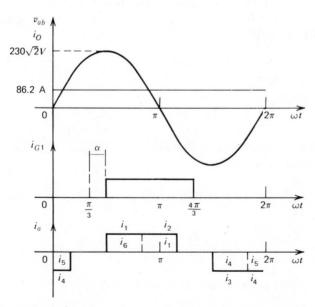

Fig. E5.5 $R = 0.8\Omega$, $L = 5$ mH, $V_C = 200$ V, $\alpha = 30°$.

Example 5.6 Repeat the problem of Example 5.5 when $V_C = -290$ V and $\alpha = 150°$; all other conditions remaining unchanged.

Solution

(i) (a)
$$\phi = 67°, \, m = \frac{-290}{\sqrt{2} \times 230} = -0.892$$

For $\alpha = 150°$, $\phi = 67°$, $m = -0.892$, Fig. 5.21 shows that current i_O is continuous and that conduction will commence at $\omega t = 210°$. From equation 5.50

$$V_O = \frac{3\sqrt{2}}{\pi} V \cos \alpha = \frac{3\sqrt{2}}{\pi} \times 230 \cos 150° = -269.0 \quad V$$

From equation 5.20

$$I_O = \frac{V_O - V_C}{R} = \frac{-269 + 290}{0.8} = 26.25 \quad A$$

(b) Since the curves of $c_n | \sqrt{2} \, V$ versus α in Fig. 5.8 are symmetrical about the line $\alpha = 90°$, I_{6R} at $\alpha = 150°$ will be the same value as that calculated in Example 5.5.

$$I_{6R} = 3.46 \quad A$$

Thus from equation 5.25

$$I_R = [26.25^2 + 3.46^2]^{1/2} = 26.5 \quad A$$

From which it may be seen that inclusion of higher current harmonics would have negligible effect.

(c)
$$I_Q = \frac{I_O}{3} = \frac{26.25}{3} = 8.75 \quad A$$

$$I_{QR} = \frac{I_R}{\sqrt{3}} = \frac{26.5}{\sqrt{3}} = 15.3 \quad A$$

(d) Output power delivered to the load circuit is

$$P_O = RI_R^2 + V_C I_O = 0.8 \times 26.5^2 + (-290) \times 26.25 = -7.05 \times 10^3$$

where the negative sign shows that power is being delivered from the dc to the ac source, that is, the system is inverting.

The three-phase apparent power is from equation 5.68

$$S = \sqrt{2}\ VI_R = \sqrt{2}\ \times 230 \times 26.5 = 8.60 \times 10^3 \quad \text{VA}$$

and power factor

$$PF = \frac{-7.05}{8.60} = -0.820$$

where the significance of the minus sign is again that the system is inverting.

(ii)   The required curves are shown in Fig. E5.6.

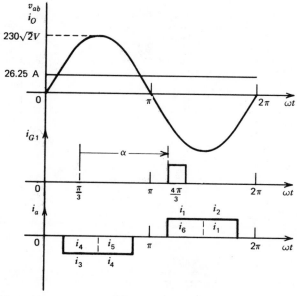

Fig. E5.6 $R = 0.8\ \Omega$, $L = 5$ mH, $V_C = -290$ V, $\alpha = 150°$.

5.4 SEPARATELY EXCITED DC MOTOR DRIVES

A basic speed-control system for a separately excited dc motor is illustrated in Fig. 5.26, where a shows the main elements of the system, and b shows the equivalent circuit of the motor and sources needed for steady-state analysis. It is assumed in a that the field current is manually controlled, although this also could be controlled by the speed feedback signal. A system such as this is frequently employed as the hoist drive for large gantry cranes, and it is convenient to use it as an example of drives in general, since the physical conditions existing in such a system at any instant are readily visualized.

A hoist system operates in all four quadrants of the speed torque diagram of the motor, and these are illustrated in Fig. 5.27. In the first quadrant, in which both speed and motor output torque at the coupling are positive, the hoist is raising a load against the action of gravity. In the second quadrant, which is entered only momentarily when a load is on the

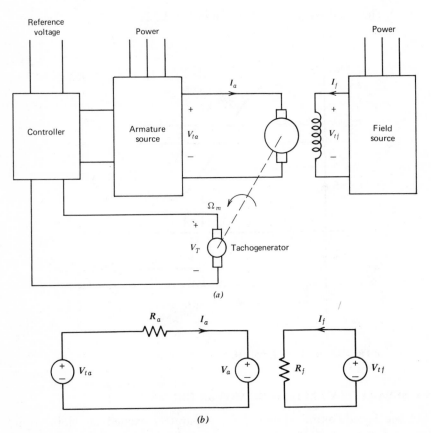

Fig. 5.26 (*a*) Speed control by separately excited d-c motor. (*b*) Equivalent circuit of motor and sources.

hook, the motor torque is reversed to decelerate the rotating parts of the mechanism that is raising the load. Operation in this quadrant naturally occurs most frequently when the light hook is being raised and there is no load on it to assist in decelerating the mechanism. In the third quadrant, both speed and torque are negative. This quadrant also is entered only momentarily when a load is on the hook, and the motor assisted by the load is accelerating the rotating parts of the mechanism at the beginning of a lowering operation. With a light hook, the motor would operate in the third quadrant throughout the greater part of the downward movement of the hook. In the fourth quadrant, the motor torque tends to raise the load, which is, however, moving downward. This means that the motor is resisting the tendency of the load to accelerate downward and run away with the system. With a light hook, this quadrant would be entered only momentarily to decelerate the mechanism at the end of operation in the third quadrant. In quadrants 1 and 3, energy must be supplied by the electrical system to the mechanical system. In quadrants 2 and 4, energy may be supplied, momentarily or continuously, by the mechanical system to the electrical system.

To correlate Fig. 5.27 with a diagram of armature induced voltage V_a versus armature current I_a, it is necessary to analyze the equivalent circuit of Fig. 5.26b.

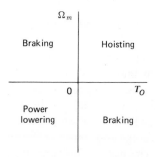

Fig. 5.27 Operation of a hoist.

5.4.1 Steady-State Model of the Motor Under steady-state operating conditions, the current in the motor field winding is

$$I_f = \frac{V_{tf}}{R_f} \quad \text{A} \tag{5.69}$$

where V_{tf} is the voltage applied at the field terminals, and R_f is the field-winding resistance. The flux per pole Φ of the motor magnetic circuit is a function of the field current

$$\Phi = f(I_f) \quad \text{Wb} \tag{5.70}$$

and this is a nonlinear relationship owing to the saturation of the magnetic circuit. In the analysis that follows, it will be assumed that V_{tf} is held constant, so that I_f and Φ are constant.

The voltage induced in the armature winding on the rotor is

$$V_a = k\Phi\Omega_m \quad \text{V} \tag{5.71}$$

where k is a constant for a particular motor and depends on its geometry and type of armature winding. Thus with constant field current, V_a is directly proportional to the speed of rotation Ω_m. If the resistance of the armature winding R_a is assumed to be constant, then from Fig. 5.26b,

$$V_{ta} = V_a + R_a I_a \quad \text{V} \tag{5.72}$$

where positive current I_a has the direction shown in the equivalent circuit and is such as to cause the motor to develop torque in a positive direction, that is, to raise the load. Under such conditions $V_{ta} > V_a$.

The internal or air-gap torque T developed in the motor is

$$T = k\Phi i_a \quad \text{n-m} \tag{5.73}$$

where k is the constant appearing in equation 5.71. The output torque T_O of the motor differs from that given in equation 5.73 due to the friction, windage, and rotor core loss. These rotational losses are often assumed to be constant at all speeds and equal to the power input to the motor armature when the motor is running on no load at rated speed. This is a considerable approximation, and more accurate methods of including rotational losses in performance calculations are possible.

The relationship between V_a and I_a for a motor in which V_{tf} and hence Φ is constant is illustrated in Fig. 5.28. The discussion of this diagram may be conducted with reference to a motor coupled to a hoist mechanism as in Fig. 5.27. The line shown in the first and second quadrants of Fig. 5.28 corresponds to a positive value of armature terminal voltage V_{ta}, shown as an intercept on the vertical axis. This line is described by equation 5.72. In the first quadrant, positive I_a produces a positive motor torque, raising the load. The torque required for this purpose will be constant, and equation 5.73 shows that the armature current will be constant also. If V_{ta} were reduced, then for I_a to remain constant, it would be necessary for V_a to decrease, since from equation 5.72

$$I_a = \frac{V_{ta} - V_a}{R_a} \quad \text{A} \tag{5.74}$$

and equation 5.71 shows that this would be brought about by a momentary reduction in I_a permitting a fall in speed. In the first quadrant, therefore, motoring takes place, and this corresponds to the first quadrant of Fig. 5.27.

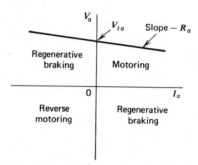

Fig. 5.28 Relationship of armature-circuit variables for the motor of Fig. 5.26 (Φ constant).

In the second quadrant of Fig. 5.28, V_a is still positive, so that from equation 5.71 the hoist hook must still be moving upward. Current I_a is negative, however, and equation 5.73 shows that a negative torque is decelerating the hoist mechanism. This condition may be reached by a reduction of V_{ta} such that I_a momentarily becomes negative. Under these conditions, the motor operates as a generator, and some of the system kinetic energy is converted to electrical energy and returned to the armature source. In the second quadrant, therefore, regenerative braking takes place.

In the third quadrant of Fig. 5.28, both V_a and I_a are negative; from equation 5.74, V_{ta} must therefore be negative, and $|V_{ta}| > |V_a|$. This quadrant of Fig. 5.28 corresponds to reverse motoring, or power lowering, as indicated in Fig. 5.27.

In the fourth quadrant of Fig. 5.28, while V_a is negative, and hence Ω_m must be negative, I_a is positive, showing that the motor is developing a torque tending to raise the load. The motor is thus driven by the load in a direction opposite to that in which the motor torque is acting: this is a condition of regenerative braking, in which kinetic energy of the mechanical system is being converted and returned to the armature source.

Correlation between the four quadrants of Figs. 5.27 and 5.28 is not exact, since while, at fixed flux Φ, V_a is directly proportional to Ω_m, the motor output torque T_O is not directly proportional to armature current I_a. Exact correlation with Fig. 5.28 is obtained if a diagram of Ω_m versus motor air-gap torque T is drawn, and such a diagram is shown in Fig. 5.29. The characteristics extending approximately horizontally across that diagram represent relationships between speed and motor air-gap torque for a series of values of applied armature terminal voltage V_{ta}, where this may be positive or negative.

The characteristics marked T_L in the first and fourth quadrants of Fig. 5.29 represent the relationship between speed and effective load torque referred to the motor shaft when a load is on the hook. This torque may be considered to be made up of two components. One is produced by the weight of the load suspended on the hook and referred through an ideal lossless mechanism to the motor shaft, and this may be given the symbol T_{DW}. The other component is due to the losses in the motor and mechanism, and this may be given the symbol T_{loss}. This component will oppose the motion of the load in whichever direction that may be and thus causes the discontinuity in the curve of T_L at the torque axis. The characteristic marked T_{LH} in the first and third quadrants of Fig. 5.29 represents the relationship between speed and the composite effective load torque for light-hook operation.

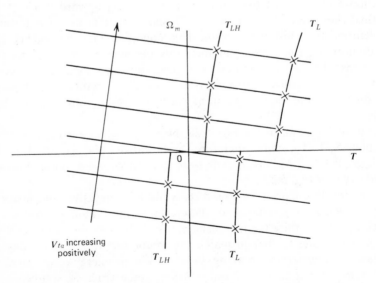

Fig. 5.29 Speed control by variation of armature terminal voltage.

5.4.2 Steady-State Model of a Hoist Drive If it is assumed that all of the losses in motor and mechanism have the nature of viscous friction, then

$$T_{\text{loss}} = F\Omega_m \quad \text{n-m} \qquad (5.75)$$

where F is the "friction factor" of the system. Then under steady-state operating conditions

$$T = T_L = T_{DW} + T_{\text{loss}} \quad \text{n-m} \qquad (5.76)$$

where T_{loss} will be positive or negative according to whether the hook is being raised or lowered.

The variable that the operator of a hoist mechanism wishes to control is the speed of the hook, and this he may do, provided that a speed-sensing device such as the tachogenerator shown in Fig. 5.26a is provided. This will usually be a permanent-magnet dc generator whose output voltage is

$$V_T = k_T\Omega_m \quad \text{V} \qquad (5.77)$$

where k_T is a constant for the tachogenerator. Normally also an overriding control will be provided in the form of a limit on the motor armature current, so that this does not become greater than the motor can commutate. The speed-sensing and current limit features may be incorporated in a block diagram showing the steady-state operation of the entire electromechanical system described by equations 5.69 to 5.77, and such a diagram is shown in Fig. 5.30.

Figure 5.29 shows that both positive and negative motor air-gap torque must be developed, since the light hook must be driven downward, and the loaded hook must also be accelerated downward at the beginning of a lowering movement. Equation 5.73 shows that negative torque will be produced if the flux per pole Φ is reversed. This could be achieved by reversing the polarity of V_{tf}. If rapid torque reversal is required, however, reversal of the flux is not suitable, since the motor field circuit has a large time constant, and thus a considerable interval is needed before I_f and consequently Φ can be reversed. In practice, therefore, the contribution to speed control made by variation of the field current is restricted to varying the magnitude of Φ, but not reversing it. This varies the "base speed" of operation of the motor, defined as the speed at which the motor will deliver rated torque at the coupling with the maximum permissible current in the field windings.

Equation 5.73 also shows that negative torque is produced if I_a is reversed, and indeed in discussing the operation of the hoist in terms of

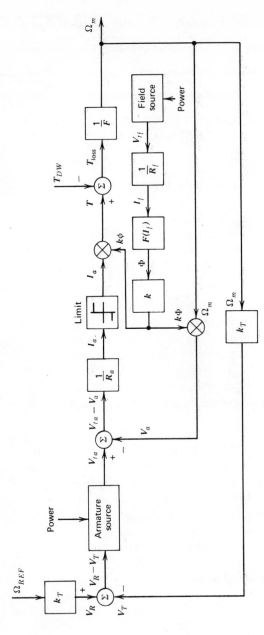

Fig. 5.30 Steady-state model of a hoist drive.

Fig. 5.28 it has been tacitly assumed that reversal of armature current could take place. This means that the motor armature source must be capable of operating with negative current as well as negative terminal voltage; that is, a four-quadrant source is required. The controlled rectifier systems that have so far been discussed form two-quadrant sources only, since they cannot accept reverse current. It is therefore necessary to discuss how the required armature source can be provided.

5.5 THE DUAL CONVERTER

Since a single controlled rectifier can provide a two-quadrant source, corresponding to quadrants 1 and 4 or 2 and 3 of Fig. 5.28, two controlled rectifiers can provide a four-quadrant source for the armature circuit of a separately excited motor. The equivalent circuit for such an arrangement is shown in Fig. 5.31, where two three-phase controlled rectifiers could equally well have been employed.

For the circuit of Fig. 5.31, it will be seen that unless

$$v_{\text{tap}} + v_{\text{tan}} = 0 \quad \text{V} \tag{5.78}$$

where v_{tap} is the instantaneous terminal voltage of the controlled rectifier giving positive armature current, and v_{tan} that of the rectifier giving

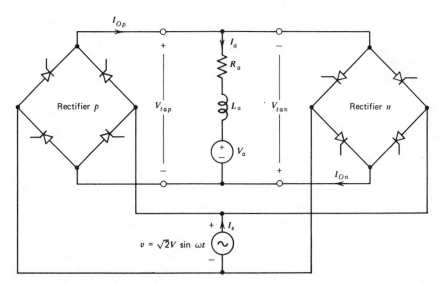

Fig. 5.31 Single-phase four-quadrant armature source.

negative armature current, the rectifier currents will be unbounded. It is possible to ensure that

$$V_{\text{tap}} + V_{\text{tan}} = 0 \quad \text{V} \tag{5.79}$$

where these are the average values of the rectifier output voltages, but it is quite impossible to ensure that equation 5.78 shall be satisfied. Thus Fig. 5.31 only represents a practical circuit if the control of the gating signals is such that those of rectifier n are blanked out whenever the armature current is positive, while those of rectifier p are blanked out whenever the armature current is negative. Under these circumstances each rectifier presents effectively infinite impedance to the output of the other, and only one rectifier acts as an energy source or sink at any one time.

With the restriction on gating signals described in the preceding paragraph in effect, the circuit of Fig. 5.31 represents a practical arrangement for a four-quadrant drive. The operating conditions for the motor and active rectifier may be determined for any quadrant of operation of Fig. 5.28 by the methods discussed in Sections 5.2.1 and 5.3 with reference to Figs. 5.10 and 5.21, respectively. When V_a and I_a are known, the operating point in Fig. 5.29 may be calculated by application of the analysis of Section 5.4.1.

A transition between quadrants 1 and 2 or 3 and 4 in Fig. 5.28, representing a transition from motoring to regenerative braking, or the converse, may be considered to take place in three stages. The current supplied by the first rectifier is reduced to zero by increase of the delay angle α_1 to the necessary value. The gating signals to the first rectifier are then blanked out, and those to the second rectifier are applied. The current supplied by the second rectifier is then increased from zero by reduction of the delay angle α_2 from its maximum value. The polarity of the voltage V_a induced in the motor armature circuit remains unchanged throughout this transition.

The block diagram in Fig. 5.32 shows how the dual controller and current limit are normally incorporated into a complete motor control circuit. It is assumed there that the motor flux per pole is constant. In this system, the speed error does not directly control the operation of the gating circuits for the dual converter. If a change of speed is called for, then the speed control unit generates a reference current signal I_R, which may be positive or negative with respect to I_a and may have any value up to the current limit of the motor. The current error signal is then applied to the gating circuits that generate the required delay angle for the appropriate bridge of the dual converter. This results in the application of an armature terminal voltage tending to produce the armature current called for by the speed control unit.

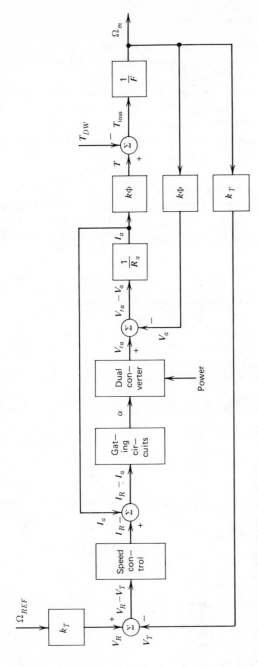

Fig. 5.32 DC motor speed control system

5.6 DC SERIES MOTOR DRIVES

For some vehicle and hoist drives, the characteristics of the dc series motor are preferable to those of the separately excited motor. A minor complication arises because the transition from motoring to regeneration demands that the armature current must reverse while the direction of the field current remains unchanged. This may be achieved by means of the arrangement shown in Fig. 5.33. The diode bridge surrounding the equivalent circuit of the motor field ensures that the field current direction remains unchanged, just as in the case of the separately excited motor.

For the determination of rectifier output rating, the important range of operation of the series motor is that surrounding zero speed where, under load, large torque is required, and large motor current flows. In this range, the magnetic circuit of the series motor is normally saturated, so that regardless of variation of current, the flux per pole Φ remains virtually

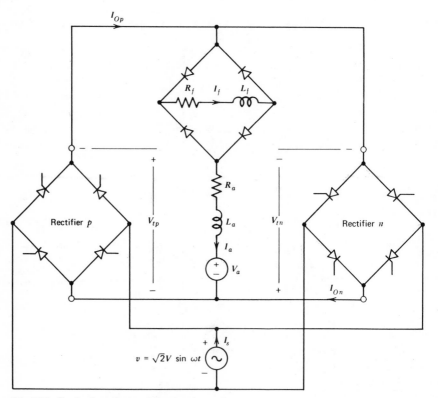

Fig. 5.33 Single-phase four-quadrant series motor source.

constant. This means that the steady-state model of the motor may be exactly the same as that for the separately excited motor described in Section 5.4.1, provided only that $R_a + R_f$ for the series motor is substituted for R_a in equations 5.72 and 5.74. The diagram of Fig. 5.32 may also be applied to a system incorporating a saturated dc series motor.

Beyond the saturated range of operation, where the vehicle is "accelerating on the speed curve" or the hoist is operating with a light hook, determination of system performance requires the use of the saturation curve for the motor.

5.7 INDUCTION MOTOR DRIVE

The conventional method of controlling the speed of a wound-rotor induction motor is by the introduction of balanced three-phase external resistances into the rotor circuit. The function of these resistances is to introduce voltages at rotor frequency which oppose the voltages induced in the rotor windings. The principal disadvantage of this method of control is that energy is dissipated in the rotor-circuit resistance, internal and external, and this energy is wasted in the form of heat. The greater the reduction in speed below the synchronous speed of the motor, the greater is the proportion of the energy supplied to the motor which is wasted.

An alternative method of introducing the required rotor-frequency voltages into the rotor circuits is illustrated in Fig. 5.34. The uncontrolled diode rectifier shown there produces a direct output voltage V_{RR} that is proportional to the alternating voltage appearing at the rotor slip rings. The controlled bridge rectifier operates as an inverter from dc to fixed-

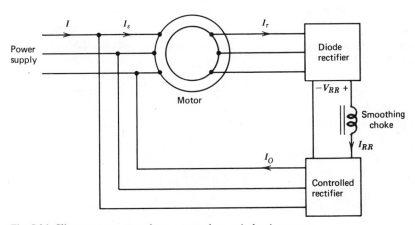

Fig. 5.34 Slip energy recovery from a wound-rotor induction motor.

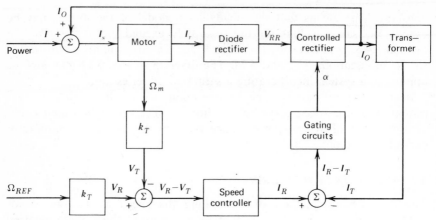

Fig. 5.35 Induction motor speed control system.

frequency ac at the voltage and frequency of the power source supplying the motor, as described in Section 5.3 in reference to Fig. 5.21. The choke between the two converters provides smoothing of the rectifier output, so that it appears to the controlled rectifier as an approximately constant direct voltage. By this means, the energy that would otherwise have been dissipated in external rotor-circuit resistances is recovered and returned to the power source.

A further advantage of the system of Fig. 5.34 is that it permits closed-loop speed control of the motor and the imposition of a current limit. A block diagram of such a speed control system is shown in Fig. 5.35. The current feedback signal in Fig. 5.35 is obtained from a current transformer in one of the ac lines from the controlled rectifier. The speed signal is obtained from a tachogenerator driven by the motor.

If an increase of speed is called for, then the speed controller in Fig. 5.35 generates a reference current signal I_R which is greater than the output current of the controlled rectifier. The gating circuits then reduce the delay angle for the controlled rectifier, calling for a lower input voltage V_{RR} for a given output current. This permits the rotor current to increase, and this is reflected in an increase of stator current I_s. The consequence is an increase in the developed torque of the motor, which accelerates.

PROBLEMS

5.1   For the circuit shown in Fig. P5.1, if $L = 0$ and $V_C = 0$:

(a)   Show that the average value of the output voltage v_O is given by

$$V_O = \frac{\sqrt{2}\, V_s}{\pi}(1 + \cos \alpha) \quad \text{V}$$

where α is the delay angle as defined in Section 5.2.1.

(b) Show that the rms value of the output current i_O is given by

$$I_R = \frac{V_s}{R}\left\{ \frac{1}{\pi}\left[(\pi-\alpha)+\frac{1}{2}\sin 2\alpha\right]\right\}^{1/2} \quad \text{A}$$

(c) Sketch to scale the time variations of v_s, i_O, i_1, and v_{AK1} for $\alpha=90°$.

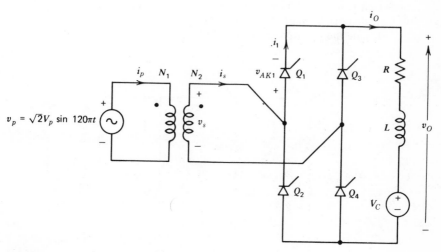

Fig. P5.1

5.2   For the circuit shown in Fig. P5.1, if $V_p=230\text{V}$, $N_1/N_2=2$, $L=10$ mH, $R=6.52\Omega$, $V_C=0$, and $\alpha=45°$, calculate the following:

(a) Average output current.
(b) RMS output current.
(c) Average and rms thyristor currents.
(d) Power factor at the ac source.

Also

(e) Sketch to scale the time variations of v_s, v_O, i_O, and v_{AK1}.

5.3   Repeat problem 5.2 with $V_C=-55\text{V}$.

5.4   For the circuit shown in Fig. P5.1, $V_s=110\text{V}$, $R=0.5\Omega$, and $I_O=10\text{A}$. If L is so large that the output current may be assumed to be constant, calculate the required delay angle α if (a) $V_C=65\text{V}$; (b) $V_C=-75\text{V}$. In each case determine also which source (ac or dc) is delivering power and the power factor at the ac source terminals. Also sketch to scale the time variations of v_s, v_O, i_O, and v_{AKi}.

5.5   For the circuit shown in Fig. P5.1, $V_p = 230$V. Calculate the transformer turns ratio required to give $V_O = 50$V when $\alpha = 0$.

5.6   For the circuit shown in Fig. P5.1, $V = 110$V, $V_C = -124$V, $\phi = 75°$. Employ Fig. 5.10 to determine the maximum delay angle α at which operation is possible if the turn-off time of the thyristors t_{off} is 20 μs.

5.7   For the circuit shown in Fig. P5.2, $V = 110$V, $R = 1.5\,\Omega$, $V_C = 30$V, and L is sufficiently large to make the current virtually constant. For $\alpha = \pi/4$ rad:

(a)   Sketch to scale the time variations of v_{ab}, v_O, i_O, v_{AK1}, and i_1.
(b)   Calculate the average values of v_O and i_O.
(c)   Calculate the input power factor.

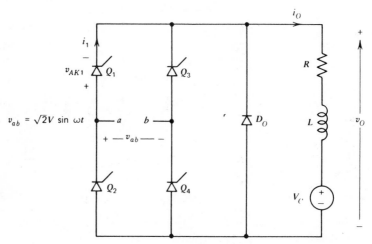

Fig. P5.2

5.8   For the circuit shown in Fig. P5.2 without diode D_O and with the circuit parameters specified in problem 5.7, assume that the values of V_O and I_O are those calculated in problem 5.7b.

(a)   Sketch to scale the time variations of v_{ab}, v_O, i_O, v_{AK1} and i_1.
(b)   Calculate the input power factor and compare it with that obtained in problem 5.7c.

5.9   For the circuit shown in Fig. P5.3, $V_p = 230$V, $N_2/N_1 = 0.75$, $R = 1\,\Omega$, $L = 0$,

$V_C = 0$. Calculate the following:

(a) Maximum average output current I_O.
(b) Control range of α required from zero to maximum output current.
(c) Maximum rms and average thyristor currents.
(d) Maximum possible positive and negative values of v_{AK1}.
(e) The value of α required to give $V_O = 50\,\text{V}$.

In addition:

(f) For the value of α determined in part a, sketch to scale the time variations of v_{s1}, i_1, v_O, v_{AK1}, and i_p.
(g) Repeat f for the value of α obtained in part e.

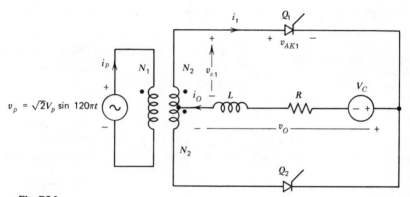

Fig. P5.3

5.10   Repeat problem 5.9 for the circuit shown in Fig. P5.1.

5.11   For the circuit shown in Fig. P5.4, if $L = 0$, and $V_C = 0$, show that the average output voltage V_O is given by

$$V_O = \frac{3}{\sqrt{2}} \frac{V}{\pi} \cos\alpha \quad \text{V:} \quad 0 < \alpha < \frac{\pi}{6}$$

$$V_O = \sqrt{\frac{3}{2}} \frac{V}{\pi} \left[\cos\left(\alpha + \frac{\pi}{6}\right) + 1 \right] \quad \text{V:} \quad \frac{\pi}{6} < \alpha < \frac{5\pi}{6}$$

(N. B. Remember the definition of α).

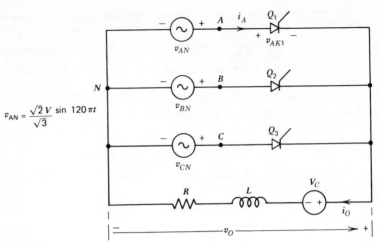

Fig. P5.4

5.12   For the circuit shown in Fig. P5.4, show that if L, R, and V_C are such that the output current i_O is continuous, then the average output voltage is given by

$$V_O = \frac{3}{\sqrt{2}} \frac{V}{\pi} \cos\alpha \quad \text{V:} \quad 0 < \alpha < \pi$$

5.13   For the circuit shown in Fig. P5.4, $V = 110$ V, $V_C = 40$ V, $R = 15\,\Omega$, and L is so large that the output current may be assumed to be constant. Calculate for $\alpha = 60°$.

(a)   The average and rms values of i_O.
(b)   The rms value of the line current i_A.

Also

(c)   Sketch to scale the time variations of v_{AN}, i_O, v_O, v_{AK1}, and i_A.

5.14   For the circuit of Fig. 5.17,

$$v_{ab} = \sqrt{2}\, V \sin\omega t \quad \text{V}$$

If $L = 0$ and $V_C = 0$, show that the average output voltage is given by

$$V_O = \frac{3\sqrt{2}\,V}{\pi} \cos\alpha \quad \text{V:} \quad 0 < \alpha < \frac{\pi}{3}$$

$$V_O = \frac{3\sqrt{2}\,V}{\pi}\left[1 + \cos\left(\alpha + \frac{\pi}{3}\right)\right] \quad \text{V:} \quad \frac{\pi}{3} < \alpha < \frac{2\pi}{3}$$

5.15   For the circuit of Fig. 5.17,

$$v_{ab} = 230\sqrt{2} \, \sin 120\pi t \quad \text{V}$$

while $L=0$, $V_C=0$, and $R=1.5\,\Omega$. For a delay angle $\alpha = \pi/3$, sketch to scale the time variations of v_{ab}, v_O, i_O, v_{AK1}, and i_a.

5.16   Repeat problem 5.15 for $\alpha = \pi/2$.

5.17   For the circuit of Fig. 5.17,

$$v_{ab} = 110\sqrt{2} \, \sin 120\pi t \quad \text{V}$$

while $L=5$ mH, $R=1.89\,\Omega$, and $V_C=0$. For a delay angle of $\alpha = \pi/2$, calculate the following:

(a)   Average output current I_O.
(b)   RMS output current I_R.
(c)   Average and rms thyristor currents I_Q and I_{QR}.
(d)   Power factor at the ac source.

Also

(e)   Sketch to scale the time variations of v_{ab}, v_O, i_O, v_{AK1}, and i_1.

5.18   Repeat problem 5.17 for $V_C = -124\,\text{V}$.

5.19   The circuit of Fig. 5.17 is employed to charge a battery with an emf of 95 V and an internal resistance of $0.25\,\Omega$. The supply voltage is 110 V line-to-line and sufficient inductance is included in the output circuit to maintain the current virtually constant at 10 A. Determine:

(a)   The delay angle α required.
(b)   The power factor at the ac source terminals.
(c)   The rate of delivery of energy to the battery.

5.20   The circuit of Fig. 5.17 is used to deliver energy to a three-phase ac system of 1500-V line-to-line. The dc source emf is 1000 V and it is delivering a current of 1000 A. The dc source resistance is $0.01\,\Omega$ and the inductance is sufficiently great to maintain a constant direct current.

Calculate the delay angle α required for this condition of operation and the power factor at the terminals of the ac system. Also calculate the efficiency of power conversion from the dc source to the ac source.

5.21   The circuit of Fig. 5.17 is operating under continuous-current conditions. The turn-off time of the thyristors is 400 μs. The ac source frequency is 60 Hz. Determine the maximum permissible value of delay angle α for this system.

SIX

DC-TO-DC
CONVERTERS (CHOPPERS)

DC-to-dc converters, commonly called choppers because of their principle of operation, are employed to vary the average value of the direct voltage applied to a load circuit by introducing one or more thyristors between the load circuit and a dc source. The function of a chopper is illustrated by Fig. 6.1a.

The manner in which the average load voltage is reduced below that of the source is illustrated in Fig. 6.1b. This shows that the chopper applies a train of unidirectional voltage pulses to the load circuit, the magnitude of these pulses being the same as that of the source voltage. Load voltage V_O may be varied in one of three different ways.

1. t_{ON} may be varied, while periodic time T is held constant—pulse-width modulation.
2. t_{ON} may be kept constant while T is varied—frequency modulation.
3. Combined pulse-width and frequency modulation.

In practice, all choppers do not maintain the ideal output voltage variation shown in Fig. 6.1b, but it is convenient to ignore this possibility while discussing the basic principles of operation of the various types of chopper that may be built. In this chapter, therefore, the converters to be analyzed are first briefly explained in terms of a highly simplified and idealized model, and they are then analyzed in the detail required by the circuit designer.

Choppers are employed in variable-speed dc drives, where it is desirable to eliminate the waste of energy in the form of heat produced in starting or control resistors, or where the operating characteristics obtained by such

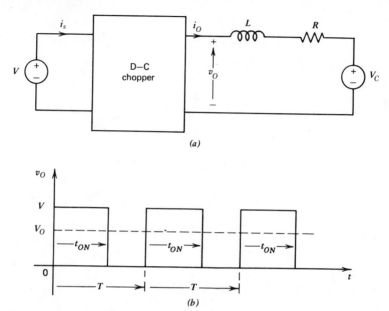

(a)

(b)

Fig. 6.1 Function of a DC chopper.

methods of control would be unsatisfactory. Thus a chopper may be employed to supply the armature of a separately-excited dc motor, or to supply a series-wound dc motor. It follows that choppers are employed in dc transportation drives of all kinds where the ability of a chopper to impose a current limit on the load circuit is of particular value.

6.1 TYPES OF CHOPPER CIRCUITS

Figure 6.2a illustrates the basic principles of a type A chopper, in which both V_O and I_O can only be positive. In that circuit diagram, the thyristor symbol enclosed in a circle represents a thyristor that may be turned on and commutated by means of circuit elements not included in the diagram; D_1 is a free-wheeling diode. Two possible conditions of operation are illustrated in Figs. 6.2b and c, where it is assumed that the control is by means of frequency modulation.

In Fig. 6.2b the load current i_O is discontinuous, so that during the interval for which i_O is zero, $v_O = V_C$. In Fig. 6.2c, the periodic time T has been reduced to such an extent that i_O has not ceased to flow before Q_1 is again turned on. As a consequence, the output voltage v_O consists of a train of rectangular pulses of magnitude V. An increase of load circuit

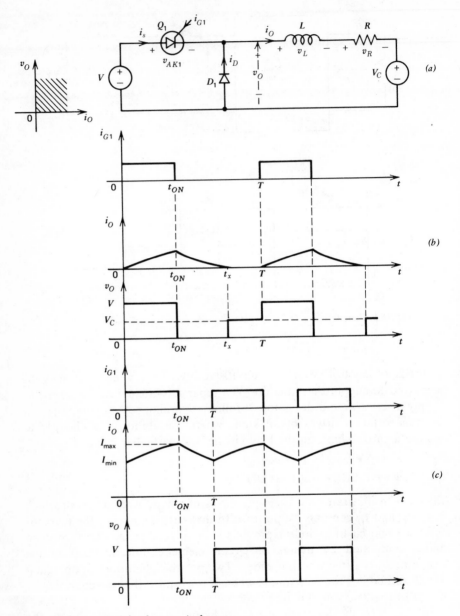

Fig. 6.2 Basic principle of a type A chopper.

inductance L or a reduction of V_C would also tend to result in a continuous output current.

Figure 6.3 illustrates a type B chopper, in which I_O may be either positive or negative, but V_O can only be positive. A converter of this type may be employed with a load circuit that is capable of regenerating and returning energy to source V. Here there are two thyristors that can be turned on and commutated. For operation with positive output current, thyristor Q_1 and diode D_1 are controlled and function in exactly the same way as do the thyristor and diode of Fig. 6.2a. For operation with negative output current elements Q_2 and D_2 are employed, while Q_1 is turned off. If $V_C > 0$, and Q_2 is turned on, then a negative i_O will flow and energy from source V_C will be stored in inductance L. If Q_2 is then commutated, a positive value of v_L will result and, in conjunction with source voltage V_C, will drive current i_O through diode D_2 and the source V, in this way supplying the energy stored in inductance L to source V.

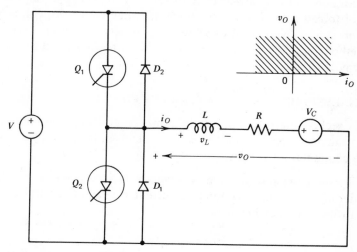

Fig. 6.3 Converter capable of regeneration.

Figure 6.4 illustrates a chopper circuit in which both v_O and i_O may be made positive or negative, separately or simultaneously. A converter of this type can provide both regeneration and reversal of the supply to the load circuit. If the control is so arranged that reversals of v_O and i_O occur cyclically at the same frequency, then the converter is in fact operating as a nonsinusoidal ac system, and functions as a single-phase bridge inverter. It is analyzed as such in Chapter 7.

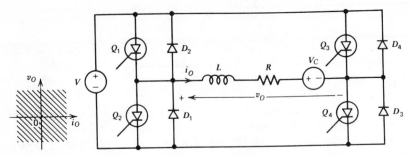

Fig. 6.4 Converter capable of reversal and regeneration.

In practical chopper circuits, both load commutation and forced commutation may be employed, although the latter is much the more common. The basic principles of forced commutation have been explained in Section 3.5.3, but forced commutation may be further subdivided into (*a*) voltage commutation and (*b*) current commutation. In this chapter both of these types of forced commutation are discussed when employed with a type A chopper. A large variety of commutation circuits have been devised, but the methods of analyzing them will be sufficiently illustrated by means of two typical circuits.

6.2 ANALYSIS OF THE TYPE A CHOPPER CIRCUIT

In this section the operation of the power circuit under continuous and discontinuous-current conditions is first discussed, followed by an analysis of the commutation circuits.

6.2.1 Power Circuit of a Type A Chopper It is convenient to start by considering the case of continuous-current operation illustrated in Fig. 6.2*c*. In the circuit of Fig. 6.2*a*,

$$-v_O + v_L + v_R + V_C = 0 \qquad V \tag{6.1}$$

from which

$$\frac{di_O}{dt} + \frac{R}{L}i_O = \frac{v_O - V_C}{L} \qquad A|s \tag{6.2}$$

when thyristor Q_1 is turned on at $t=0$, then at $t=0^+$, $v_O = V$, and $i_O = I_{min}$. From equation 6.2 and these initial conditions,

$$i_O = \frac{V - V_C}{R}(1 - \epsilon^{-t/\tau}) + I_{min}\epsilon^{-t/\tau} \qquad A: \qquad 0 \leqslant t < t_{ON} \quad s \tag{6.3}$$

where

$$\tau = \frac{L}{R} \quad \text{s} \tag{6.4}$$

At $t = t_{\text{ON}}$, when Q_1 is commutated,

$$i_O = I_{\max} = \frac{V - V_C}{R}(1 - \epsilon^{-t_{\text{ON}}/\tau}) + I_{\min}\epsilon^{-t_{\text{ON}}/\tau} \quad \text{A} \tag{6.5}$$

and since v_O then becomes zero, due to conduction of the free-wheeling diode D_1, from equation 6.2

$$\frac{di_O}{dt'} + \frac{R}{L}i_O = -\frac{V_C}{L} \quad \text{A|s} \tag{6.6}$$

where

$$t' = t - t_{\text{ON}} \quad \text{s} \tag{6.7}$$

At $t' = 0^+$, $i_O = I_{\max}$, and from equation (6.6)

$$i_O = -\frac{V_C}{R}(1 - \epsilon^{-t'/\tau}) + I_{\max}\epsilon^{-t'/\tau} \quad \text{A:} \qquad t_{\text{ON}} < t \leqslant T \tag{6.8}$$

At $t' = T - t_{\text{ON}}$, or $t = T$, $i_O = I_{\min}$, and from equation 6.8,

$$i_O = I_{\min} = -\frac{V_C}{R}(1 - \epsilon^{-(T - t_{\text{ON}})/\tau}) + I_{\max}\epsilon^{-(T - t_{\text{ON}})/\tau} \quad \text{A} \tag{6.9}$$

Solution of equations 6.5 and 6.9 for $I_{\max}$ and $I_{\min}$ yields

$$I_{\max} = \frac{V}{R}\frac{(1 - \epsilon^{-t_{\text{ON}}/\tau})}{(1 - \epsilon^{-T/\tau})} - \frac{V_C}{R} \quad \text{A} \tag{6.10}$$

$$I_{\min} = \frac{V}{R}\frac{(\epsilon^{t_{\text{ON}}/\tau} - 1)}{(\epsilon^{T/\tau} - 1)} - \frac{V_C}{R} \quad \text{A} \tag{6.11}$$

From equations 6.10 and 6.11 it will be noted that when Q_1 is continuously turned on, so that $t_{\text{ON}} = T$, then

$$I_{\max} = I_{\min} = \frac{V - V_C}{R} \quad \text{A} \tag{6.12}$$

If t_{ON} is decreased to the value t_{ON}^x at which $I_{min} = 0$, then the converter is operating at the point of changeover from continuous-current operation, illustrated in Fig. 6.2c, to discontinuous-current operation, illustrated in Fig. 6.2b. For this boundary condition, from equation 6.11

$$\frac{V_C}{V} = \frac{\epsilon^{(t_{ON}^x/T)(T/\tau)} - 1}{\epsilon^{T/\tau} - 1} \tag{6.13}$$

or

$$m = \frac{\epsilon^{\rho\sigma} - 1}{\epsilon^{\sigma} - 1} \tag{6.14}$$

where

$$m = \frac{V_C}{V} \tag{6.15}$$

$$\rho = \frac{t_{ON}^x}{T} \tag{6.16}$$

$$\sigma = \frac{T}{\tau} \tag{6.17}$$

From equation 6.14, a family of curves of m versus ρ may be plotted with σ as a parameter, and these are shown in Fig. 6.5. A point $[\rho, m]$ lying below the curve for a particular value of σ signifies continuous-current operation for a circuit operating with that ratio of T/τ. A point above the curve signifies discontinuous-current operation. Operation at a point $[\rho, m]$ above the line marked $\sigma = 0$ is impossible, since this line corresponds to a purely inductive load circuit.

For discontinuous-current operation at and beyond the boundary condition defined by the curves of Fig. 6.5, $I_{min} = 0$, and from equation 6.5,

$$I_{max} = \frac{V - V_C}{R}[1 - \epsilon^{-(t_{ON}/\tau)}] \quad \text{A}: \qquad 0 < t_{ON} < t_{ON}^x \quad \text{s} \tag{6.18}$$

and from equations 6.8 and 6.18

$$i_O = -\frac{V_C}{R}(1 - \epsilon^{-t'/\tau}) + \frac{V - V_C}{R}(1 - \epsilon^{-t_{ON}/\tau})\epsilon^{-t'/\tau} \quad \text{A}: \qquad 0 < t_{ON} < t_{ON}^x \quad \text{s}$$

$$\tag{6.19}$$

This current will become zero at time $t = t_x$, or $t' = t_x - t_{ON}$, and substitu-

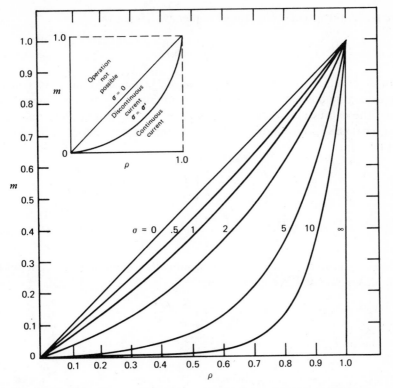

Fig. 6.5 Boundary between discontinuous and continuous current operation.

tion of these end conditions in equation 6.19 yields

$$t_x = \tau \ln \left\{ \epsilon^{t_{\text{ON}}/\tau} \left[1 + \frac{V - V_C}{V_C} (1 - \epsilon^{-t_{\text{ON}}/\tau}) \right] \right\} \quad \text{s} \qquad (6.20)$$

In many cases it will be found that the time variation of load voltage v_O is a very close approximation to one of the ideal waveforms shown in Fig. 6.2b and c. When this is so, i_O may readily be determined by Fourier analysis. Thus the time variation of v_O may be represented by the series

$$v_O = V_O + \sum_{n=1}^{\infty} a_n \sin n\omega t + \sum_{n=1}^{\infty} b_n \cos n\omega t \quad \text{V}$$

$$= V_O + \sum_{n=1}^{\infty} c_n \sin (n\omega t + \theta_n) \quad \text{V} \qquad (6.21)$$

where ω is the angular chopping frequency defined by

$$\omega = \frac{2\pi}{T} \quad \text{rad/s} \tag{6.22}$$

In determining the series in equation 6.21, it is convenient to consider the discontinuous-current mode shown in Fig. 6.2b as the general case. From the expressions obtained for V_O, c_n, and θ_n it may then readily be seen what the corresponding expressions for the continuous-current mode will be. Thus

$$V_O = \frac{1}{T} \int_0^T v_O \, dt = \frac{1}{T} \left[\int_0^{t_{ON}} V \, dt + \int_{t_x}^T V_C \, dt \right]$$

$$= \frac{t_{ON}}{T} V + \frac{(T - t_x)}{T} V_C \quad \text{V} \tag{6.23}$$

When $t_x = T$, that is, at the commencement of continuous-current operation, this expression becomes simply

$$V_O = \frac{t_{ON}}{T} V \quad \text{V} \tag{6.24}$$

In equation 6.21

$$a_n = \frac{2}{T} \int_0^T v_O \sin n\omega t \, dt$$

$$= \frac{2}{T} \left[\int_0^{t_{ON}} V \sin \frac{2n\pi t}{T} \, dt + \int_{t_x}^T V_C \sin \frac{2n\pi t}{T} \, dt \right]$$

$$= \frac{V}{n\pi} [1 - \cos n\omega t_{ON}] - \frac{V_C}{n\pi} [1 - \cos n\omega t_x] \quad \text{V} \tag{6.25}$$

and when $t_x = T$,

$$a_n = \frac{V}{n\pi} [1 - \cos n\omega t_{ON}] \quad \text{V} \tag{6.26}$$

Similarly in equation 6.21

$$b_n = \frac{2}{T} \int_0^T v_O \cos n\omega t \, dt$$

$$= \frac{V}{n\pi} \sin n\omega t_{ON} - \frac{V_C}{n\pi} \sin n\omega t_x \quad \text{V} \tag{6.27}$$

and when $t_x = T$

$$b_n = \frac{V}{n\pi} \sin n\omega t_{ON} \quad V \tag{6.28}$$

Also

$$c_n = [a_n^2 + b_n^2]^{1/2} \quad V \tag{6.29}$$

and

$$\theta_n = \tan^{-1} \frac{b_n}{a_n} \tag{6.30}$$

and when $t_x = T$, substitution from equations 6.26 and 6.28 in these last two expressions yields

$$c_n = \frac{\sqrt{2}\,V}{n\pi} (1 - \cos n\omega t_{ON})^{1/2} \quad V \tag{6.31}$$

$$\theta_n = \tan^{-1} \frac{\sin n\omega t_{ON}}{1 - \cos n\omega t_{ON}} \tag{6.32}$$

The rms values of the output voltage v_O and current i_O as well as the rms harmonic values and ripple factors may now be determined as explained in Section 2.6.1.

The maximum values of the average and rms thyristor currents are one and the same and occur when $t_{ON} = T$, giving

$$I_{Q\,max} = I_{QR\,max} = \frac{V - V_C}{R} \quad A \tag{6.33}$$

By substitution from equation 6.10 in equation 6.8 an expression for the diode current i_D may be obtained and may be employed to determine the average and rms diode currents. This is a cumbersome method of obtaining the rms diode current, but somewhat less so when employed to obtain the average diode current. Since in rating the main thyristor and diode the average rather than the rms current is usually the determining factor, it is perhaps worthwhile to determine the average diode current in this way. However this average current I_D is zero when $t_{ON} = 0$ and also when $t_{ON} = T$. Between these two limits there will be a maximum value which will decide the required diode rating. For a given value of V_C, several values of I_D could be calculated for a series of values of t_{ON} and this maximum value obtained graphically.

An approximate diode rating may be obtained by assuming that the load-circuit inductance is great enough to maintain i_0 at a constant value

$$I_0 = \frac{V_0 - V_C}{R} = \frac{t_{ON}}{T} \times \frac{V}{R} - \frac{V_C}{R} \qquad (6.34)$$

The approximate average diode current is then

$$I_D = \frac{T - t_{ON}}{T} I_0 = \frac{T - t_{ON}}{RT}[\frac{t_{ON}}{T}V - V_C] \qquad (6.35)$$

and this will have its maximum value when

$$\frac{dI_D}{dt_{ON}} = \frac{1}{RT}(1 - \frac{2t_{ON}}{T})V + V_C = 0 \qquad (6.36)$$

from which

$$\frac{t_{ON}}{T} = \frac{V + V_C}{2V} \qquad (6.37)$$

and substitution in equation 6.35 gives

$$I_{D\,max} = \frac{V}{4R}\left[1 - (\frac{V_C}{V})\right]^2 \quad A \qquad (6.38)$$

The worst case will occur when $V_C = 0$, and from equations 6.37 and 6.38

$$\frac{t_{ON}}{T} = \frac{1}{2} \quad : \quad I_{D\,max} = \frac{V}{4R} \quad A \qquad (6.39)$$

The rms diode current corresponding to the conditions of equation 6.39 will then be

$$I_{DR\,max} = \left[\frac{1}{T}\int_{T/2}^{T} (\frac{V}{2R})^2 dt\right]^{1/2} = \frac{V}{2\sqrt{2R}} \quad A \qquad (6.40)$$

In deriving these expressions it has not been possible to take into account external constraints placed on the operation of the converter by the system in which it and the load circuit are embodied. It is very common practice, for example, to include a feedback loop in such a system that would limit the output current to a certain specified value much lower than the maximum values obtained from the equations in the foregoing analysis of the power circuit. Such external constraints must be taken into account when choosing the devices for the system.

On the basis of the power circuit analysis, and in the light of external constraints on the system variables, the devices are chosen. The next stage in converter design is that of devising a commutation circuit on the basis of the manufacturers' specifications of the devices.

Example 6.1 In the type A chopper circuit of Fig. 6.2a, $V = 110$ V, $L = 1$ mH, $R = 0.25\,\Omega$, $V_C = 11$ V, $T = 2500$ μs, $t_{ON} = 1000$ μs.

(a) Calculate the average output current I_O and the average output voltage V_O.

(b) Calculate the maximum and minimum values of instantaneous output current I_{max} and I_{min}.

(c) Sketch to scale the time variations of i_{G1}, v_O, i_O, i_D, i_s, and v_{AK1}.

(d) Calculate the rms values of the first harmonic (fundamental) output voltage and current.

Solution

(a) First determine whether the current is continuous or not. From equations 6.13 to 6.17,

$$m = \frac{V_C}{V} = \frac{11}{110} = 0.1$$

$$\sigma = \frac{T}{\tau} = 2.5 \times 10^{-3} \times \frac{0.25}{1 \times 10^{-3}} = 0.625$$

The value of $\rho = t_{ON}^x/T$ at which the output current changes from continuous to discontinuous mode is obtained by equation 6.14 from

$$0.1 = \frac{\epsilon^{0.625\rho} - 1}{\epsilon^{0.625} - 1}$$

From which

$$\rho = \frac{t_{ON}^x}{T} = 0.133$$

The actual value of t_{ON}/T is

$$\frac{t_{ON}}{T} = \frac{1}{2.5} = 0.4$$

Since this is greater than ρ, the current is continuous. From equation 6.24

$$V_O = 0.4 \times 110 = 44 \quad V$$

also

$$I_O = \frac{V_O - V_C}{R} = \frac{44 - 11}{0.25} = 132 \quad A$$

(b) Expressions for I_{max} and I_{min} are given in equations 6.10 and 6.11. In this system

$$\frac{t_{ON}}{\tau} = 10^{-3}\frac{0.25}{10^{-3}} = 0.25$$

and

$$\frac{T}{\tau} = \sigma = 0.625$$

thus

$$I_{max} = \frac{110}{0.25}\frac{(1 - \epsilon^{-0.25})}{(1 - \epsilon^{-0.625})} - \frac{11}{0.25} = 165 \quad A$$

$$I_{min} = \frac{110}{0.25}\frac{(\epsilon^{0.25} - 1)}{(\epsilon^{0.625} - 1)} - \frac{11}{0.25} = 99.9 \quad A$$

(c) Time variations of the variables are shown in Fig. E6.1.
(d) From equation 6.22 the angular chopping frequency is

$$\omega = \frac{2\pi}{T} = \frac{2\pi}{2.5 \times 10^{-3}} = 2513 \quad rad/s$$

From equation 6.31, the rms value of the first harmonic of the output voltage is

$$V_{1R} = \frac{c_1}{\sqrt{2}} = \frac{V}{\pi}(1 - \cos \omega t_{ON})^{1/2}$$

$$= \frac{110}{\pi}\left(1 - \cos \frac{2513}{10^3}\right)^{1/2} = 47.1 \quad V$$

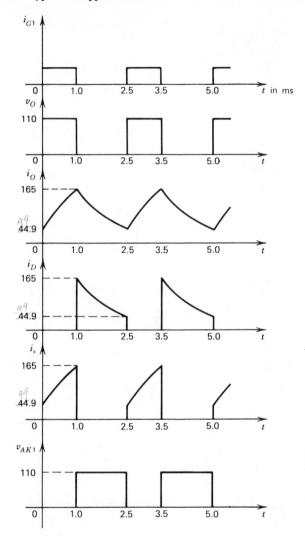

Fig. E6.1 Type A chopper.

The first harmonic of the output current is then

$$I_{1R} = \frac{V_1}{\left[R^2 + (\omega L)^2\right]^{1/2}} = \frac{47.1}{\left[0.25^2 + 2.513^2\right]^{1/2}} = 18.7 \quad A$$

Example 6.2 Repeat example 6.1 when $V = 110\,V$, $L = 0.2$ mH, $R = 0.25\,\Omega$, $V_C = 40\,V$, $T = 2500\,\mu s$, $t_{ON} = 1250\,\mu s$.

Solution

(a)
$$m = \frac{V_C}{V} = \frac{40}{110} = 0.364$$

$$\sigma = \frac{T}{\tau} = 2.5 \times 10^{-3} \times \frac{0.25}{0.20 \times 10^{-3}} = 3.125$$

$$0.364 = \frac{\epsilon^{3.125\rho} - 1}{\epsilon^{3.125} - 1}$$

From which

$$\rho = \frac{t_{ON}{}^x}{T} = 0.700$$

The actual value of t_{ON}/T is

$$\frac{t_{ON}}{T} = \frac{1250}{2500} = 0.5$$

Since this is less than ρ, the current is discontinuous. In this system

$$\tau = \frac{0.2 \times 10^{-3}}{0.25} = 0.8 \times 10^{-3} \quad \text{s}$$

$$\frac{t_{ON}}{\tau} = \frac{1.25 \times 10^{-3}}{0.8 \times 10^{-3}} = 1.56$$

From equation 6.20

$$t_x = 0.8 \times 10^{-3} \ln \left\{ \epsilon^{1.56} \left[1 + \frac{110 - 40}{40} (1 - \epsilon^{-1.56}) \right] \right\} \quad \text{s}$$

$$= 1.94 \times 10^{-3} \quad \text{s}$$

From equation 6.23

$$V_O = 0.5 \times 110 + \frac{(2.5 - 1.94)10^{-3}}{2.5 \times 10^{-3}} \times 40 = 64.0 \quad \text{V}$$

thus

$$I_O = \frac{64.0 - 40}{0.25} = 96 \quad \text{A}$$

(b)   Since the current is discontinuous, $I_{min} = 0$. From equation 6.18

$$I_{max} = \frac{110-40}{0.25}(1 - \epsilon^{-1.56}) = 221 \quad A$$

(c)   Time variations of the variables are shown in Fig. E6.2.

(d)   As in the preceding example, the angular chopping frequency is

$$\omega = 2513 \quad rad/s$$

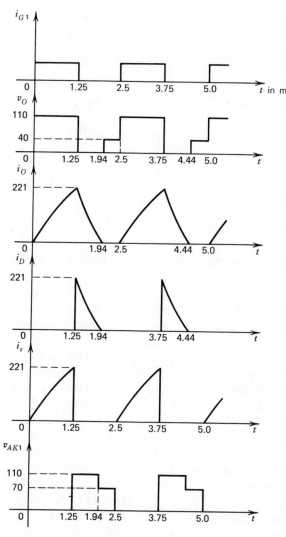

Fig. E6.2 Type A chopper.

From equations 6.25 and 6.27,

$$a_1 = \frac{110}{\pi}[1 - \cos 2.513 \times 1.25] - \frac{40}{\pi}[1 - \cos 2.513 \times 1.94]$$

$$= 70.0 - 10.7 = 59.3 \quad V$$

$$b_1 = \frac{110}{\pi}\sin 2.513 \times 1.25 - \frac{40}{\pi}\sin 2.513 \times 1.94$$

$$= 0.012 - 12.6 \cong -12.6$$

From equation 6.29

$$V_{1R} = \frac{c_1}{\sqrt{2}} = \frac{1}{\sqrt{2}}[59.3^2 + 12.6^2]^{1/2} = 42.9 \quad V$$

and

$$I_{1R} = \frac{42.9}{\left[0.25^2 + (2.513 \times 0.20)^2\right]^{1/2}} = 76.4 \quad A$$

6.2.2 Voltage Commutation of a Type A Chopper The principles of voltage commutation have already been explained in Section 3.5.3, and a rereading of that section would be helpful at this point. The equivalent circuit of a Type A chopper with voltage commutation is shown in Fig. 6.6a. This is an idealization of the actual circuit, not only because the elements required to protect the thyristors are omitted, but also because the transformer composed of coupled linear inductances L_1 and L_2 is assumed to be ideal apart from requiring a magnetizing current. In other words, inductances L_1 and L_2 are ideally coupled and of negligible resistance. The main devices that carry load current may be seen by comparison with the basic circuit of Fig. 6.2a to be thyristor Q_1 and diode D_1. The additional elements not shown in Fig. 6.2a are required for voltage commutation. In particular, capacitor C is charged and employed to commutate the main thyristor Q_1 via diode D_1 by causing v_{AK1} to become negative. Thus if $v_C > 0$, thyristors Q_A and Q_B must be turned on to commutate Q_1. Conversely if $v_C < 0$, then Q_C and Q_D must be turned on. The purpose of the bridge arrangement of thyristors surrounding capacitor C is therefore that of alternately charging and discharging it, and thereby producing an alternating voltage at its terminals.

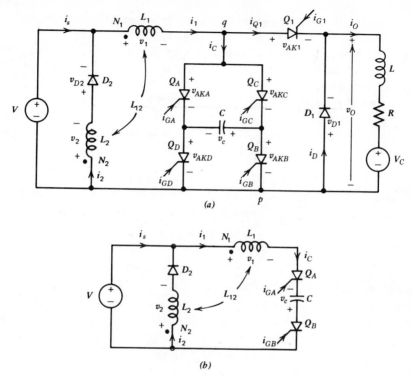

Fig. 6.6 Type A converter—voltage commutation.

The sequence of operations for the entire circuit is as follows:

1. The commutating circuit is switched on for several cycles of alternation of v_C, so that a steady-state alternating condition is achieved.
2. At $t = 0$, when it will be assumed that v_C has the polarity shown in Fig. 6.6a, thyristor Q_1 is turned on, and load current i_O increases exponentially from zero.
3. At $t = t_{ON}$, thyristors Q_A and Q_B are turned on, commutating Q_1 and diverting current i_{Q1} through C as i_C.
4. During interval $t_{ON} < t < T$, the two parts of the circuit separated by Q_1 operate independently as follows:

    a. i_O decays exponentially through diode D_1.
    b. The energy stored in L_1 and C results in an oscillation which leaves $|v_C| > V$, after which that part of the initial circuit energy not finally stored in C is returned to source V by inductance L_2.

5. When all the energy stored in the transformer has been returned to source V by L_2, thyristor Q_1 is again turned on at $t = T$.

6. Load current i_o may be continuous or discontinuous, depending on the values of T, t_{ON}, L, R, and V_C. If i_o is discontinuous, then it again increases during interval $T < t < T + t_{ON}$ exponentially from zero. If i_o is continuous, then it will increase from some initial value I_{min1}. After several cycles of operation have taken place the steady-state value I_{min}, shown in Fig. 6.2c will be achieved, and that value will be maintained until T, t_{ON}, or both are changed.

The detailed analysis of the commutation circuit of Fig. 6.6a may now be carried out. In the transformer of winding self inductances L_1 and L_2, the numbers of turns are N_1 and N_2 respectively. Thus

$$\frac{N_1}{N_2} = \frac{v_1}{v_2} = n \tag{6.41}$$

It will be assumed that, at the end of the preliminary settling period described in Step 1, an alternating voltage has been produced at the terminals of capacitor C, such that at the end of each cycle of operation

$$v_C = \pm V(1+n) \quad \text{V} \tag{6.42}$$

The variables during a typical steady-state cycle of the commutation circuit may now be determined, first for operation on no load, then for operation on load.

Commutation on No-Load   In Fig. 6.6a, let

$$v_C = V(1+n) \quad \text{V}: \quad t = 0 \quad \text{s} \tag{6.43}$$

so that at the end of one cycle, a capacitor voltage of this same magnitude but of opposite sign may be anticipated. Let thyristor Q_A and Q_B be turned on at $t = 0$. Then for the outer loop of the equivalent circuit shown in Fig. 6.6b,

$$V = v_1 - v_C = L_1 \frac{di_C}{dt} + \frac{1}{C} \int i_C \, dt - V(1+n) \quad \text{V} \tag{6.44}$$

From which

$$\frac{d^2 i_C}{dt^2} + \frac{1}{L_1 C} i_C = 0 \quad \text{A/s}^2 \tag{6.45}$$

Also from equation 6.44,

$$v_1 = L_1 \frac{di_C}{dt} = V + V(1+n) \quad \text{V}: \qquad t=0 \quad \text{s} \qquad (6.46)$$

so that the initial conditions for the solution of equation 6.45 are

$$i_C = 0 \quad \text{A}: \qquad \frac{di_C}{dt} = \frac{V(2+n)}{L_1} \quad \text{A/s}: \qquad t=0 \quad \text{s} \qquad (6.47)$$

The solution of equation 6.45 is then

$$i_C = \frac{V(2+n)}{\omega_r L_1} \sin \omega_r t \quad \text{A} \qquad (6.48)$$

where

$$\omega_r = \frac{1}{\sqrt{L_1 C}} \quad \text{rad/s} \qquad (6.49)$$

and is the ringing frequency of the $L_1 C$ circuit. Also from equation 6.48,

$$v_1 = L_1 \frac{di_C}{dt} = V(2+n) \cos \omega_r t \quad \text{V} \qquad (6.50)$$

and

$$v_2 = \frac{v_1}{n} = \frac{V(2+n)}{n} \cos \omega_r t \quad \text{V} \qquad (6.51)$$

while from equations 6.44 and 6.50,

$$v_C = v_1 - V = V(2+n) \cos \omega_r t - V \quad \text{V} \qquad (6.52)$$

The variations of current and voltages described in equations 6.48 and 6.50 to 6.52 continue during period I of the commutation interval and are shown in Fig. 6.7, where it is assumed that $n=1$. Period I ends at some instant $t=t_1$, when v_2 becomes negative and equal in magnitude to V. When this situation is reached, D_2 begins to conduct, v_2, v_1, and v_C are clamped, i_C becomes zero, and thyristors Q_A and Q_B turn off. At this instant

$$v_1 = nv_2 = -nV \quad \text{V}: \qquad t=t_1 \quad \text{s} \qquad (6.53)$$

and

$$v_C = v_1 - V = -V(n+1) \quad \text{V}: \qquad t=t_1 \quad \text{s} \qquad (6.54)$$

The value of v_C given in equation 6.54 justifies the assumption made in equation 6.42, since the capacitor voltage has become equal in magnitude but opposite in polarity to that at $t = 0$. Substitution for v_2 and t in equation 6.51 yields

$$t_1 = \frac{1}{\omega_r} \cos^{-1} \frac{-n}{2+n} \quad \text{s} \qquad (6.55)$$

and from Fig. 6.7 it may be seen that $\omega_r t_1$ is a second-quadrant angle.

Period II of the commutation interval now commences at $t' = 0$, where

$$t' = t - t_1 \quad \text{s} \qquad (6.56)$$

At $t' = 0^-$

$$i_1 = i_C = \frac{V(2+n)}{\omega_r L_1} \sin \omega_r t_1 = I_{O2} \quad \text{A} \qquad (6.57)$$

At $t' = 0^+$, i_C becomes zero, current I_{O2} in inductance L_1 becomes current $n I_{O2}$ in inductance L_2, and diode D_2 continues to conduct until at $t' = t'_1$ the energy that was stored in the transformer at $t' = 0$ has been returned to source V. During period II

$$V = -L_2 \frac{di_2}{dt} \quad \text{V}: \qquad 0 < t' < t'_1 \quad \text{s} \qquad (6.58)$$

and by employing the initial condition

$$i_2 = n I_{O2} \quad \text{A}: \qquad t' = 0^+ \quad \text{s} \qquad (6.59)$$

the solution of equation 6.58 is seen to be

$$i_2 = n I_{O2} - \frac{V}{L_2} t' \quad \text{A} \qquad (6.60)$$

Thus when all the transformer energy has been restored to source V and i_2 becomes zero,

$$t' = t'_1 = \frac{n I_{O2} L_2}{V} \quad \text{s} \qquad (6.61)$$

Period II is also shown in Fig. 6.7, ending at

$$t_2 = t_1 + t'_1 \quad \text{s} \qquad (6.62)$$

The commutation interval is now complete, and a new cycle of the commutation circuit variables may be initiated by turning on thyristors Q_C

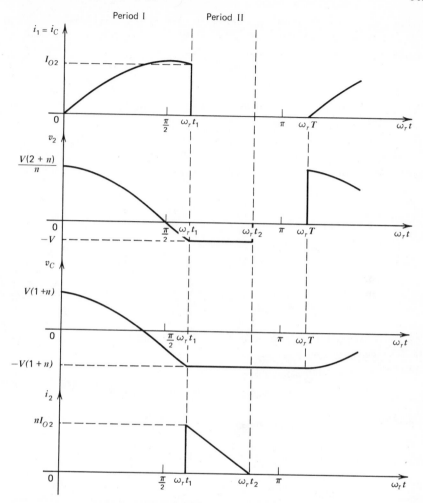

Fig. 6.7 No-load operation of the voltage-commutation circuit ($n=1$).

and Q_D. At the end of this second cycle, it will be found that v_C again has the value given in equation 6.43.

Commutation on Load   In Fig. 6.6a, as for no-load operation, let

$$v_C = V(1+n) \quad \text{V}: \qquad t=0 \quad \text{s} \tag{6.63}$$

Period I of the commutation interval then starts at $t=0$ with the removal of the gating signal from thyristor Q_1 and the simultaneous application of gating signals to thyristors Q_A and Q_B. The capacitor voltage appears between nodes p and q in Fig. 6.6a, so that Q_1 is commutated, and D_1 begins to conduct. At this instant,

$$i_O = I_{O1} \quad \text{A}: \quad t = 0 \quad \text{s} \tag{6.64}$$

and the load current decays through diode D_1 with I_{O1} as its initial value. In this particular case it happens that $I_{O1} = I_{\max}$; however since this is not true for all possible load circuits, it is better to employ another symbol for the initial capacitor current. Equations 6.44 and 6.45 again apply to the operation of the commutation circuit, however the initial conditions are now

$$i_C = i_1 = I_{O1} \quad \text{A}; \qquad \frac{di_C}{dt} = \frac{V(2+n)}{L_1} \quad \text{A/s}; \qquad t = 0 \quad \text{s} \tag{6.65}$$

and the solution of equation 6.45 in this case is

$$i_C = I_{O1}\cos\omega_r t + \frac{V(2+n)}{\omega_r L_1}\sin\omega_r t \quad \text{A} \tag{6.66}$$

This expression for i_C may be compared with that in equation 6.48. Also, from equation 6.66,

$$v_1 = L_1\frac{di_C}{dt} = -\omega_r L_1 I_{O1}\sin\omega_r t + V(2+n)\cos\omega_r t \quad \text{V} \tag{6.67}$$

and

$$v_2 = \frac{v_1}{n} = \frac{-\omega_r L_1 I_{O1}}{n}\sin\omega_r t + \frac{V(2+n)}{n}\cos\omega_r t \quad \text{V} \tag{6.68}$$

while from equation 6.44,

$$v_C = v_1 - V = -\omega_r L_1 I_{O1}\sin\omega_r t + V(2+n)\cos\omega_r t - V \quad \text{V} \tag{6.69}$$

The variations of current and voltages described in equations 6.66 to 6.69 continue during period I of the commutation interval and are shown in Fig. 6.9, where it is again assumed that $n = 1$. Period I ends, as in the no-load case, at $t = t_1$ when $v_2 = -V$ with the consequences already described in the no-load case. Equations 6.53 and 6.54 again apply, and substitution for v_2 and t in equation 6.68 yields

$$t_1 = \frac{1}{\omega_r}[\cos^{-1}A - \tan^{-1}B] \quad \text{s} \tag{6.70}$$

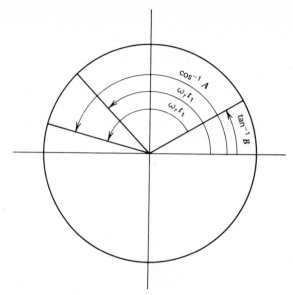

Fig. 6.8 Possible angles in equation 6.70.

where in equation 6.70,

$$A = \frac{-nV}{\left[(\omega_r L_1 I_{O1})^2 + V^2(2+n)^2\right]^{1/2}} \tag{6.71}$$

$$B = \frac{\omega_r L_1 I_{O1}}{V(2+n)} \tag{6.72}$$

Moreover since $\tan^{-1} B$ is a first-quadrant angle, then $\cos^{-1} A$ must be a second-quadrant angle. Possible values of these angles are indicated in Fig. 6.8.

Period II of the commutation interval commences at $t'=0$, where equation 6.56 again applies. At $t'=0^-$, from equation 6.66

$$i_1 = i_C = I_{O1} \cos \omega_r t_1 + \frac{V(2+n)}{\omega_r L_1} \sin \omega_r t_1 = I_{O2} \quad \text{A} \tag{6.73}$$

and this equation may be compared with equation 6.57, where the value of I_{O2} on no load is given. The discussion and equations following equation 6.57 are again applicable and lead to equation 6.61 which is repeated here for convenience, that is,

$$t_1' = \frac{nI_{O2}L_2}{V} \quad \text{s} \tag{6.74}$$

where t_1' is the instant at which all of the energy stored in the transformer core at $t' = 0$ has been returned to source V. The value of I_{O2} to be substituted in equation 6.74 is, of course, that obtained from equation 6.73. Once again, the commutation interval is given by

$$t_2 = t_1 + t_1' \quad \text{s} \tag{6.75}$$

The remaining circuit variable of importance is v_{AK1}. When thyristor Q_1 is conducting, necessarily

$$v_{AK1} = 0 \quad \text{V}: \qquad i_{Q1} \neq 0 \quad \text{A} \tag{6.76}$$

Also, during period I of the commutation interval,

$$v_{AK1} = -v_C \quad \text{V}: \qquad 0 < t < t_1 \quad \text{s} \tag{6.77}$$

During period II of the commutation interval, while D_1 and D_2 are both conducting

$$v_O = 0 \quad \text{V}: \qquad v_1 = nv_2 = -nV \quad \text{V}: \qquad t_1 < t < t_2 \tag{6.78}$$

and

$$v_{AK1} = V - v_1 = V(1+n) \quad \text{V}: \qquad t_1 < t < t_2 \tag{6.79}$$

From the end of the commutation interval to the instant at which Q_1 is again turned on, $v_1 = 0$, and from equation 6.79

$$v_{AK1} = V \quad \text{V}: \qquad t_2 < t < T - t_{ON} \quad \text{s} \tag{6.80}$$

The time variations of the load-circuit and commutation-circuit variables are shown in Fig. 6.9, and with the assistance of this diagram certain critically important time intervals may be determined.

From Fig. 6.9, the angular interval $\omega_r t_q$ during which $v_{AK1} < 0$ is shown. The time available for turn off of thyristor Q_1 is therefore t_q. The interval $\omega_r t_q$ is also that for which $v_C > 0$. The components of v_C expressed in equation 6.69 are shown in Fig. 6.10, where as previously it is assumed that $n = 1$. From this diagram, it may be seen that t_q has a maximum value when $I_{O1} = 0$, and the converter is operating on no load. For any value of

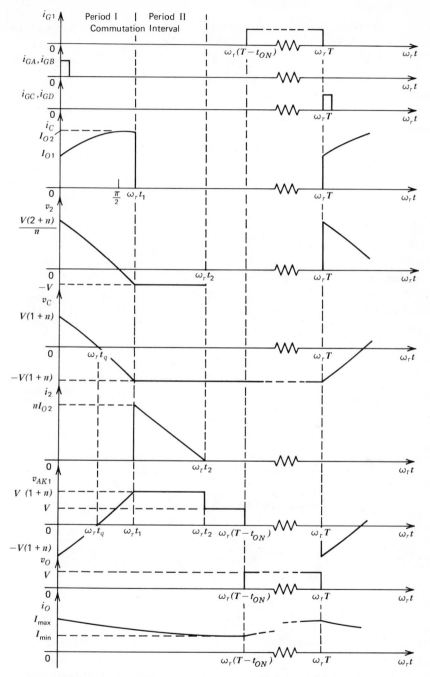

Fig. 6.9 Time variation of the currents and voltages in the circuit of Fig. 6.6a ($n = 1$).

307

I_{O1}, t_q may readily be obtained graphically from a diagram such as that of Fig. 6.10 or alternatively it may be obtained by setting $v_C = 0$ and $t = t_q$ in equation 6.69, giving

$$t_q = \frac{1}{\omega_r} [\cos^{-1} C - \tan^{-1} D] \quad \text{s} \tag{6.81}$$

where in equation 6.81

$$C = \frac{V}{\left[(\omega_r L_1 I_{O1})^2 + V^2 (2 + n)^2 \right]^{1/2}} \tag{6.82}$$

$$D = \frac{\omega_r L_1 I_{O1}}{V(2 + n)} \tag{6.83}$$

Moreover since $\tan^{-1} D$ and $\cos^{-1} C$ are both first-quadrant angles, then $\omega_r t_q$ must also be a first-quadrant angle.

An inspection of Fig. 6.10 shows that $\omega_r t_q$ approaches its maximum value as I_{O1} approaches zero, that is, as the converter nears the no-load

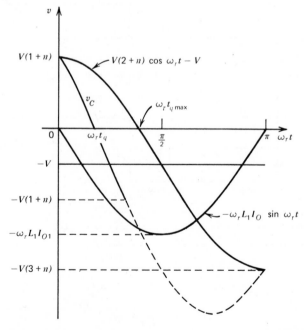

Fig. 6.10 Time available for turn off of thyristor Q_1 in Fig. 6.6.

condition. Substitution in equations 6.81 to 6.83 then yields

$$t_{q\,max} = \frac{1}{\omega_r} \cos^{-1} \frac{1}{2+n} = \sqrt{LC} \cos^{-1} \frac{1}{2+n} \quad \text{s} \qquad (6.84)$$

Thus the time available for turn off decreases as I_{O1} increases, and a permissible limit is reached when $\omega_r t_q = \omega_r t_{off}$ for thyristor Q_1. The limiting value of I_{O1} may be determined by setting $v_C = 0$ and $t = t_{off}$ in equation 6.69, giving the limiting value of I_{O1} as

$$\hat{I}_{O1} = \frac{V(2+n)\cos \omega_r t_{off} - V}{\omega_r L_1 \sin \omega_r t_{off}} \quad \text{A} \qquad (6.85)$$

If the transformer energy is not returned to the source before thyristor Q_1 is again turned on, and if the load circuit has little inductance, there is a possibility that the dissipation of the remaining part of the transformer energy in the load circuit may be accompanied by a peak of current sufficiently great to damage thyristor Q_1. Thus the pulse of i_{G1} must not appear before i_2 has fallen to zero. A necessary operating condition is therefore

$$T > t_{ON} + t_2 \quad \text{s} \qquad (6.86)$$

As a consequence of the limitation expressed in equation 6.86, the average output voltage must always be less than the source voltage, and its maximum possible value is given by

$$\hat{V}_O = \frac{T - t_2}{T} V \quad \text{V} \qquad (6.87)$$

6.2.3 Rating of Circuit Components—Voltage Commutation. Now that the entire circuit of Fig. 6.6a has been analyzed, it is possible to determine the required current and voltage ratings of the various components of the converter.

Main Thyristor Q_1   The current ratings of thyristor Q_1 have already been obtained in Section 6.2.1 and are given in equation 6.33. From the curve of v_{AK1} shown in Fig. 6.9, it may be seen that the peak repetitive voltages applied to thyristor Q_1 are given by

$$V_{FB} = V_{RB} = V(1+n) \quad \text{V} \qquad (6.88)$$

in which the symbols of Fig. 3.24 have been employed.

Free-Wheeling Diode D_1 The current ratings of diode D_1 also have been obtained in Section 6.2.1 and are given in equations 6.35 and 6.38. The peak repetitive voltage applied to diode D_1 is seen from Fig. 6.6a to be

$$V_{RB} = V \quad \text{V} \tag{6.89}$$

Commutating Thyristors The commutating thyristors conduct only on alternate cycles of operation of the converter. Thus the average current is

$$I_{\text{AVEC}} = \frac{1}{2T} \int_0^{t_1} i_C \, dt \quad \text{A} \tag{6.90}$$

and the rms current is

$$I_{\text{RMSC}} = \left[\frac{1}{2T} \int_0^{t_1} i_C^2 \, dt \right]^{1/2} \quad \text{A} \tag{6.91}$$

The expression for i_C given in equation 6.66 must be substituted in equations 6.90 and 6.91, and the value of I_{O1} to be employed in equation 6.66 must be the greatest anticipated value, $\hat{I}_{O1}$. However since t_1 is normally much smaller than T, it is probable that the choice of thyristor will be determined by the peak current, given from equation 6.66 by

$$I_{\text{PEAK}} = \left\{ I_{O1}^2 + \left[\frac{V(2+n)}{\omega_r L_1} \right]^2 \right\}^{1/2} \quad \text{A} \tag{6.92}$$

Data for permissible values of I_{PEAK} do not appear in the data sheet of Fig. 3.24, which refers to a series of thyristors that would not normally be applied in chopper commutation circuits. For thyristors suitable for this application, curves of permissible peak current for sinusoidal and rectangular current pulses are published. The time variation of i_C shown in Fig. 6.9 approximates more closely to a rectangle than to a pulse consisting of the first part of a sine wave, so that this determines the type of published data to be employed in this instance.

In Fig. 6.11 is shown a family of curves of I_{PEAK} versus di/dt for different frequencies. The value of di/dt with which this diagram is entered depends on the amount of protective inductance placed in series with the commutating thyristors and not shown in Fig. 6.6. The diagram in Fig. 6.11 applies to situations in which the thyristor is conducting the peak current for a stated proportion of the duty cycle, given by the maximum value of the ratio $t_1/2T$, and in which the thyristor case temperature T_C does not exceed a stated value. Families of curves for various combinations of $t_1/2T$ and T_C are published by manufacturers, so that a diagram approximating to any practical set of conditions is available. The frequency of the curve in Fig. 6.11 which must be used is given by

$$f = \frac{1}{2T} \quad \text{Hz} \tag{6.93}$$

so that it will normally be necessary to interpolate between the published curves.

From Fig. 6.6a, the greatest voltages applied to the commutating thyristors may be determined by considering one mesh of the circuit comprising thyristors Q_A and Q_C and the capacitor C. For this mesh,

$$v_{AKA} - v_C - v_{AKC} = 0 \quad \text{V} \tag{6.94}$$

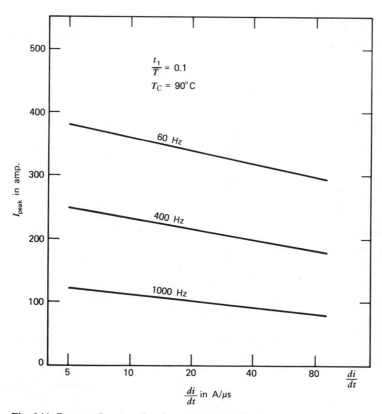

Fig. 6.11 Rectangular wave data for commutating thyristors.

The limiting values of the capacitor voltage are given by

$$v_C = \pm V(1+n) \quad \text{V} \tag{6.95}$$

and since when either of the commutating thyristors is conducting the voltage across its terminals will be zero, it follows that for these thyristors

$$V_{FBC} = V_{RBC} = V(1+n) \quad \text{V} \tag{6.96}$$

Energy-Recovery Diode D_2   From equations 6.56 and 6.60,

$$i_2 = nI_{O2} - \frac{V}{L_2}(t - t_1) \quad \text{A} \tag{6.97}$$

in which I_{O2} is obtained from equation 6.73, and the value of I_{O1} to be substituted in equation 6.73 must again be the greatest anticipated value, $\hat{I}_{O1}$. Since D_2 conducts in each cycle of operation of the converter, then

$$I_{AVE2} = \frac{1}{T}\int_{t_1}^{t_2} i_2 \, dt \quad \text{A} \tag{6.98}$$

$$I_{RMS2} = \left[\frac{1}{T}\int_{t_1}^{t_2} i_2^2 \, dt\right]^{1/2} \quad \text{A} \tag{6.99}$$

In this case it is probable that the choice of the component will be determined by the surge current obtained from

$$I_{FM2} = \left[\frac{1}{t_2 - t_1}\int_{t_1}^{t_2} i_2^2 \, dt\right]^{1/2} \quad \text{A} \tag{6.100}$$

Commutating Capacitor C   The peak voltage applied to the capacitor has been shown to be

$$v_{C\max} = \pm V(1+n) \quad \text{V} \tag{6.101}$$

Since the capacitor is charged each cycle, its average current will be twice that for the commutating thyristors, and its rms current will be $\sqrt{2}$ times that for the commutating thyristors. These values may therefore be obtained from equations 6.90, 6.91, and 6.92.

Coupled Inductors L_1 and L_2   These inductors must be linear; that is, their common core must not saturate at the peak value of i_1, which is also given by equation 6.92.

The rms current rating of the L_1 winding is given by the rms value of i_1, and this is

$$I_1 = [I_{RMS1}^2 + I_{RMSC}^2]^{1/2} \quad A \qquad (6.102)$$

where I_{RMS1} is the maximum value of the main thyristor current obtained from equation 6.33.

The rms current rating of the L_2 winding is the same as that for diode D_2 and is given by equation 6.99.

6.2.4 Design of the Voltage Commutation Circuit The current and voltage ratings of the various circuit components have been specified in Section 6.2.3 in terms of undetermined circuit parameters L_1, L_2, n, and C. A number of optimisation procedures arising out of theoretical criteria have been proposed for the determination of these parameters, but the problem is so complicated that it appears preferable to employ a cut-and-try procedure based on accumulated design experience. For any one set of proposed parameters, the weight, size, and heat generation of the converter may then be determined, and from a number of such sets the most suitable converter design may be selected.

As an initial approximation, it may be assumed that in equation 6.87 $T \gg t_2$. The source voltage V is then either determined by the maximum average output voltage required, or by the value of a dc source already conveniently available. When V is known, the value of n may be chosen to suit the voltage range of thyristors and diodes that it will be economical to employ. From this point on, design may be conducted in terms of a dimensionless quantity.

$$x = \frac{V}{\omega_r L_1 I_{O1}} = \frac{V}{(L_1/C)^{1/2} I_{O1}} \qquad (6.103)$$

and design experience has shown that when I_{O1} is the greatest anticipated value, $\hat{I}_{O1}$, then $0.5 < x < 1.5$.

From equations 6.81 to 6.83, let

$$G(x) = \omega_r t_q = \cos^{-1} C - \tan^{-1} D \quad \text{rad} \qquad (6.104)$$

where

$$C = \frac{1}{\left[x^{-2} + (2+n)^2 \right]^{1/2}} \qquad (6.105)$$

$$D = \frac{1}{x(2+n)} \qquad (6.106)$$

so that the value of $G(x)$ may be calculated from chosen values of x and n. Moreover since the value of I_{O1} employed in determining x is the greatest anticipated value, $\hat{I}_{O1}$, then the value of t_q shown in equation 6.104 must be the minimum value of time available for turn-off of Q_1.

Since an essential condition of operation is that $t_q \geqslant t_{off}$, where t_{off} is the turn-off time for the range of thyristors that it will be economical to employ, let t_q in equation (6.104) be

$$t_q = t_{off} + \Delta t \quad \text{s} \tag{6.107}$$

where Δt is a margin to allow for design approximations and tolerances in the building of the converter. From equations 6.104 and 6.107

$$\omega_r = \frac{G(x)}{t_{off} + \Delta t} \quad \text{rad/s} \tag{6.108}$$

From equations 6.103 and 6.108,

$$L_1 = \frac{V(t_{off} + \Delta t)}{xG(x)I_{O1}} \quad \text{H} \tag{6.109}$$

and

$$C = \frac{1}{\omega_r^2 L_1} = \frac{xI_{O1}(t_{off} + \Delta t)}{G(x)V} \quad \text{F} \tag{6.110}$$

All of the necessary parameters are now available for an evaluation of the converter design and for a second iteration with new values of n and x. It should be noted that under no circumstances may the instantaneous value of the load current exceed the limit specified in equation 6.85.

Example 6.3 In Example 6.1, the following Type A chopper parameters were given: $V = 110$ V, $L = 1$ mH, $R = 0.25\,\Omega$, $V_C = 11$ V, $T = 2500\ \mu\text{s}$, $t_{ON} = 1000\ \mu\text{s}$.

If additionally, in the circuit of Fig. 6.6 $L_1 = 10\ \mu\text{H}$, $C = 50\ \mu\text{F}$, $n = 1$:

(a)   Determine capacitor current I_{O1} at the instant of commutation.
(b)   Calculate the time t_q available for turn-off of thyristor Q_1.
(c)   Calculate the total commutation interval.
(d)   Determine the peak current in the commutation thyristors and the instant at which it occurs.
(e)   Sketch to scale the time variations of i_C, i_2, v_1, v_C and v_{AK1} during the commutation interval.

Solution

(a) From part *b* of Example 6.1, the capacitor current at the instant of commutation is

$$I_{O1} = 165 \quad A$$

(b) From equation 6.103,

$$x = \frac{V}{(L_1/C)^{1/2} I_{O1}} = \frac{110}{\left(\dfrac{10}{50}\right)^{1/2} \times 165} = 1.49$$

From equations 6.105 and 6.106,

$$C = \frac{1}{\left[x^{-2} + (2+n)^2\right]^{1/2}} = \frac{1}{\left[1.49^{-2} + 9\right]^{1/2}} = 0.325$$

$$D = \frac{1}{x(2+n)} = \frac{1}{1.49 \times 3} = 0.224$$

From equation 6.104,

$$\omega_r t_q = \cos^{-1} C - \tan^{-1} D = 1.02^c$$

$$\omega_r = (L_1 C)^{-1/2} = \frac{10^6}{\sqrt{500}}$$

$$\therefore t_q = \frac{1.02\sqrt{500}}{10^6} = 22.8 \quad \mu s$$

(c) The commutation interval is

$$t_2 = t_1 + t_1' \quad s \tag{6.75}$$

The quantities A and B defined in equations 6.71 and 6.72 may be expressed in terms of the dimensionless quantity x as follows:

$$A = \frac{-n}{\left[x^{-2} + (2+n)^2\right]^{1/2}} = \frac{-1}{(1.49^{-2} + 9)^{1/2}} = -0.325$$

$$B = \frac{1}{x(2+n)} = D = 0.224$$

Thus from equation 6.70

$$t_1 = \frac{1}{\omega_r}[\cos^{-1}A - \tan^{-1}B] = 37.6 \quad \mu s$$

The current I_{O2} in equation 6.74 is obtained from equation 6.73. Thus

$$\omega_r t_1 = \frac{41.2}{\sqrt{500}} = 1.84^c = 105.6°$$

$$\omega_r L_1 = \frac{10}{\sqrt{500}} = 0.447 \quad \Omega$$

$$I_{O2} = 165\cos 105.6° + \frac{110 \times 3}{0.447}\sin 105.6°$$

$$= -44.4 + 711.1 = 667 \quad A$$

Then from equation 6.74,

$$t_1' = \frac{nI_{O2}}{V}\frac{L_1}{n^2} = \frac{667}{110} \times 10 \times 10^{-6} = 60.6 \quad \mu s$$

and

$$t_2 = 41.2 + 60.6 = 101.8 \quad \mu s$$

(d)   From equation 6.66

$$i_C = I_{O1}\left[\cos\omega_r t + \frac{V(2+n)}{\omega_r L_1 I_{O1}}\sin\omega_r t\right]$$

$$= I_{O1}[\cos\omega_r t + x(2+n)\sin\omega_r t] \tag{1}$$

$$\therefore I_{PEAK} = I_{O1}\left[1 + x^2(2+n)^2\right]^{1/2}$$

$$= 165[1 + 1.49^2 \times 9]^{1/2} = 756 \quad A$$

and this value is reached when $di_C/dt = 0$, that is, when from equation 1

$$-\sin\omega_r t + x(2+n)\cos\omega_r t = 0$$

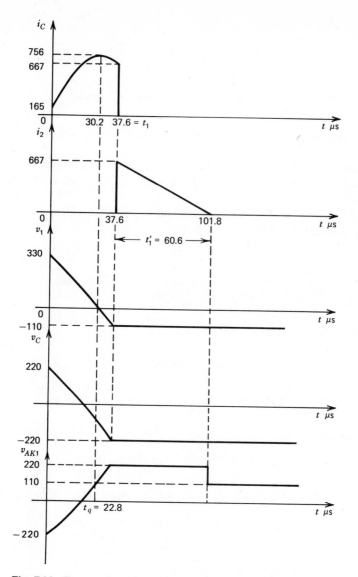

Fig. E6.3 Type A chopper.

From which

$$t = \frac{1}{\omega_r} \tan^{-1} x(2+n)$$

$$= \frac{\sqrt{500}}{10^6} \tan^{-1} 1.49 \times 3 = 30.2 \quad \mu s$$

e)   Time variations of the variables are shown in Fig. E6.3.

Example 6.4 In Example 6.2, the following type A chopper parameters were given: $V = 110$ V, $L = 0.2$ mH, $R = 0.25$ Ω, $V_C = 40$ V, $T = 2500$ μs, $t_{ON} = 1250$ μs.
  If additionally in Fig. 6.6, $n = 0.5$ and $t_{off} = 35$ μs, calculate L_1 and C when the selected value of x is 0.8.

Solution

From Example 6.2, part *b*

$$I_{O1} = I_{max} = 221 \quad A$$

From equations 6.105 and 6.106

$$C = \frac{1}{[0.8^{-2} + 2.5^2]^{1/2}} = 0.358$$

$$D = \frac{1}{0.8 \times 2.5} = 0.5$$

From equation 6.104

$$G(x) = \cos^{-1} 0.358 - \tan^{-1} 0.5 = 0.741^c$$

In equation 6.107, let $\Delta t = 5 \mu s$. Then from equations 6.109 and 6.110,

$$L_1 = \frac{V(t_{off} + \Delta t)}{x G(x) I_{O1}} = \frac{110(35+5)10^{-6}}{0.8 \times 0.741 \times 221} = 33.6 \quad \mu H$$

$$C = \frac{x I_{O1}(t_{off} + \Delta t)}{G(x) V} = \frac{0.8 \times 221(35+5)10^{-6}}{0.741 \times 110} = 86.8 \quad \mu F$$

6.2.5 Current Commutation of a Type A Chopper   The circuit of a type A chopper with current commutation is shown in Fig. 6.12*a*. Protective circuit components are again omitted, and inductor L_1 is linear. The main devices that carry load current are, as before, thyristor Q_1 and free-wheeling diode D_1. The additional elements not shown in Fig. 6.2a are required for current commutation. In particular, capacitor C is charged and employed to initiate the commutation of the main thyristor Q_1.

The sequence of operation for the entire circuit is as follows:

1.   The converter is connected to the source by closing switch SW, and capacitor C is charged up to $v_C = V$ volts via resistor R_1.

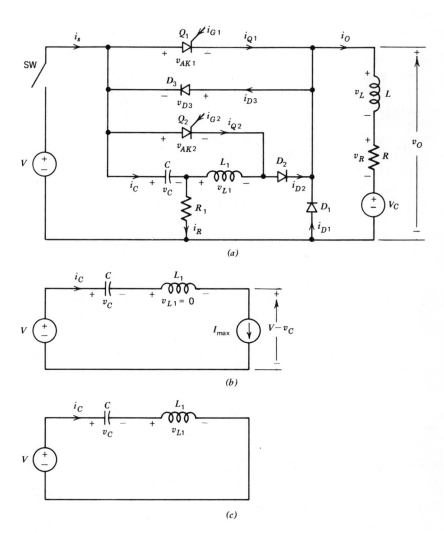

(a)

(b)

(c)

Fig. 6.12   Type A converter—current commutation.

2. At $t = 0$, when the capacitor is fully charged, thyristor Q_1 is turned on, and load current i_O increases exponentially from zero to I_{max}.

3. At $t = t_{ON}$, thyristor Q_2 is turned on, initiating the commutation cycle, and an oscillatory current flows in the ringing circuit comprising C, L_1, and Q_2; i_C is initially negative. N.B. It is assumed that the commutation interval is so short that i_O remains sensibly constant at the value I_{max} throughout the interval. It is also assumed that R_1 is sufficiently large to permit i_R to be neglected in the analysis of the commutation circuit, but is yet small enough to permit v_C to decay to the value V before the next commutation cycle is initiated.

4. When i_C becomes positive, diode D_2 conducts, Q_2 turns off, and since i_O is assumed constant, i_C reduces i_{Q1}.

5. When i_{Q1} is reduced to zero by the increasing value of i_C, diode D_3 begins to conduct, and the forward voltage drop across this diode commutates thyristor Q_1. Current $i_C - I_{max}$ then flows through diode D_3.

6. After i_C has passed its maximum positive value and again become less than I_{max}, diode D_1 conducts. A new oscillatory circuit then exists, comprising C, L_1, D_2, D_1, and source V.

7. The oscillatory cycle of i_C is completed, and i_C becomes zero, leaving $v_C > V$.

8. i_O decays exponentially through D_1 from the value I_{max}, and simultaneously v_C decays through R_1 to the value $v_C = V$.

9. At $t = T$, when $i_O = I_{min}$, Q_1 is again turned on.

The detailed analysis of the commutation circuit of Fig. 6.12 may now be carried out, employing the assumptions made at the end of step 3 in the preceding paragraph. A new time scale is employed, such that $t = 0$ when the commutating cycle is initiated by the application of the gating signal to thyristor Q_2. Thus for the circuit mesh comprising C, L_1, and Q_2,

$$\frac{1}{C} \int i_C \, dt + L_1 \frac{di_C}{dt} = 0 \quad \text{V} \tag{6.111}$$

from which, by differentiation,

$$\frac{d^2 i_C}{dt^2} + \frac{1}{L_1 C} i_C = 0 \quad \text{A/s}^2 \tag{6.112}$$

The initial conditions required for the solution of equation 6.112 are

$$i_C = 0 \quad \text{A}: \quad t = 0^+ \quad \text{s} \tag{6.113}$$

$$v_{AK2} = 0 \quad \text{V}; v_C = V \quad \text{V}; \quad t = 0^+ \quad \text{s} \tag{6.114}$$

$$v_C + v_{L1} = 0 \quad \text{V}; \quad t = 0^+ \quad \text{s} \tag{6.115}$$

so that from equations 6.114 and 6.115

$$\frac{di_C}{dt} = -\frac{V}{L_1} \quad \text{A/s}; \qquad t = 0^+ \quad \text{s} \tag{6.116}$$

From equations 6.112 and 6.116

$$i_C = -\frac{V}{\omega_r L_1} \sin \omega_r t \quad \text{A} \tag{6.117}$$

and from equations 6.114 and 6.117,

$$v_C = V \cos \omega_r t \quad \text{V} \tag{6.118}$$

Equations 6.117 and 6.118 apply throughout period I of the commutation interval, during which time the oscillatory elements C and L_1 are short-circuited through a succession of thyristors and diodes. The time variations of i_C and v_C over the interval $0 < t < t_2$ are therefore purely sinusoidal, and are as shown in Fig. 6.13.

At instant $t = \pi | \omega_r, i_C$ reverses and Q_2 turns off; i_C now flows through diode D_2, and since the main thyristor voltage $v_{AK1} = 0$, the oscillatory elements continue to be short-circuited. For this part of period I, i_C may be considered to flow in a negative direction through Q_1, or alternatively through the load circuit and source, where the equal and opposite voltages, $v_O = V$ and source voltage V, again constitute a short circuit. From either point of view,

$$i_{Q1} = I_{max} - i_C = I_{O1} - i_C \quad \text{A:} \qquad t > \frac{\pi}{\omega_r} \quad \text{s} \tag{6.119}$$

At instant $t = t_1$, $i_C = I_{max}$, $i_{Q1} = 0$, and diode D_3 begins to conduct, so that

$$i_{D3} = i_C - I_{O1} \quad \text{A} \tag{6.120}$$

Also

$$v_{AK1} = -v_{D3} \quad \text{V} \tag{6.121}$$

and Q_1 is commutated; that is, use is made here of the nonideal nature of the devices, in that they do possess an internal resistance, but the resistance is of so low a value that for the purpose of analysis it may be neglected. When Q_1 is commutated

$$i_C = I_{O1} = -\frac{V}{\omega_r L_1} \sin \omega_r t_1 \quad \text{A} \tag{6.122}$$

from which

$$t_1 = \frac{1}{\omega_r} \sin^{-1}\left(-\frac{\omega_r L_1 I_{O1}}{V}\right) = \frac{\pi}{\omega_r} + \frac{1}{\omega_r} \sin^{-1}\frac{\omega_r L_1 I_{O1}}{V} \quad \text{s} \tag{6.123}$$

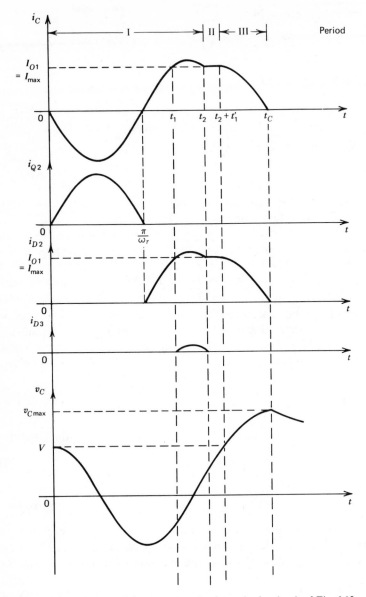

Fig. 6.13 Time variations of the currents and voltages in the circuit of Fig. 6.12.

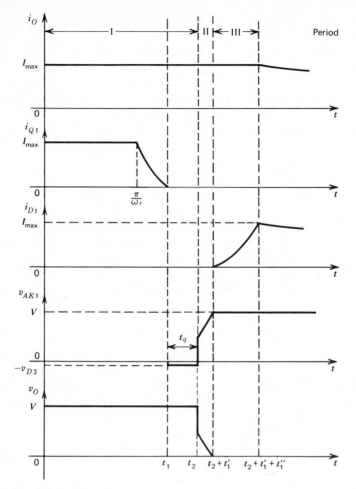

Fig. 6.13 (continued).

From Fig. 6.13 it may be seen that $\omega_r t_1$ is a third-quadrant angle. At $t = t_2$, i_C has passed its positive maximum and fallen to the value I_{O1}. Thus

$$t_2 = \frac{2\pi}{\omega_r} - \frac{1}{\omega_r} \sin^{-1} \frac{\omega_r L_1 I_{O1}}{V}$$

$$= \frac{2\pi}{\omega_r} - t_1 + \frac{\pi}{\omega_r} = \frac{3\pi}{\omega_r} - t_1 \quad \text{s} \qquad (6.124)$$

At this instant, which ends period I of the commutation interval, diode D_1 tends to begin to conduct, since i_O is constant and i_C is falling. The result of conduction of D_1, however, would be that v_O became zero, and since $v_C < V, i_C$ would tend to rise, turning off D_1. This apparent contradiction arises out of the original simplifying assumption that throughout the commutation period i_O remains constant at I_{max}. It must therefore be considered that a short interval ensues during which capacitor C is charged at constant current $i_C = I_{O1}$, until $v_C = V$. During this charging interval, which is called period II of the commutating interval,

$$v_O = V - v_C \quad \text{V} \tag{6.125}$$

This implies that while a voltage appears across inductance L such that $v_L = -v_C$, yet i_O does not vary appreciably from the value I_{max}.

Also at instant $t = t_2, D_3$ ceases to conduct, so that the reverse voltage is removed from thyristor Q_1, and for the outer loop of Fig. 6.12a,

$$-V + v_{AK1} + v_O = 0 \quad \text{V} \tag{6.126}$$

and substitution for v_O from equation 6.125 yields

$$v_{AK1} = v_C \quad \text{V} \tag{6.127}$$

that is, a forward voltage is applied to thyristor Q_1 as shown in Fig. 6.13.

The equivalent circuit for period II of the commutation interval is shown in Fig. 6.12b where, since i_C is constant, $v_{L1} = 0$. Thus

$$i_C = I_{O1} = C\frac{dv_C}{dt'} \quad \text{A} \tag{6.128}$$

where

$$t' = t - t_2 \quad \text{s} \tag{6.129}$$

At $t = t_2$, from equation 6.118,

$$v_C = V_{C2} = V \cos \omega_r t_2 \quad \text{V} \tag{6.130}$$

Thus from equations 6.128 and 6.130,

$$v_C = \frac{I_{O1}}{C} t' + V_{C2} \quad \text{V} \tag{6.131}$$

At $t' = t_1'$,

$$v_C = V = \frac{I_{O1}}{C} t_1' + V_{C2} \quad \text{V} \tag{6.132}$$

so that from equations 6.130 and 6.132

$$t_1' = \frac{CV}{I_{O1}} (1 - \cos \omega_r t_2) \quad \text{s} \tag{6.133}$$

At this instant v_O becomes zero, and diode D_1 begins to conduct; $i_C = I_{O1}$ now becomes the initial current in an oscillatory circuit comprising C, L, D_1, and source V shown in Fig. 6.12c. The current in this circuit may be considered to flow in the reverse direction through D_1, that is

$$i_{D1} = i_O - i_C > 0 \quad \text{A} \tag{6.134}$$

For this new oscillatory circuit, which exists during period III of the commutation interval,

$$v_C + v_{L1} - V = 0 \quad \text{V} \tag{6.135}$$

or

$$\frac{1}{C} \int i_C \, dt'' + L_1 \frac{di_C}{dt''} - V = 0 \quad \text{V} \tag{6.136}$$

where

$$t'' = t' - t_1' = t - t_2 - t_1' \quad \text{s} \tag{6.137}$$

From equation 6.136,

$$\frac{d^2 i_C}{(dt'')^2} + \frac{1}{L_1 C} i_C = 0 \quad \text{A/s}^2 \tag{6.138}$$

The initial conditions for the solution of this equation are

$$i_C = I_{O1} \quad \text{A:} \qquad \frac{di_C}{dt''} = 0 \quad \text{A/s:} \qquad t'' = 0 \quad \text{s} \tag{6.139}$$

and from equations 6.138 and 6.139,

$$i_C = I_{O1} \cos \omega_r t'' \quad \text{A} \tag{6.140}$$

i_C becomes zero when $t'' = t_1''$ and

$$t_1'' = \frac{\pi}{2\omega_r} \quad \text{s} \tag{6.141}$$

At this instant, all of the energy stored in inductance L_1 at $t'' = 0$ has been transferred to capacitor C, and as a consequence $v_C > V$. It is necessary to determine this maximum capacitor voltage. Thus from equation 6.140

$$v_C = \frac{1}{C} \int I_{O1} \cos \omega_r t'' \, dt'' \quad \text{V} \tag{6.142}$$

also

$$v_C = V \quad \text{V}: \qquad t'' = 0 \quad \text{s} \tag{6.143}$$

so that from equations 6.142 and 6.143

$$v_C = \frac{I_{O1}}{\omega_r C} \sin \omega_r t'' + V \quad \text{V} \tag{6.144}$$

and substitution from equation 6.141 yields

$$v_{C\max} = V + \frac{I_{O1}}{\omega_r C} \quad \text{V} \tag{6.145}$$

This capacitor voltage then decays through R_1 to $v_C = V$, and simultaneously i_O decays exponentially from the value $I_{\max}$.

The length of the entire commutation interval may now be determined. It is

$$t_C = t_2 + t_1' + t_1'' \quad \text{s} \tag{6.146}$$

and from the expressions derived for the components of t_C it would be possible to express the commutation interval as a function of $I_{\max}$ and the circuit parameters. It would be a very cumbersome expression, however, and not particularly useful or enlightening when obtained. The length of the commutation interval is not an important consideration in the design of type A choppers. In the design of type B choppers on the other hand its effect on the operation of the converter must be considered.

From the curve of v_{AK1} in Fig. 6.13 the time available for commutation of thyristor Q_1 is seen to be

$$t_q = t_2 - t_1 \quad \text{s} \tag{6.147}$$

and substitution for t_1 and t_2 from equations 6.123 and 6.124 yields

$$t_q = \frac{\pi}{\omega_r} - \frac{2}{\omega_r} \sin^{-1} \frac{\omega_r L_1 I_{O1}}{V} \quad \text{s} \tag{6.148}$$

Equation 6.148 shows that the time available for turn-off of the main thyristor decreases with increase of load current.

6.2.6 Rating of Circuit Components—Current Commutation The method of determining the required ratings of the circuit components has already been very fully illustrated for the voltage commutation circuit. There is therefore no need for a detailed discussion of this topic here. The important point to bear in mind is that each component must be rated for that condition of operation of the converter that makes the heaviest demands on it. What, for example, is the operating condition that results in the highest average and rms values of the current i_{D3} in diode D_3?

Only one situation exists in the circuit of Fig. 6.12 that has not already been met elsewhere, and that involves thyristor Q_1, which is in inverse-parallel connection with diode D_3. The turn-off time for a thyristor is somewhat reduced with increase of reverse voltage applied to it. In this circuit the reverse voltage during commutation is virtually zero, being simply the forward voltage drop across diode D_3. For the circuit of Fig. 6.12 the turn-off time to be taken from the manufacturers' data sheets (not shown in Fig. 3.24) is that listed as "Turn-Off Time (with Feedback Diode)," or some corresponding expression. The presence of the feedback diode may in extreme cases increase the turn-off time by as much as 50% of that required with reverse voltage.

6.2.7 Design of the Current Commutation Circuit Once again a design procedure is described such that, when the main devices have been chosen on the basis of the power-circuit analysis, the commutation-circuit design may be carried out in terms of a dimensionless quantity

$$x = \frac{V}{\omega_r L_1 I_{O1}} = \frac{V}{(L_1/C)^{1/2} I_{O1}} \tag{6.149}$$

Typically for a current-commutated converter $1.4 < x < 3.0$. An initial value of x must be chosen that yields circuit parameters giving acceptable values of:

1. Maximum capacitor voltage $v_{C\max}$.
2. Peak commutating thyristor current I_{PEAK}.
3. Charging-circuit losses in resistor R_1.

From equations 6.145 and 6.149,

$$\frac{v_{C\max}}{V} = 1 + \frac{1}{x} \tag{6.150}$$

so that $1/x$ is the factor by which the maximum capacitor voltage overshoots the source voltage, which must also be the capacitor voltage at the start of the commutation cycle. If now a function is defined from equations 6.148 and 6.149

$$G(x) = \omega_r t_q = \pi - 2\sin^{-1}\frac{1}{x} \quad \text{rad} \tag{6.151}$$

where

$$t_q = t_{\text{off}} + \Delta t \quad \text{s} \tag{6.152}$$

then this function may be employed to determine L_1 and C. In equation 6.152, t_{off} is the specified turn-off time for the chosen main thyristor, and Δt is a margin to allow for design approximations and tolerances in the building of the converter.

The expressions for L_1 and C are similar to those derived from the voltage commutation circuit, so that by analogy with equations 6.109 and 6.110

$$L_1 = \frac{V(t_{\text{off}} + \Delta t)}{xG(x)I_{O1}} \quad \text{H} \tag{6.153}$$

$$C = \frac{xI_{O1}(t_{\text{off}} + \Delta t)}{G(x)V} \quad \text{F} \tag{6.154}$$

If the analysis of Section 6.2.5 is to give reliable predictions of system performance, then the value of resistance R_1 must be chosen such that,

$$T \geqslant 3R_1C = 3\tau \quad \text{s} \tag{6.155}$$

since if this were not the case, there would be insufficient time for the capacitor voltage to fall from $v_{C\max}$ to V before another commutation cycle was initiated.

Of the energy stored in the capacitor at the end of the commutation interval, some is dissipated in resistance R_1, some is returned to source V, and some remains stored in the capacitor for the initiation of the next commutation cycle. The amount dissipated in R_1 should be minimised so far as is compatible with other design criteria. The amount of energy which remains stored in the capacitor is

$$W_C = \tfrac{1}{2} C V^2 \quad \text{J} \tag{6.156}$$

During the discharge period, while the capacitor voltage is falling from $v_{C\max}$ to V, the capacitor current is

$$i_C = \frac{-(v_{C\max} - V)\epsilon^{-t/\tau}}{R} \quad \text{A} \tag{6.157}$$

On the assumption that inequality 6.155 is satisfied, and $T \geqslant 3 R_1 C$, the energy returned to source V will be, to a close approximation

$$W_s = \int_0^\infty -Vi_C \, dt = \int_0^\infty \frac{V}{R}(v_{C\max} - V)\epsilon^{-t/\tau} dt$$

$$= CV(v_{C\max} - V) \quad \text{J} \tag{6.158}$$

The change in energy stored in the capacitor during the discharge period is

$$W_{C\max} - W_C = \tfrac{1}{2} C [v_C{}^2{}_{\max} - V^2] \quad \text{J} \tag{6.159}$$

The energy dissipated in resistor R_1 is therefore

$$W_R = (W_{C\max} - W_C) - W_s = \tfrac{1}{2} C [v_{C\max} - V]^2 \quad \text{J} \tag{6.160}$$

Substitution for $v_{C\max}$ from equation 6.150 in equation 6.160 yields

$$W_R = \frac{1}{2} C \frac{V^2}{x^2} \quad \text{J} \tag{6.161}$$

and further substitution for C from equation 6.154 in equation 6.161 gives

$$W_R = \frac{V I_{O1}(t_{\text{off}} + \Delta t)}{2x[\pi - 2\sin^{-1}(1/x)]} \quad \text{J} \tag{6.162}$$

From equation 6.162, it is evident that x should be made as large as possible; that is, the overshoot of capacitor voltage should be made as small as possible, consistent with satisfaction of the peak current criterion. The permissible peak current will, of course, determine the choice of commutating thyristor.

From equation 6.161, the average power dissipated in resistor R_1 is

$$P_R = \frac{1}{2} \frac{CV^2}{x^2 T} \quad \text{W} \tag{6.163}$$

Example 6.5 In Example 6.1, the following Type A chopper parameters were given: $V = 110$ V, $L = 1$ mH, $R = 0.25\,\Omega$, $V_C = 11$ V, $T = 2500\,\mu\text{s}$, $t_{ON} = 1000\,\mu\text{s}$. If the chopper is current commutated as in Fig. 6.12, with $L_1 = 4\,\mu\text{H}$ and $C = 40\,\mu\text{F}$:

    (a)   Calculate the time available for turn-off of thyristor Q_1.
    (b)   Calculate the commutation interval.
    (c)   Sketch to scale the time variations of i_C, i_{Q2}, i_{D2}, i_{D3}, i_{Q1}, i_{D1}, v_{AK1}, and v_{AK2}.
    (d)   Determine the time available for turn-off of thyristor Q_2.

Solution

    (a)   From part b of Example 6.1

$$I_{O1} = I_{max} = 165 \quad \text{A}$$

From equation 6.149

$$x = \frac{V}{I_{O1}} \left(\frac{C}{L_1} \right)^{1/2} = \frac{110}{165} \left(\frac{40}{4} \right)^{1/2} = 2.11$$

Also

$$\omega_r = \frac{1}{\sqrt{L_1 C}} = \frac{10^6}{\sqrt{40 \times 4}} = \frac{10^6}{12.65} \quad \text{rad/s}$$

From equation 6.151

$$t_{q1} = \frac{1}{\omega_r} \left(\pi - 2\sin^{-1} \frac{1}{x} \right) = \frac{12.65}{10^6} (\pi - 2 \times 0.494) = 27.2 \quad \mu\text{s}$$

    (b)   From equation 6.146,

$$t_C = t_2 + t_1' + t_1'' \quad \text{s}$$

From equation 6.124,

$$t_2 = \frac{1}{\omega_r}\left(2\pi - \sin^{-1}\frac{1}{x}\right) = \frac{10.65}{10^6}(2\pi - 0.494) = 73.2 \quad \mu s$$

From equation 6.133,

$$t_1' = \frac{x}{\omega_r}(1 - \cos\omega_r t_2) = \frac{2.11 \times 12.65}{10^6}\left(1 - \cos\frac{73.2}{12.65}\right)$$

$$= 3.22 \quad \mu s$$

From equation 6.141

$$t_1'' = \frac{\pi}{2\omega_r} = \frac{\pi}{2} \times \frac{12.65}{10^6} = 19.9 \quad \mu s$$

Thus

$$t_C = 73.2 + 3.2 + 19.9 = 96.3 \quad \mu s$$

(c) The time variations of the variables are shown in Fig. E6.5.

(d) Thyristor Q_2 ceases to conduct at instant $t = \pi/\omega_r$, and a forward voltage is again applied to it at the end of period I of the commutation interval. From Fig. E6.5 therefore,

$$t_{q2} = t_2 - \frac{\pi}{\omega_r} = (73.2 - 12.65\pi)10^{-6} = 33.5 \quad \mu s$$

Example 6.6 In example 6.2, the following Type A chopper parameters were given: $V = 110$ v, $L = 0.2$ mH, $R = 0.25\Omega$, $V_c = 40$ V, $T = 2500\mu s$, $t_{ON} = 1250\mu s$. If for the chosen main thyristor of the chopper of Fig. 6.12, t_q as defined in equation 6.152 is 50 μs, then for $x = 1.5$ determine:

(a) The commutating inductance L_1 and capacitance C.
(b) The maximum capacitor voltage $v_{C\max}$.
(c) The losses in the commutating resistor R_1.
(d) The peak current in the commutating thyristor I_{PEAK}.

Also

(e) to (h) Repeat a to d for $x = 2.5$.
(i) Compare the two sets of results.

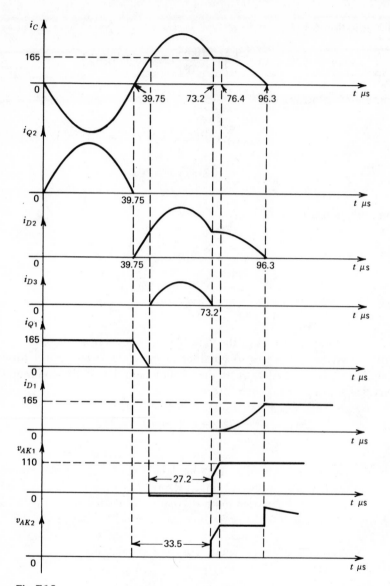

Fig. E6.5

Solution

  $x = 1.5$

(a) From equation 6.151,

$$G(x) = \pi - 2\sin^{-1}\frac{1}{1.5} = 1.68$$

From example 6.2,

$$I_{O1} = I_{max} = 221 \quad A$$

From equations 6.153 and 6.154,

$$L_1 = \frac{110 \times 50 \times 10^{-6}}{1.5 \times 1.68 \times 221} = 9.88 \times 10^{-6} \quad H$$

$$C = \frac{1.5 \times 221 \times 50 \times 10^{-6}}{1.68 \times 110} = 89.7 \times 10^{-6} \quad F$$

(b) From equation 6.150,

$$v_{C\,max} = 110\left(1 + \frac{1}{1.5}\right) = 183 \quad V$$

(c) From equation 6.163,

$$P_R = \frac{1}{2} \times \frac{89.7 \times 10^{-6} \times 110^2}{1.5^2 \times 2500 \times 10^{-6}} = 96.5 \quad W$$

(d) From equation 6.122 and 6.149,

$$I_{PEAK} = \frac{V}{\omega_r L_1} = xI_{O1} = 1.5 \times 221 = 332 \quad A$$

  $x = 2.5$
(e)

$$G(x) = \pi - 2\sin^{-1}\frac{1}{2.5} = 2.32$$

$$L_1 = \frac{110 \times 50 \times 10^{-6}}{2.5 \times 2.32 \times 221} = 4.29 \times 10^{-6} \quad H$$

$$C = \frac{2.5 \times 221 \times 50 \times 10^{-6}}{2.32 \times 110} = 108 \times 10^{-6} \quad F$$

(f)

$$v_{C\max} = 110\left(1 + \frac{1}{2.5}\right) = 154 \quad V$$

(g)

$$P_R = \frac{1}{2} \times \frac{108 \times 10^{-6} \times 110^2}{2.5^2 \times 2500 \times 10^{-6}} = 41.8 \quad W$$

(h)

$$I_{PEAK} = 2.5 \times 221 = 552 \quad A$$

(i)   From two or more such sets of figures as have been obtained in *a* to *h* the designer must attempt to choose the best design.

There is little to choose between the two sets of parameters obtained in *a* and *e*. As one component increases, the other decreases in rating, and the total cost of the two will not vary greatly.

Since this is a low-voltage converter, the capacitor voltage rating is no problem, so that the voltages obtained in *b* and *f* are equally acceptable.

Steps *c* and *g* show that as a result of increasing *x*, the heat to be dissipated from the converter due to the commutation circuit has been halved. On the other hand *d* and *h* show that this reduction in heat dissipation will be gained at the expense of an increased commutation thyristor rating and no doubt also an increase in the heat to be dissipated from that thyristor.

6.3 ANALYSIS OF THE TYPE B CHOPPER CIRCUIT

In this section, the operation of the power circuit is first discussed. Two methods of current commutation are then described.

6.3.1 Power Circuit of a Type B Chopper The power circuit of a type B chopper shown in Fig. 6.3 is reproduced in Fig. 6.14a. Gating signals for the two thyristors are shown in Fig. 6.14b, and it is assumed that with these gating signals the values of V and V_C are such that the average output current I_O is positive, and that the converter is therefore operating in the first quadrant of a diagram of V_O versus I_O.

After the converter has been switched on for a short time, it may be found to be operating with time variations of the circuit variables such as are shown in Fig. 6.14b. The waveforms of Fig. 6.14b should be compared with those of Fig. 6.2c. When this is done, it will be seen that the essential difference between the two converters is that, for part of the cycle of the type B chopper, $i_O < 0$, and this occurs because of the presence of thyristor Q_2 and diode D_2. In other words, the type B chopper cannot operate with discontinuous load current. The type B waveform of i_O may be broken up into four segments, and in Fig. 6.14b the semiconductor device current identical with i_O for any interval is indicated on the i_O waveform.

It is now necessary to determine relationships between the independent variables T, t_{ON}, V, V_C, and the dependent variables in the circuit, of which the most significant are V_O, I_O, and I_s, the average source current.

It should be noted at this stage that, since discontinuous-current operation of this converter is not possible, it may be assumed for the purpose of the power-circuit analysis that there is no interval between the end of one gating signal and the beginning of the other. It is however shown later that the operation of the commutating circuit requires such an interval, albeit a very small one.

The analysis developed in Section 6.2.1 for continuous-current operation of the Type A chopper may be directly applied to the circuit of Fig. 6.14a. Thus from equations 6.3, 6.4, 6.8, and 6.7,

$$i_O = \frac{V - V_C}{R}(1 - \epsilon^{-t/\tau}) + I_{min}\epsilon^{-t/\tau} \quad A: \quad 0 \leqslant t < t_{ON} \quad s \quad (6.164)$$

where

$$\tau = \frac{L}{R} \quad s \quad (6.165)$$

$$i_O = -\frac{V_C}{R}(1 - \epsilon^{-t'/\tau}) + I_{max}\epsilon^{-t'/\tau} \quad A: \quad t_{ON} < t \leqslant T \quad s \quad (6.166)$$

where

$$t' = t - t_{ON} \quad s \quad (6.167)$$

Substitution of the end conditions in equations 6.164 and 6.166 and solving for I_{max} and I_{min} then yields expressions for these two currents, which are, from equations 6.10 and 6.11,

$$I_{max} = \frac{V}{R}\frac{[1 - \epsilon^{-t_{ON}/\tau}]}{[1 - \epsilon^{-T/\tau}]} - \frac{V_C}{R} \quad A \quad (6.168)$$

$$I_{min} = \frac{V}{R}\frac{(\epsilon^{t_{ON}/\tau} - 1)}{(\epsilon^{T/\tau} - 1)} - \frac{V_C}{R} \quad A \quad (6.169)$$

The independent variables in equations 6.168 and 6.169 may be such that $I_{min} > 0$. In this case, the converter is operating simply as a Type A chopper with continuous output current. Thyristor Q_2 and diode D_2 do not

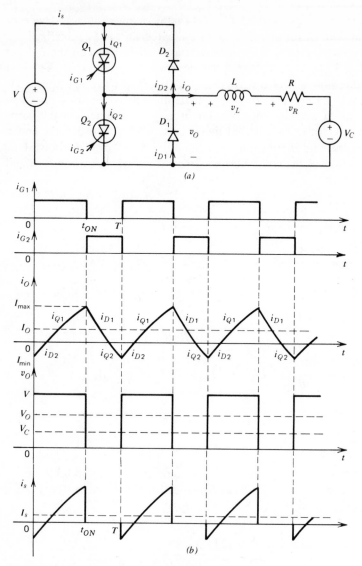

Fig. 6.14 Basic principles of a type B chopper.

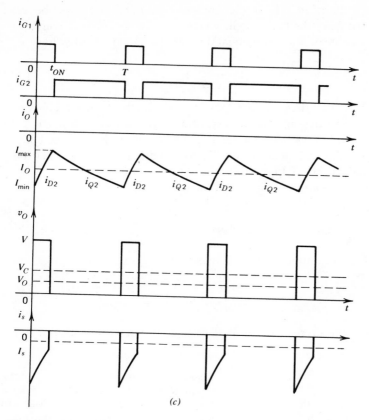

Fig. 6.14 (continued).

conduct during any part of the cycle. This is first-quadrant operation, as illustrated in Fig. 6.2c.

If the independent variables are such that $I_{max} > 0$, and $I_{min} < 0$, then the result may be first-quadrant operation, as illustrated in Fig. 6.14b, where $I_O > 0$. If the independent variables are such that I_{max} has a smaller positive value and I_{min} a larger negative value than are shown in Fig. 6.14b, so that $I_O < 0$, then the result will be second-quadrant operation.

Finally, if the independent variables are such that $I_{max} < 0$, then $I_O < 0$, and the result is second-quadrant operation. Under these conditions, thyristor Q_1 and diode D_1 do not conduct during any part of the cycle. This condition is illustrated in Fig. 6.14c.

The quadrant in which the converter is operating may be readily determined from the load-circuit voltages. As in the Type A converter, v_O is always defined, and from equation 6.24 the average load voltage is

$$V_O = \frac{t_{ON}}{T} V \quad V \tag{6.170}$$

If now $V_O > V_C$, then average power flows to the load circuit, and the converter operates in the first-quadrant. If, however, $V_O < V_C$, then average power flows from the load circuit to source V, and the converter operates in the second quadrant. These conditions are illustrated on the voltage waveforms of Fig. 6.14.

When Q_1 is continuously turned on, so that $t_{ON} = T$, then

$$I_{max} = I_{min} = \frac{V - V_C}{R} \quad A \tag{6.171}$$

When Q_2 is continuously turned on, so that $t_{ON} = 0$, then

$$I_{max} = I_{min} = -\frac{V_C}{R} \quad A \tag{6.172}$$

Also, since the thyristors and diodes are considered to be ideal, it follows that

$$V I_s = V_O I_O \quad W \tag{6.173}$$

The results of the Fourier analysis of the Type A chopper operating under continuous-current conditions may be applied directly to the Type B chopper. Thus from equations 6.21, 6.22, 6.24, 6.31, and 6.32

$$v_O = V_O + \sum_{n=1}^{\infty} c_n \sin(n\omega t + \theta_n) \quad V \tag{6.174}$$

when

$$\omega = \frac{2\pi}{T} \quad \text{rad/s} \tag{6.175}$$

$$V_O = \frac{t_{ON}}{T} V \quad V \tag{6.176}$$

$$c_n = \frac{\sqrt{2} V}{n\pi} (1 - \cos n\omega t_{ON})^{1/2} \quad V \tag{6.177}$$

$$\theta_n = \tan^{-1} \frac{\sin n\omega t_{ON}}{1 - \cos n\omega t_{ON}} \quad \text{rad} \tag{6.178}$$

From these expressions, the rms values of the output voltage v_O and current i_O as well as the rms harmonic currents and ripple factors may be determined as explained in Section 2.6.1.

The maximum average and rms values of the current in thyristor Q_1 are identical and occur during first-quadrant operation, when $t_{ON} = T$, giving as in equation 6.33

$$I_{Q1\,max} = I_{Q1R\,max} = \frac{V - V_C}{R} \quad \text{A} \tag{6.179}$$

Correspondingly, the maximum values for diode D_1 also occur during first-quadrant operation and from equations 6.38 and 6.40 are approximately

$$I_{D1\,max} = \frac{V}{4R}\left[1 - \frac{V_C}{V}\right]^2 \quad \text{A} \tag{6.180}$$

$$I_{D1R\,max} = \frac{V}{2\sqrt{2}\,R} \quad \text{A} \tag{6.181}$$

The maximum average and rms values of the current in thyristor Q_2 are identical and occur during second-quadrant operation, when $t_{ON} = 0$, giving

$$I_{Q2\,max} = I_{Q2R\,max} = \frac{V_C}{R} \quad \text{A} \tag{6.182}$$

For diode D_2, approximate ratings may be obtained by again assuming that the load-circuit inductance is sufficiently great to maintain i_O at a constant value given by equation 6.34,

$$I_O = \frac{t_{ON}}{T} \times \frac{V}{R} - \frac{V_C}{R} \quad \text{A} \tag{6.183}$$

The average diode current is then

$$I_{D2} = \frac{t_{ON}}{T}\left[\frac{t_{ON}}{T}\frac{V}{R} - \frac{V_C}{R}\right] \quad \text{A} \tag{6.184}$$

Differentiation of equation 6.184 with respect to t_{ON} and equating to zero yields,

$$t_{ON} = \frac{1}{2}\frac{V_C}{V}T \quad \text{S} \tag{6.185}$$

and

$$I_{D2\,max} = \frac{1}{4}\frac{V_C^2}{VR} \quad \text{A} \tag{6.186}$$

The warning already given at the end of Section 6.2.1, that in fact the maximum currents in the devices will in many cases be constrained by current-limiting feedback loops in the overall system, must be borne in mind here also.

Example 6.7 In the type B chopper circuit of Fig. 6.14a, $V = 110$ V, $L = 0.2$ mH, $R = 0.25\,\Omega$, $V_C = 40$ V, $T = 2500$ μs, $t_{ON} = 1250$ μs.

(a) Calculate the average output current I_O and the average output voltage V_O.

(b) Calculate the maximum and minimum value of instantaneous output current I_{max} and I_{min}.

(c) Sketch to scale the time variations of v_O, i_O, i_{Q1}, i_{Q2}, i_{D1}, i_{D2}, and i_s.

Solution

(a) From equation 6.170

$$V_O = \frac{1250}{2500} \times 110 = 55 \quad V$$

$$I_O = \frac{V_O - V_C}{R} = \frac{55 - 40}{0.25} = 60 \quad A$$

(b) From equation 6.165

$$\tau = \frac{0.2 \times 10^{-3}}{0.25} = 800 \quad \mu s$$

Also

$$\frac{t_{ON}}{\tau} = \frac{1250}{800} = 1.562$$

$$\frac{T}{\tau} = \frac{2500}{800} = 3.125$$

From equations 6.168 and 6.169,

$$I_{max} = \frac{110}{0.25} \frac{(1 - \epsilon^{-1.562})}{(1 - \epsilon^{-3.125})} - \frac{40}{0.25} = 204 \quad A$$

$$I_{min} = \frac{110}{0.25} \frac{(\epsilon^{1.562} - 1)}{(\epsilon^{3.125} - 1)} - \frac{40}{0.25} = -83.8 \quad A$$

(c) Time variations of the variables are shown in Fig. E6.7.

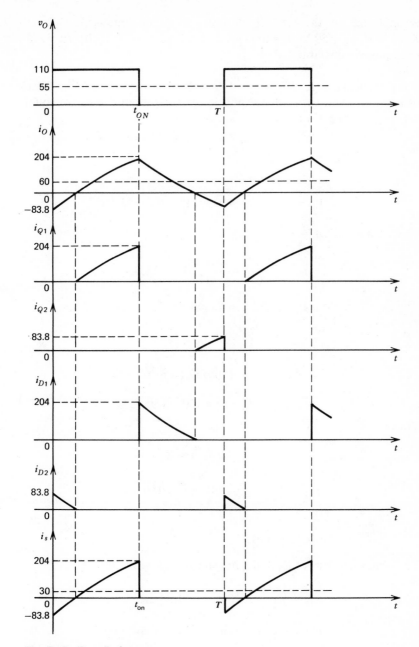

Fig. E6.7 Type B chopper.

Example 6.8 In the type B chopper circuit of Fig. 6.14a, $V = 110$ V, $R = 0.15\,\Omega$, $V_C = 50$ V, $T = 4000$ μs, and L is so large that effectively i_O may be assumed constant.

   (a)   Calculate t_{ON} if $i_O = I_O = 100$ A. Sketch to scale the time variations of v_O and i_s, and verify that input power and output power are equal.
   (b)   Repeat a if $i_O = I_O = -100$ A.

Solution

   (a)

$$V_O = V_C + RI_O = 50 + 0.15 \times 100 = 65 \quad V$$

From equation 6.170

$$t_{ON} = \frac{V_O T}{V} = \frac{65 \times 4000}{110} = 2364 \quad \mu s$$

The time variations of v_O and i_s are shown in Fig. E6.8a.

$$I_s = \frac{2364}{4000} \times 100 = 59.1 \quad A$$

$$P_{IN} = V_s I_s = 110 \times 59.1 = 6.50 \quad kW$$

$$P_O = V_O I_O = 65 \times 100 = 6.50 \quad kW$$

Thus

$$P_O = P_{IN}$$

   (b)

$$V_O = V_C + RI_O = 50 - 0.15 \times 100 = 35 \quad V$$

$$t_{ON} = \frac{35 \times 4000}{110} = 1273 \quad \mu s$$

The time variations of v_O and i_s are shown in Fig. E6.8b.

$$I_s = \frac{1273}{4000}(-100) = -31.8 \quad A$$

$$P_{IN} = V_s I_s = -110 \times 31.8 = -3.50 \quad kW$$

$$P_O = V_O I_O = 35 \times (-100) = -3.50 \quad kW$$

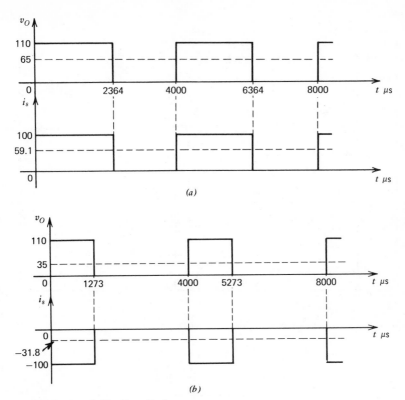

Fig. E6.8 (*a*) and (*b*)   Type B chopper.

Thus

$$P_O = P_{\text{IN}}$$

6.3.2 Commutation of a Type B Chopper Two methods of current commutation of a type B chopper are described. As has already been remarked, many commutation circuits have been devised. Those described here are typical.

Type 1 Current Commutation The circuit of a type B chopper with type 1 current commutation is shown in Fig. 6.15*a*. The circuit employed for first-quadrant operation is identical with that shown in Fig. 6.12*a*, pro-

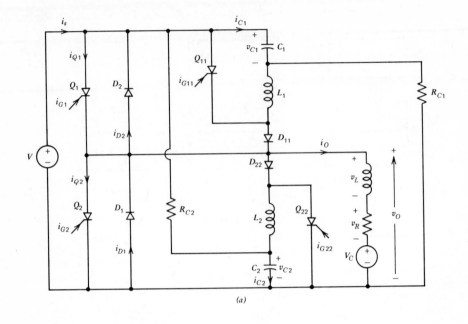

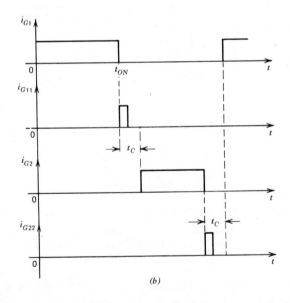

Fig. 6.15 Type B chopper—type 1 current commutation.

vided that the following correspondences are observed:

Fig. 6.15a		Fig. 6.12a
Q_1	corresponds to	Q_1
D_1	corresponds to	D_1
Q_{11}	corresponds to	Q_2
D_{11}	corresponds to	D_2
D_2	corresponds to	D_3
C_1	corresponds to	C
L_1	corresponds to	L_1
R_{C1}	corresponds to	R_1

If the Type B chopper is operating in the first quadrant, and i_O is always positive, then the explanation of the commutation circuit given in Section 6.2.5 may be applied directly to the circuit of Fig. 6.15a.

The commutation circuit employed for second-quadrant operation is identical with that shown in Fig. 6.12a, provided that the following correspondences are observed:

Fig. 6.15a		Fig. 6.12a
Q_2	corresponds to	Q_1
D_2	corresponds to	D_1
Q_{22}	corresponds to	Q_2
D_{22}	corresponds to	D_2
D_1	corresponds to	D_3
C_2	corresponds to	C
L_2	corresponds to	L_1
R_{C2}	corresponds to	R_1

If the type B chopper is operating in the second quadrant, and i_O is always negative, then the explanation of Section 6.2.5 may again be applied. The fact that the power circuit configuration for this quadrant of operation differs from that of the type A chopper does not in any way affect the operation of the commutation circuit.

If the type B chopper is operating either in the first or second quadrant under such conditions that $I_{max} > 0$, and $I_{min} < 0$, as illustrated in Fig. 6.14b, then an interval must be allowed to elapse between the end of the gating signal for one main thyristor and the beginning of the gating signal for the other. The reason for this may be seen by considering the operation of the circuit at the end of a current pulse from thyristor Q_1. When the gating signal is removed from Q_1, and Q_{11} is turned on to initiate the commutation cycle, it is not permissible to turn on Q_2 until Q_1 has turned off, otherwise the source V would be short-circuited. The Q_1 commutation cycle must therefore be completed, at least to the instant $t = t_2$ shown in Fig. 6.13, before Q_2 may be turned on. It is therefore normal practice to delay the gating signal on each of the main thyristors until the entire commutation interval of the other, as defined in equation 6.146, has elapsed. The necessary arrangement of the gating signals is therefore that illustrated in Fig. 6.15b.

Type 2 Current Commutation   The circuit of a type B chopper with type 2 current commutation is shown in Fig. 6.16a. In analyzing the behavior of this curcuit, it is convenient to make assumptions almost identical with those made in the analysis of Type 1 current commutation. The assumptions are that i_O remains constant throughout the commutation interval at the value I_{max}, and that R_1 is of sufficiently great resistance to damp out any tendency to oscillation on the part of the $R_1 L_1 C$ series circuit; consequently, i_R is small enough to be ignored during the commutation interval.

Operation of the converter in the first quadrant of the V_O versus I_O diagram with i_O always positive is as follows:

1.   At $t = 0$, thyristors Q_1 and Q_{22} are turned on, as a consequence of which, simultaneously:

   a. i_O increases exponentially to the value I_{max}, during which $v_O = V$.
   b.   An oscillatory current i_C flows in the ringing circuit comprising L_1, C, and source V, with the result that, when i_C falls to zero, and Q_{22} turns off, capacitor C is charged to voltage $v_C = -2V$.

2.   The capacitor charge decays through the circuit comprising R_1, D_{22}, source V, and D_2, leaving $v_C = -V$.

.3.   At $t = t_{ON}$, thyristor Q_{11} is turned on; i_C then flows in the ringing circuit comprising Q_{11}, source V, and the load circuit, thus decreasing current i_{Q1} below the value of $i_O = I_{max}$.

4.   When i_{Q1} is reduced to zero by the increasing value of i_C, that is, when $i_C = I_{O1} = I_{max}$, diode D_2 begins to conduct, and the forward voltage across D_2 results in a negative value of v_{AK1}, so that thyristor Q_1 is commutated.

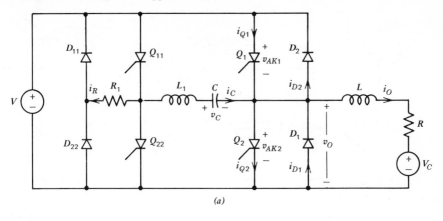

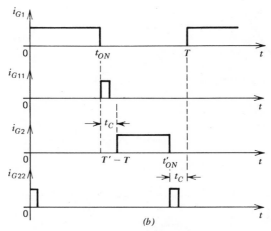

Fig. 6.16 Type B chopper—type 2 current commutation.

5. When i_C again falls to the value I_{O1}, period I of the commutation interval is completed, as illustrated in Fig. 6.17. (Note that the time scale begins at $t = t_{ON}$). Diode D_1 now tends to begin to conduct and period II of the commutation interval ensues, during which capacitor C is charged at constant current $i_C = I_{O1}$. The reason for this condition of operation has been explained in Section 6.2.5

6. When capacitor C has been charged to voltage $v_C = V$, D_1 begins to conduct, and $v_O = 0$. This instant marks the beginning of period III of the commutation interval. Simultaneously:

a. i_O decays exponentially through D_1.

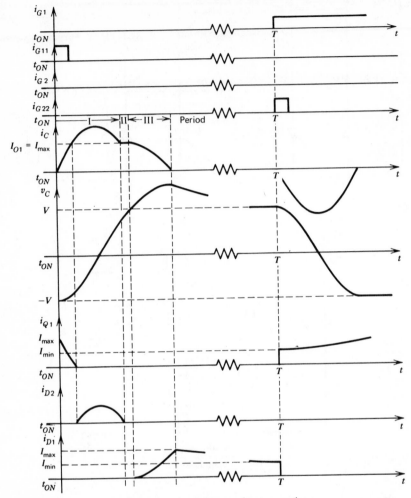

Fig. 6.17   Type 2 current commutation—first quadrant operation.

   b.   An oscillatory current i_C flows in the ringing circuit comprising Q_{11}, source V, and the load circuit, for which now $v_O = 0$.

   7. i_C falls to zero, and thyristor Q_{11} turns off, leaving $v_C > V$ at the end of period III of the commutation interval.
   8.   The capacitor charge decays through the circuit comprising R_1, D_{11}, source V, and D_1 to give $v_C = V$.
   9.   At $t = T$, when i_O has fallen to value I_{min}, Q_1 and Q_{22} are again

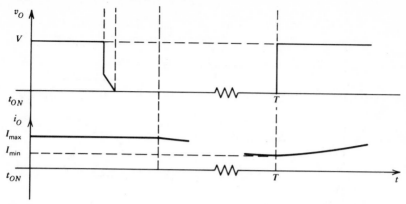

Fig. 6.17 (*continued*).

turned on, and the resulting current i_C in the ringing circuit produces a capacitor voltage $v_C = -V$, ready for the next commutation cycle.

The time variations of some of the principal circuit variables are illustrated in Fig. 6.17.

Operation of the converter in the second quadrant of the V_O versus I_O diagram with i_O always negative is as follows:

1. At $t = 0$, thyristors Q_2 and Q_{11} are turned on, as a consequence of which, simultaneously:

a. i_O increases in a negative direction to the value I_{min}, during which $v_O = 0$.
b. An oscillatory current i_C flows in the ringing circuit until, when $i_C = 0$, thyristor Q_{11} turns off, and $v_C = 2V$.

2. The capacitor charge decays through D_{11}, R_1, and D_1, leaving $v_C = V$.
3. At $t = t'_{ON}$, thyristor Q_{22} is turned on; i_C then flows in a negative direction, thus decreasing the current i_{Q2} below the value of $-i_O = -I_{min}$.
4. When i_{Q2} is reduced to zero, that is, when $i_C = I_{O1} = I_{min}$, diode D_1 begins to conduct, and the forward voltage across D_1 results in a negative value of v_{AK2}, so that thyristor Q_2 is commutated.
5. When i_C again becomes equal to I_{O1}, period I of the commutation interval is completed, as illustrated in Fig. 6.18. Diode D_2 now tends to begin to conduct, and period II of the commutation interval ensues during which capacitor C is charged at constant current $i_C = I_{O1}$.

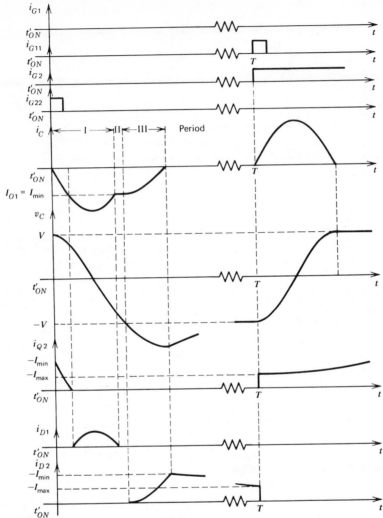

Fig. 6.18 Type 2 current commutation—second quadrant operation.

6.   When capacitor C has been charged to voltage $v_C = -V$, D_2 begins to conduct, and $v_O = V$. This instant marks the beginning of period III of the commutation interval. Simultaneously:

a. i_O decays exponentially through D_2 and source V.

b.   An oscillatory current i_C flows in the ringing circuit comprising Q_{22} and the load circuit, for which now $v_O = V$.

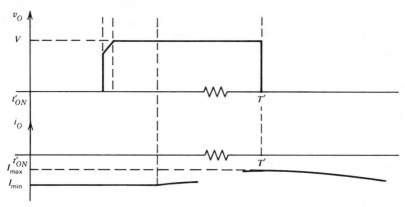

Fig. 6.18   (*continued*).

7. i_C falls to zero, and Q_{22} turns off, leaving $v_C < -V$ at the end of period III of the commutation interval.

8.   The capacitor charge decays through D_2, source V, D_{22}, and R_1 to give $v_C = -V$.

9.   At $t = T'$, when $i_O = I_{max}$, Q_2 and Q_{11} are again turned on, and the resulting current i_C in the ringing circuit produces a capacitor voltage $v_C = V$, ready for the next commutation cycle.

The time variations of some of the principal circuit variables are illustrated in Fig. 6.18.

Operation of the converter in the first or second quadrant of the V_O versus I_O diagram under conditions such that $I_{max} > 0$, and $I_{min} < 0$, may now be visualized in terms of the two conditions already discussed. If at $t = 0$ thyristor Q_1 is turned on, then Q_{22} will already have been conducting at $t < 0$ in order to commutate Q_2. As a consequence $v_C = -V$, ready for the commutation cycle of Q_1. Conversely, when Q_2 is turned on, then Q_{11} will already have been conducting to commutate Q_1. As a consequence $v_C = +V$, ready for the commutation cycle of Q_2.

As in the case of the type B converter with type 1 current commutation, an interval equal in length to the commutation interval t_C must elapse between the end of the gating signal for one main thyristor and the beginning of the gating signal for the other; otherwise, source V will be short-circuited through Q_1 and Q_2. However when the converter is switched on, before any steady-state cycle of the circuit variables has been established, it is initially necessary to have Q_1 and Q_{22} or Q_2 and Q_{11} turned on simultaneously in order that C may be charged. This can be arranged by allowing a small overlap of the gating signals, as illustrated in Fig. 6.16b. The gating signals shown there will also result in operation of the converter when i_O is always positive, or when i_O is always negative.

The circuits of Figs. 6.15a and 6.16a may be analyzed in detail employing the assumptions and procedure used for the type A converter. If this is done, it will be found that equations 6.145, 6.146, and 6.148, giving the maximum capacitor voltage, commutation interval, and time available for turn-off respectively, apply to these circuits also. In addition, the rating of the devices and design of the commutation circuits may be carried out along the lines prescribed for the type A converter in Sections 6.2.6 and 6.2.7.

6.4 THE FOUR-QUADRANT CHOPPER

The power circuit of a dc-to-dc converter capable of operating in any of the four quadrants of the V_O versus I_O diagram is shown in Fig. 6.19a. If thyristor Q_4 is turned on continuously, then the anti-parallel-connected pair of devices Q_4 and D_3 constitute a short circuit of nodes a and b in Fig. 6.19a. Clearly Q_3 may not be turned on, since this would short circuit source V. The anti-parallel-connected pair of devices Q_3 and D_4 thus constitute an open circuit between nodes a and c. The remainder of the circuit now forms a Type B converter operating in the first and second quadrant with positive values of V_O.

If thyristor Q_2 in the circuit of Fig. 6.19a is turned on continuously, then Q_2 and D_1 short-circuit nodes p and q. Q_1 may not be turned on; thus Q_1 and D_2 constitute an open circuit between nodes p and r. The remainder of the circuit may now be rearranged, as shown in Fig. 6.19b where it is seen to constitute a Type B converter operating in the third and fourth quadrants with negative values of V_O.

The methods of commutation described for the Type A and B choppers may be applied to this circuit also, and the devices may be rated and the circuit designed by similar procedures.

6.5 DC MOTOR DRIVES

Choppers find a particularly rich field of application in electric railways, where the advantages of dc power distribution are great and—at least in underground systems—the elimination of heating from motor control resistors is of first importance. As explained in Sections 5.4 and 5.6, a variable-voltage dc source may be employed to excite the armature circuit of a separately excited motor or the armature and field of a series motor.

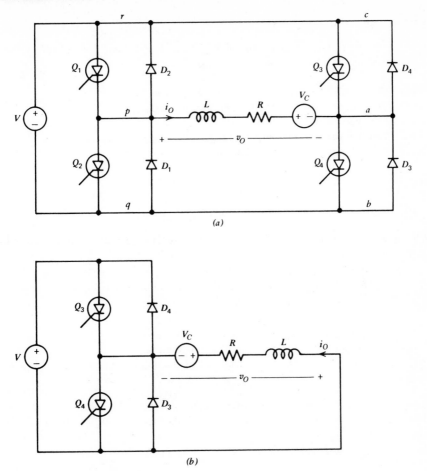

Fig. 6.19 Four-quadrant chopper.

The chopper of Fig. 6.3 is suitable for transportation drives, since it provides for operation in the first and second quadrant of the V_O versus I_O diagram, and therefore gives the required transition between motoring and regenerative braking. The analysis of Section 6.3 may be combined with that of Section 5.4.1 to determine the performance of an electric train drive. Series motors may be employed in this application also, provided that the diode bridge shown in Fig. 5.33 is used to give unidirectional field current. Once again, the operation of the motor in magnetic saturation determines the converter rating. Four-quadrant operation of either type of motor may be obtained by means of the chopper circuit of Fig. 6.4.

PROBLEMS

6.1   In the type A chopper circuit of Fig. 6.2a, $V_C=0$, and $L=0$. Express the following variables as functions of V, R, t_{ON}, and T:

- (a)   The average output voltage and current, V_O and I_O.
- (b)   The output current at the instant of commutation.
- (c)   The average and rms diode currents.
- (d)   The rms value of the output voltage.
- (e)   The rms and average thyristor currents.

also

- (f)   Sketch to scale the time variations of i_{G1}, v_O, and i_O.

6.2   Repeat problem 6.1 for $V_C=0$ and L of so large a value that the output current may be assumed constant or ripple-free at the value V_O/R.

6.3   In the type A chopper circuit of Fig. 6.2a, $V=600$ V, $V_C=200$ V, $L=4$ mH, $R=1.5\,\Omega$, $T=4000\ \mu$s, $t_{ON}=2500\ \mu$s. Show that the output current i_O is continuous.

6.4   For the chopper circuit of problem 6.3, determine:

- (a)   The average output voltage and current, V_O and I_O.
- (b)   The output current at the instant of commutation.
- (c)   The rms values of the first harmonics of the output voltage and the output current.

6.5   In the type A chopper circuit of Fig. 6.2a, $V=600$ V, $V_C=200$ V, $L=1$ mH, $R=1.5\,\Omega$, $T=4000\ \mu$s, and $t_{ON}=2500\ \mu$s. Show that the output current i_O is discontinuous.

6.6   For the chopper circuit of problem 6.5 determine:

- (a)   The average output voltage and current, V_O and I_O.
- (b)   The output current at the instant of commutation.
- (c)   The rms values of the first harmonics of the output voltage and the output current.

6.7   In the type A chopper circuit of Fig. 6.2a, $V=600$ V, $V_C=350$ V, $R=0.1\,\Omega$, $T=1800\ \mu$s, and L is of so large a value that the output current may be assumed constant or ripple-free. If the output current is to be $I_O=100$ A:

- (a)   Calculate the required value of t_{ON}.
- (b)   Sketch to scale the time variations of i_{G1}, v_O, i_O, i_D, and i_s.

6.8   A type A chopper has a load circuit consisting of $R=1.5\,\Omega$, $L=1.59$ mH, and $C=99.5\ \mu$F connected in series. In addition $V=230$ V, $T=2500\ \mu$s, and $t_{ON}=1100\ \mu$s.

(a)   Determine the steady-state values of average output voltage V_O and average output current I_O.

(b)   Calculate the rms values of the first harmonics of the output voltage and the output current.

6.9   In the voltage commutation circuit of Fig. 6.6a, $i_{G1}=0$, so that Q_1 does not turn on. Verify that the following statements apply to this condition:

(a)   If $n \leqslant 1$ and Q_A and Q_B are turned on, the final capacitor voltage will be $v_C = -V(1+n)$ V.

(b)   If $n > 1$ and Q_A and Q_B are turned on, then subsequently Q_C and Q_D must be turned on before the magnitude of the final capacitor voltage becomes $V(1+n)$ V.

6.10   In the type A chopper circuit of Fig. 6.6a, $V = 300$ V, $L_1 = 40\,\mu$H, $C = 10\,\mu$F, $n = 0.5$, and $T = 4000$ μs. If the load current at commutation i_O is 120 A:

(a)   Calculate the time t_q available for turn-off of thyristor Q_1.

(b)   Calculate the total commutation interval t_C.

(c)   Determine the peak current in the commutation thyristors and the peak value of voltage v_{AKA}.

(d)   Sketch to scale the time variations of i_C, i_2, v_1, v_C, and v_{AK1} during the commutation interval.

6.11   In the type A chopper circuit of Fig. 6.6a, $V = 300$ V, $n = 0.4$, $t_{\text{off}} = 25$ μs, $\Delta t = 5$ μs, and $I_{O1} = 200$ A. Calculate the values of L_1 and C when the selected value of x is 1.

6.12   In the Type A chopper circuit of Fig. 6.6a, $V = 300$ V, $L_1 = 40$ μH, $C = 10$ μF, $n = 0.5$, $t_{\text{off}} = 20$ μs, and $\Delta t = 10$ μs.

(a)   Determine the permissible value of the load current at commutation.

(b)   Determine the losses in the circuit if the thyristors and capacitor are ideal and the resistance of the inductor is negligible.

6.13   In the type A chopper circuit of Fig. 6.6a, explain clearly what would happen if the inductance L_2 were open circuited.

6.14   In the type A chopper circuit of Fig. 6.12a, $V = 600$ V, $L_1 = 42$ μF, $C = 6$ μF, and $T = 2500$ μs. If the load current at commutation i_O is 150 A:

(a)   Calculate the time t_q available for turn-off of thyristor Q_1.

(b)   Calculate the total commutation interval t_C.

(c)   Sketch to scale the time variations of i_C, i_{Q2}, i_{D2}, i_{D3}, i_{Q1}, i_{D1}, v_C, v_{AK1}, and v_{AK2}.

(d)   Determine the time available for turn-off of thyristor Q_2.

6.15   In the type A chopper circuit of Fig. 6.12a, $V = 600$ V, $I_{\text{max}} = 150$ A, $t_{\text{off}} =$

25 μs, $\Delta t = 6$ μs. Calculate the values of L_1 and C if the maximum capacitor voltage is not to exceed 1000 V and the losses in resistor R_1 are to be as small as possible.

6.16   In the type A chopper circuit of Fig. 6.12a, $V = 600$ V, $L_1 = 42$ μH, $C = 6$ μF, $T = 2500$ μs, $t_{off} = 15$ μs, $\Delta t = 5$ μs. Calculate the maximum load current that can be commutated and the corresponding losses in the resistor R_1.

6.17   In the type B chopper circuit of Fig. 6.14a, the data are identical with those of problem 6.3.

(a)   Calculate the average output current I_O and the average output voltage V_O.
(b)   Calculate I_{max} and I_{min}.
(c)   Calculate the average value of the source current i_s.
(d)   Sketch to scale the time variations of v_O, i_O, i_{Q1}, i_{Q2}, i_{D1}, i_{D2}, i_s.
(e)   Explain in what way the time variations of v_O and i_O differ from those that would be obtained with the type A chopper of problem 6.3.

6.18   In the type B chopper circuit of Fig. 6.14a, the data are identical with those of problem 6.5. Repeat problem 6.17 for this circuit.

6.19   In the type B chopper circuit of Fig. 6.14a, the data are identical with those of problem 6.8. Repeat problem 6.8 for this circuit.

SEVEN

INVERTERS

Inverters convert dc power to ac power at some desired output voltage and frequency. Applications of inverters include the following:

1. Stand-by power supplies.
2. Uninterruptible power supplies for computers.
3. Variable-speed ac motor drives.
4. Aircraft power supplies.
5. Induction heating.
6. Output of dc transmission lines.

In most inverter applications, it is necessary to be able to control both the output voltage and the output frequency. The controllable-voltage requirement may arise out of the need to overcome regulation in the connected ac equipment or to maintain constant flux in ac motors driven at variable speed by variation of their supply frequency. If the dc input voltage is controllable, then an inverter with a fixed ratio of dc input voltage to ac output voltage may be satisfactory. If the dc input voltage is not controllable, then control of the output voltage must be obtained by employing pulse-width modulation, already discussed in connection with dc-to-dc converters at the beginning of Chapter 6.

The output-voltage waveform of an inverter is nonsinusoidal, and in most applications the voltage harmonics have a significant effect on the overall system performance. These harmonics may be reduced at the cost of increasing the complexity of the inverter circuit, and an economic decision must be made on the degree to which this should be done. Some harmonic-reduction techniques are briefly discussed in this chapter.

In the following sections the simple single-phase half-bridge inverter and the methods of commutation employed in it are first discussed in detail.

357

This is followed by a description of the single-phase bridge inverter and the methods of controlling voltage and reducing output harmonics. Three-phase and series inverters are then discussed, and finally the use of inverters in motor control is briefly surveyed.

7.1 TYPES OF INVERTER CIRCUITS

Figure 7.1 illustrates the basic principles of a single-phase half-bridge inverter. As before, the thyristor symbol enclosed in a circle represents a thyristor that may be turned on and, if necessary, commutated by means of circuit elements not included in the diagram. The sequence of gating signals and the resulting output-voltage waveform are shown in Fig. 7.1b,

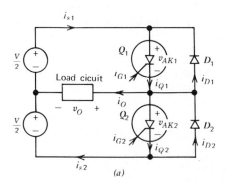

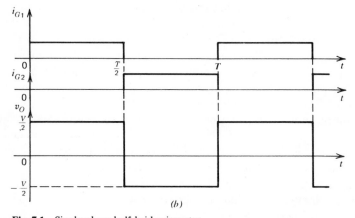

Fig. 7.1 Single-phase half-bridge inverter.

where the output angular frequency is given by

$$\omega = \frac{2\pi}{T} \quad \text{rad/s} \tag{7.1}$$

It might appear that the thyristors alone in Fig. 7.1 would be sufficient to provide the voltage waveform shown there. However, provision must be made for the fact that the load circuit will possess some inductance or capacitance, as a result of which the load current i_O will not necessarily reverse at the same instants as does the load voltage v_O. The diode in antiparallel with each thyristor will then permit load current to flow.

A serious disadvantage of the half-bridge inverter is that it requires a three-wire dc source. Consequently the full-bridge inverter circuit shown in Fig. 7.2 is commonly employed when a single-phase ac output is required from a dc source. The circuit of Fig. 7.2 will be recognized as being identical in configuration with that of the four-quadrant chopper of Fig. 6.19a. The necessary gating signals for the thyristors and the resultant output-voltage waveform are shown in Fig. 7.2b.

Figure 7.3a shows the circuit of a three-phase bridge inverter of which the three-phase load circuit would be connected to terminals A, B, and C. The gating signals required to produce three-phase output voltages are shown in Fig. 7.3b, as also are the resulting line-to-line voltages applied to the load circuit. It may be seen that the fundamental components of these line-to-line voltages will form a balanced three-phase voltage source. The diodes in antiparallel with the thyristors permit currents to flow that are out of phase with these voltages.

7.2 ANALYSIS OF THE HALF-BRIDGE INVERTER

The operation of the power circuit of this inverter is first discussed, and this is followed by an explanation of methods of commutation. The method of rating the circuit devices and of designing the commutation circuit is then considered.

7.2.1 Power Circuit of the Half-Bridge Inverter
If the load circuit of Fig. 7.1a is an RLC series circuit to which a rectangular voltage wave of amplitude $V/2$ V and periodic time T is applied, then this system may be represented by the equivalent circuit of Fig. 7.4a, where

$$
\begin{aligned}
v_s &= \frac{V}{2} \quad \text{V:} \qquad 0 < t < \frac{T}{2} \quad \text{s} \\[2mm]
v_s &= -\frac{V}{2} \quad \text{V:} \qquad \frac{T}{2} < t < T \quad \text{s}
\end{aligned}
\tag{7.2}
$$

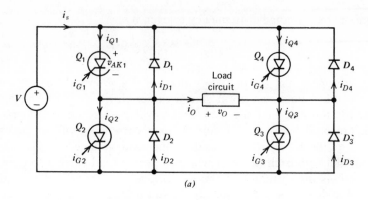

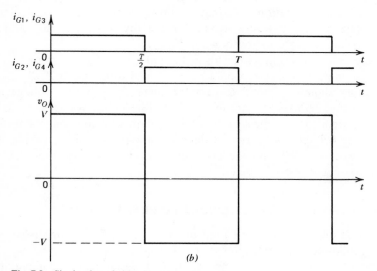

Fig. 7.2 Single-phase bridge inverter.

When several cycles of source voltage v_s have elapsed, the time variation of the current will have settled down to a periodic form such that

$$i_O = I_{O1} \quad \text{A:} \quad t = \frac{T}{2} \quad \text{s}$$

$$i_O = -I_{O1} \quad \text{A:} \quad t = 0, T \quad \text{s}$$

(7.3)

For the circuit of Fig. 7.1a, during the interval $0 < t < T/2$ s,

$$v_O = \frac{V}{2} = v_R + v_L + v_C \quad \text{V}$$

(7.4)

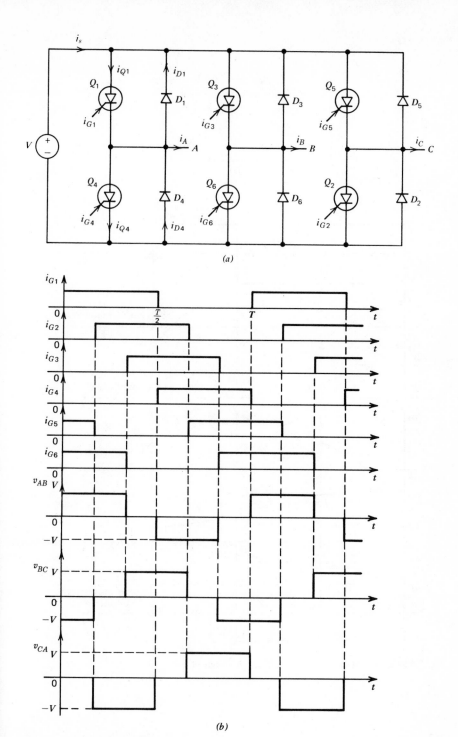

(a)

(b)

Fig. 7.3 Three-phase bridge inverter.

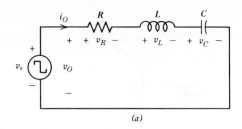

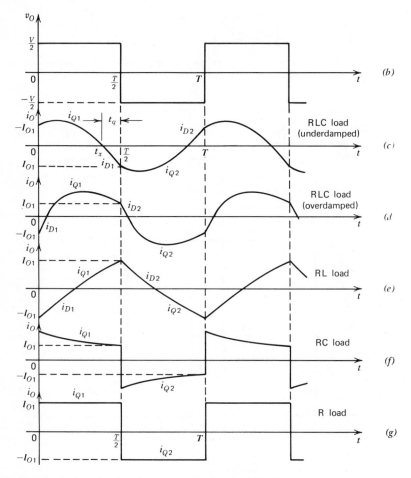

Fig. 7.4 Equivalent circuit and responses for circuit of Fig. 7.1a.

or

$$\frac{V}{2} = Ri_O + \frac{L\, di_O}{dt} + \frac{1}{C}\int_0^t i_O\, dt + V_{C0} \quad \text{V} \tag{7.5}$$

where V_{C0} is the voltage across the capacitive element of the load circuit at $t = 0$. By differentiation of equation 7.5

$$\frac{d^2 i_O}{dt^2} + \frac{R}{L}\frac{di_O}{dt} + \frac{1}{LC}i_O = 0 \quad \text{A/s}^2 \tag{7.6}$$

The solution of equation 7.6 employing the initial conditions of equations 7.3 gives a general expression describing the load current i_O as a function of time. However, since the load voltage v_O is defined, the current values necessary for rating the devices in the inverter circuit can readily be obtained by Fourier analysis. Thus determination of an analytical expression for i_O in any particular case is unnecessary. A qualitative understanding of how i_O varies as a function of time is helpful, however, and possible waveforms of i_O are shown in Fig. 7.4c to g. An underdamped oscillatory load circuit could have a waveform such as that shown in c, whereas d illustrates an overdamped case with increased load-circuit resistance. Waveforms e to g represent ideal cases, since in practice any circuit will possess all three properties of resistance, inductance, and capacitance in some degree. On the first cycle of each of the current waveforms the device current of Fig. 7.1a that is identical with the load current during a particular part of the cycle is indicated. From this, it would appear as if an RC or resistive load circuit would require no diodes in the inverter circuit. Since any circuit possesses inductance, even if only the make-up inductance due to the circuit wiring, waveforms f and g must simply be considered as limiting cases of waveform d.

From the curves of Fig. 7.4, it may be seen that the thyristors may start to conduct at different instants in the half cycle, depending on the nature of the load. To ensure that the thyristors will begin to conduct when required, each must be gated continuously throughout the half cycle, as indicated in Fig. 7.1b.

The time variation of the load-circuit voltage v_O shown in Fig. 7.4b may be described by the Fourier series

$$v_O = \sum_{n=1}^{\infty} \frac{2V}{n\pi}\sin n\,\omega t \quad \text{V:} \qquad n = 1, 3, 5\ldots \tag{7.7}$$

The load current i_O may therefore be described by the series

$$i_O = \sum_{n=1}^{\infty} \frac{2V}{n\pi Z_n} \sin(n\omega t - \phi_n) \quad \text{A:} \qquad n = 1, 3, 5 \ldots \qquad (7.8)$$

where

$$Z_n = \left[R^2 + \left(n\omega L - \frac{1}{n\omega C} \right)^2 \right]^{1/2} \quad \Omega \qquad (7.9)$$

$$\phi_n = \tan^{-1} \frac{n\omega L - \dfrac{1}{n\omega C}}{R} \quad \text{rad} \qquad (7.10)$$

The load current at the instant of commutation may be obtained by substituting $\omega t = \pi$ in equation 7.8, giving

$$i_O = I_{O1} \quad \text{A:} \qquad \omega t = \pi \quad \text{rad} \qquad (7.11)$$

If $I_{O1} > 0$, forced commutation must be employed. If $I_{O1} < 0$, then it may be possible to rely on load commutation of the thyristors, as explained in the following section.

7.2.2 Load Commutation of the Half-Bridge Inverter

If the parameters of the RLC load circuit are such that the response to the rectangular wave of applied voltage is an oscillatory load current, then the time variation of that current may be a waveform such as illustrated in Fig. 7.4c. In such a case, i_O reverses before v_O reverses, so that at the end of each half cycle the load current is carried by the diodes. If, during the first half-cycle of the voltage in Fig. 7.4, the current i_O goes negative at $t = t_x$, and

$$t_q = \frac{T}{2} - t_x > t_{\text{off}} \quad \text{s} \qquad (7.12)$$

where t_{off} is the turn-off time of thyristors Q_1 and Q_2, then when the gating signal is removed from Q_1, and Q_2 is turned on at $t = T/2$, Q_1 will not conduct. Under these circumstances, the inverter is load-commutated. If inequality 7.12 is not satisfied, or if any of the conditions represented in Fig. 7.4d to g apply, then the thyristors must be forced-commutated.

Example 7.1 In the circuit of Fig. 7.1a, the load circuit is an RLC series circuit for which $R = 0.8\,\Omega$, $\omega L = 10\,\Omega$, $1/\omega C = 10\,\Omega$, $V/2 = 25\,\text{V}$. The gating signals are as shown in Fig. 7.1b.

(a)   Sketch to scale the waveforms of v_O, i_O, i_{Q1}, i_{Q2}, i_{D1}, i_{D2}, i_{s1}, i_{s2}, and v_{AK1}. Higher current harmonics than the first or fundamental component may be neglected.

(b)   Show that the power delivered by the two sources is equal to the power delivered to the load circuit.

(c)   State whether this circuit will require forced commutation.

Solution

(a)   Time variations of the variables are shown in Fig. E7.1.

(b)   The rms value of the fundamental component of the output current is, from equations 7.8 and 7.9

$$I_{1R} = \frac{2 \times 50}{\sqrt{2}\,\pi \times 0.8} = 28.13 \quad \text{A}$$

Power delivered to the load is

$$P_O = R I_{1R}{}^2 = 0.8 \times 28.13^2 = 633 \quad \text{W}$$

Power taken from each source is

$$P_s = \frac{V}{2} I_s \quad \text{W}$$

where I_s is the average value of the source current over one cycle. Thus

$$P_s = \frac{V}{2} \times \frac{\sqrt{2}\,I_{1R}}{\pi} = \frac{25\sqrt{2} \times 28.13}{\pi} = 317 \quad \text{W}$$

Power delivered by the two sources is

$$2P_s = 634 = P_O \quad \text{W}$$

(c)   The curves of Fig. E7.1 show that, from inequality 7.12

$$\frac{T}{2} - t_x = 0 \quad \text{s}$$

Thus no time is available for turn off, and forced commutation is required.

Example 7.2   Repeat Example 7.1 with $1/\omega C = 9.2$ and all other data unchanged.

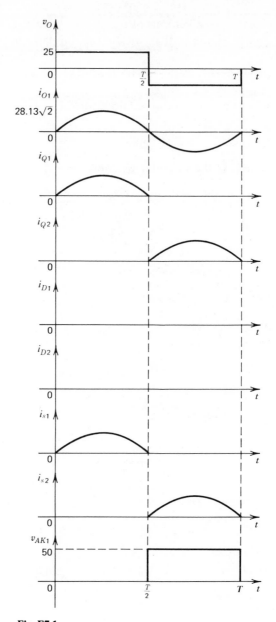

Fig. E7.1

Solution

(a) Time variations of the variables are shown in Fig. E7.2.

(b)
$$I_{1R} = \frac{2 \times 50}{\sqrt{2}\, \pi (0.8^2 + 0.8^2)^{1/2}} = 19.89 \quad A$$

$$P_O = 0.8 \times 19.89^2 = 317 \quad W$$

$$\phi_1 = \tan^{-1} \frac{10 - 9.2}{0.8} = \frac{\pi}{4} \text{rad}$$

From equation 7.8,

$$i_{O1} = \sqrt{2}\, I_{1R} \sin\left(\omega t - \frac{\pi}{4}\right) \quad A$$

and as may be seen from Fig. E7.2, during the first half-cycle of the supply voltage, $i_{O1} = i_{s1}$, and during the second half-cycle, $i_{O1} = -i_{s2}$. Thus the average power from the sources is

$$P_s = \frac{1}{\pi} \int_0^\pi \frac{V}{2} i_O \, d(\omega t) = \frac{50}{2\pi} \times 19.89\sqrt{2} \int_0^\pi \sin\left(\omega t - \frac{\pi}{4}\right) d(\omega t)$$

$$= 223.8 \left(2 \cos \frac{\pi}{4}\right) = 317 \quad W$$

(c) Since $t_x > T/2$, $t_q < 0$ in inequality 7.12. Thus there is no time available for load commutation, and forced commutation must be employed.

Example 7.3 Repeat Example 7.1 with $\omega L = 9.2\,\Omega$ and $T = 2500\ \mu s$. All other data are unchanged.

(a) Time variations of the variables are shown in Fig. E7.3.

(b) As in Example 7.2,

$$I_{1R} = 19.89 \quad A: \qquad P_O = 317 \quad W$$

$$\phi_1 = \tan^{-1} \frac{9.2 - 10}{0.8} = -\frac{\pi^c}{4}$$

Average power from the two sources is

$$P_s = \frac{50}{2\pi} \times 19.89\sqrt{2} \int_0^\pi \sin\left(\omega t + \frac{\pi}{4}\right) d(\omega t) = 317 \quad W$$

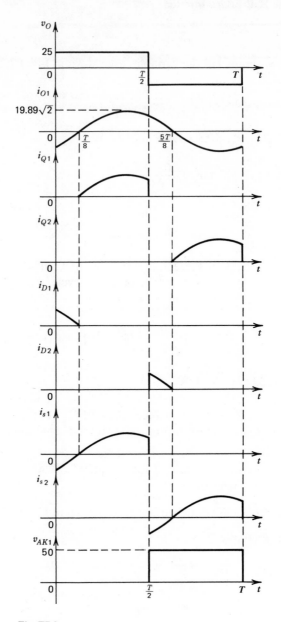

Fig. E7.2

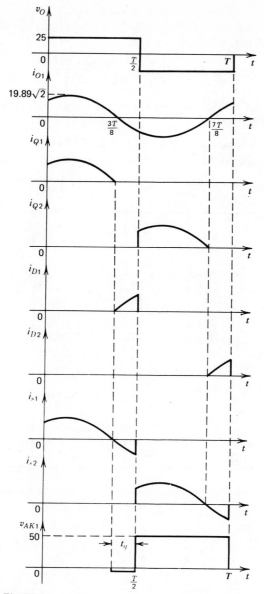

Fig. E7.3

The current in thyristor Q_1 falls to zero before Q_2 is turned on. Thus load commutation is adequate if inequality 7.11 is satisfied. The angular interval ωt_q is shown in Fig. E7.3, from which

$$t_q = \frac{\pi}{4\omega} = \frac{\pi}{4}\frac{T}{2\pi} = \frac{2500}{8} = 312.5 \quad \mu s$$

And since t_{off} for a thyristor rarely exceeds $150\,\mu s$, this system requires no forced commutation.

7.2.3 Voltage Commutation of the Half-Bridge Inverter

The circuit in Fig. 7.5a illustrates the McMurray-Bedford method of complementary voltage

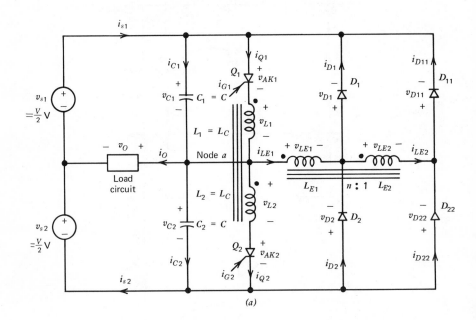

(a)

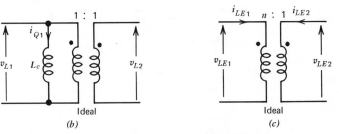

Ideal   Ideal
(b) (c)

Fig. 7.5 Complementary voltage commutation of the half-bridge inverter.

commutation. In this circuit, the turning on of thyristor Q_2 turns off thyristor Q_1, if it is conducting, and conversely. This technique has the advantage of not requiring additional thyristors for commutation, but the advantage is gained at the expense of additional inductive and capacitive components.

In Fig. 7.5a, D_1 and D_2 are the diodes shown in the power circuit of Fig. 7.1a. The two commutating capacitors are of equal magnitude, so that

$$C_1 = C_2 = C \quad \text{F} \tag{7.13}$$

The two commutating inductors are also of equal magnitude

$$L_1 = L_2 = L_C \quad \text{H} \tag{7.14}$$

These two inductors are wound on one core, which contains an air gap to ensure linearity. They may therefore be regarded as constituting a transformer that is ideal, apart from requiring a magnetising current, and represented by the equivalent circuit shown in Fig. 7.5b. Thus at all times

$$v_{L1} = v_{L2} \quad \text{V} \tag{7.15}$$

Since the two windings have the same number of turns, the magnetizing inductance L_C may be considered to be connected across either winding of the ideal transformer, as convenient. It should also be noted that these windings are connected one in series with each thyristor. Since the two thyristors may not conduct simultaneously, it follows that only one winding of this transformer will carry current at any instant. Furthermore, the statement that "the current in an inductive circuit cannot change instantaneously," may in this situation be replaced by the more fundamental statement that "the flux linkage of any circuit cannot change instantaneously." This means that, while the *mmf* producing the flux in this transformer core cannot change instantaneously, the currents in the circuits providing it can do so, as long as one exactly replaces the *mmf* formerly provided by the other.

Inductors L_{E1} and L_{E2} are also wound on one core, but this contains no air gap, and the windings and core therefore constitute a transformer that may be regarded as ideal. This transformer may be represented by the equivalent circuit shown in Fig. 7.5c. Its turns ratio is

$$n = \frac{N_1}{N_2} < 1 \tag{7.16}$$

so that at all times

$$\frac{v_{LE1}}{v_{LE2}} = n = -\frac{i_{LE2}}{i_{LE1}} \tag{7.17}$$

The nature of this transformer is such that current cannot flow in one winding unless current also flows in the other, since the net *mmf* on the core must be zero. Diodes D_{11} and D_{22}, in conjunction with this transformer, form an energy-recovery circuit that returns to the voltage sources v_{s1} and v_{s2} the energy stored at the instant of commutation in inductors L_1 and L_2. It may be noted in passing that while the inductance L_C of the commutating transformer may be of the order of microhenries, the inductances L_{E1} and L_{E2} of the energy-recovery transformer may be of the order of millihenries.

A brief verbal description of the operation of the circuit during the commutation interval will be of assistance in following the course of the detailed analysis that follows.

When thyristor Q_2 is turned on at the beginning of period I of the commutation interval, the voltage across capacitor C_2 is applied to inductor L_2, and hence appears at the terminals of inductor L_1. This commutates thyristor Q_1, which turns off instantaneously. The *mmf* on the core of inductors L_1 and L_2 cannot change instantaneously; consequently the current formerly in inductor L_1 is instantaneously replaced by an equal current in inductor L_2. To this current is added an oscillatory component flowing in the ringing circuit formed by capacitor C_2 and inductor L_2. Due to the oscillatory current, the voltages of inductors L_{E1} and L_{E2} vary, and when v_{LE2} acquires a negative value equal in magnitude to the combined source voltages, diodes D_2 and D_{11} conduct. This ends period I of the commutation interval.

During period II of the commutation interval, the voltages across all four inductors are clamped, due to the flow of current through diodes D_2 and D_{11} into the constant-voltage sources. The current in inductor L_2 and thyristor Q_2 therefore falls at a uniform rate to zero, and this ends period II of the commutation interval. During period II, the energy stored in the core of inductor L_2 is recovered and returned to the sources. A detailed analysis of the circuit may now be undertaken.

When thyristor Q_1 is conducting, $i_{Q1} = i_o$, and as may be seen from the current waveforms of Fig. 7.4, di_{Q1}/dt may be large at the beginning of the half cycle. However, shortly after Q_1 has begun to conduct the rate of change of current will have decreased considerably, so that since L_C is a small inductance

$$L_C \frac{di_{Q1}}{dt} = v_{L1} \ll \frac{V}{2} \quad \text{V} \tag{7.18}$$

Thus for the circuit mesh comprising C_1, Q_1, and L_1,

$$v_{C1} = v_{L1} \cong 0 \quad \text{V} \tag{7.19}$$

and for the loop comprising Q_1, L_1, the load circuit and source v_{s1},

$$v_O = \frac{V}{2} \quad \text{V} \tag{7.20}$$

For the loop comprising C_2, the load circuit and source v_{s2},

$$v_{C2} = v_O + v_{s2} = V \quad \text{V} \tag{7.21}$$

The commutation interval begins when thyristor Q_2 is turned on. This instant, corresponding to $t = T/2$ in Fig. 7.4, will be taken as the origin, $t = 0$, of the curves showing the time variations of the circuit variables during the second half cycle of v_O in Fig. 7.7 and in the discussion that follows. Thus

$$v_{C1} = 0 \quad \text{V:} \qquad v_{C2} = V \quad \text{V:} \qquad t = 0^- \quad \text{s} \tag{7.22}$$

and since the capacitor voltages have reached the values given in equation 7.22 some time before Q_2 is turned on, it follows that

$$i_{C1} = i_{C2} = 0 \quad \text{A:} \qquad t = 0^- \quad \text{s} \tag{7.23}$$

At this instant also the load current has reached the value defined in equation 7.11, so that

$$i_O = I_{O1} \quad \text{A:} \qquad t = 0^- \quad \text{s} \tag{7.24}$$

For node a of the circuit of Fig. 7.4a,

$$i_O - i_{C1} + i_{C2} - i_{Q1} + i_{Q2} + i_{LE1} = 0 \quad \text{A} \tag{7.25}$$

Since $i_{Q1} = i_O$ and $i_{Q2} = 0$, it follows from equation 7.23 that

$$i_{LE1} = 0 \quad \text{A:} \qquad t = 0^- \quad \text{s} \tag{7.26}$$

and consequently

$$i_{LE2} = 0 \quad \text{A:} \qquad t = 0^- \quad \text{s} \tag{7.27}$$

For the circuit mesh comprising C_2, L_2 and Q_2, from equation 7.19,

$$v_{L2} = v_{L1} \cong 0 \quad \text{V:} \qquad t = 0^- \quad \text{s} \tag{7.28}$$

and consequently

$$v_{AK2} = v_{C2} = V \quad \text{V:} \qquad t = 0^- \quad \text{s} \tag{7.29}$$

The conditions immediately before the beginning of the commutation interval have now been established and are shown in Table 7.1.

Table 7.1 Conditions at $t = 0^-$ for Circuit of Fig. 7.5

$i_O = I_{O1}$ A	:	$v_O = \dfrac{V}{2}$ V	
$v_{C1} = 0$ V	:	$v_{C2} = V$ V	
$i_{C1} = 0$ A	:	$i_{C2} = 0$ A	
	$v_{L1} = v_{L2} = 0$ V		
$i_{Q1} = I_{O1}$ A	:	$i_{Q2} = 0$ A	
$v_{AK1} = 0$ V	:	$v_{AK2} = V$ V	
$i_{LE1} = 0$ A	:	$i_{LE2} = 0$ A	

Period I   At $t = 0$, thyristor Q_2 is turned on, and voltage v_{AK2} falls to zero. For the mesh comprising Q_2, C_2, and L_2

$$v_{L2} = v_{C2} = V \quad \text{V}: \qquad t = 0^+ \quad \text{s} \tag{7.30}$$

Thus for the mesh comprising Q_1, C_1, and L_1,

$$v_{AK1} = -v_{L1} = -V \quad \text{V}: \qquad t = 0^+ \quad \text{s} \tag{7.31}$$

and Q_1 is commutated. Instantaneously i_{Q1} becomes zero, and i_{Q2} rises to the value

$$i_{Q2} = I_{O1} \quad \text{A}: \qquad t = 0^+ \quad \text{s} \tag{7.32}$$

Since v_{L1} and v_{L2} both have the positive value V, and since no currents were flowing in L_{E1} and L_{E2} at $t = 0^-$, it follows that no currents are flowing in any of the four diodes at $t = 0^+$. The circuit to be considered during period I of the commutation interval is therefore that shown in Fig. 7.6a. The initial conditions for this circuit are given in Table 7.2.

Table 7.2 Conditions at $t = 0^+$
for Circuit of Fig. 7.6a

$$i_O = I_{O1} \ \text{A} \quad : \quad v_O = \frac{V}{2} \ \text{V}$$

$$v_{C1} = 0 \ \text{V} \quad : \quad v_{C2} = V \ \text{V}$$

$$i_{C1} = I_{O1} \ \text{A} \quad : \quad i_{C2} = -I_{O1} \ \text{A}$$

$$v_{L1} = v_{L2} = V \ \text{V}$$

$$i_{Q1} = 0 \ \text{A} \quad : \quad i_{Q2} = I_{O1} \ \text{A}$$

$$v_{AK1} = -V \ \text{V} \quad : \quad v_{AK2} = 0 \ \text{V}$$

For the circuit loop in Fig. 7.6a comprising C_1, C_2, v_{s2}, and v_{s1},

$$\frac{1}{C} \int_0^t i_{C1} \, dt + \frac{1}{C} \int_0^t i_{C2} \, dt + V = V \quad \text{V} \tag{7.33}$$

where the initial conditions of equation 7.22 have been employed. From equation 7.33,

$$i_{C1} = -i_{C2} \quad \text{A} \tag{7.34}$$

For node a of the circuit

$$i_O - i_{C1} + i_{C2} + i_{Q2} = 0 \quad \text{A} \tag{7.35}$$

or

$$I_{O1} + i_{Q2} = i_{C1} - i_{C2} \quad \text{A} \tag{7.36}$$

From equations 7.34 and 7.36,

$$i_{C1} = \frac{I_{O1} + i_{Q2}}{2} \quad \text{A}: \quad i_{C2} = -\frac{(I_{O1} + i_{Q2})}{2} \quad \text{A} \tag{7.37}$$

The source currents are therefore

$$i_{s1} = i_{Q1} + i_{C1} = \frac{i_{Q2} + I_{O1}}{2} \quad \text{A} \tag{7.38}$$

$$i_{s2} = i_{Q2} + i_{C2} = \frac{i_{Q2} - I_{O1}}{2} \quad \text{A} \tag{7.39}$$

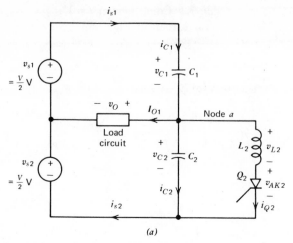

$$(a)$$

Fig. 7.6 Circuits applying during the commutation interval of thyristor Q, in Fig. 7.5a.

From equation 7.30

$$L_C \frac{di_{Q2}}{dt} = \frac{1}{C} \int_0^t i_{C2}\, dt + V \quad \text{V} \tag{7.40}$$

Substitution from equation 7.37 in equation 7.40 and differentiation yields

$$\frac{d^2 i_{Q2}}{dt^2} + \frac{i_{Q2}}{2L_C C} = -\frac{I_{O1}}{2L_C C} \quad \text{A/s}^2 \tag{7.41}$$

Solution of equation 7.41 gives the time variation of i_{Q2} during period I of the commutation interval. This is

$$i_{Q2} = 2I_{O1}\cos\omega_r t + \frac{V}{\omega_r L_C}\sin\omega_r t - I_{O1} \quad \text{A} \tag{7.42}$$

where

$$\omega_r = \frac{1}{[2L_C C]^{1/2}} \quad \text{rad/s} \tag{7.43}$$

During period I, therefore

$$v_{L2} = L_C \frac{di_{Q2}}{dt} = -2\omega_r L_C I_{O1}\sin\omega_r t + V\cos\omega_r t \quad \text{V} \tag{7.44}$$

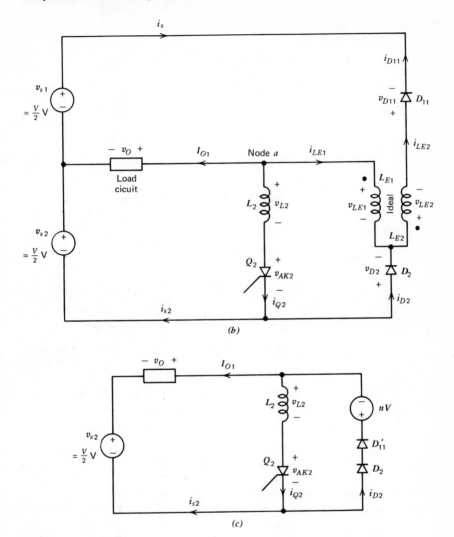

Fig. 7.6 (*Continued*)

and since $v_{AK2} = 0$

$$v_{C2} = v_{L2} \quad \text{V} \tag{7.45}$$

For the loop comprising C_1, C_2, v_{s2}, and v_{s1},

$$v_{C1} = V - v_{C2} = V - v_{L2} \quad \text{V} \tag{7.46}$$

and for the loop comprising source v_{s2}, the load circuit, and C_2,

$$v_{s2} + v_O = v_{C2} \quad \text{V} \tag{7.47}$$

so that from equation 7.44, 7.45, and 7.47,

$$v_O = -2\omega_r L_C I_{O1} \sin \omega_r t + V \cos \omega_r t - \frac{V}{2} \quad \text{V} \tag{7.48}$$

For the circuit loop of Fig. 7.5a comprising Q_1, L_1, L_2, Q_2, v_{s2}, and v_{s1},

$$v_{AK1} + v_{L1} + v_{L2} + v_{AK2} - V = 0 \quad \text{V} \tag{7.49}$$

from which, since $v_{AK2} = 0$, and $v_{L1} = v_{L2}$,

$$v_{AK1} = V - 2v_{L2} \quad \text{V} \tag{7.50}$$

Substitution for v_{L2} from equation 7.44 then yields

$$v_{AK1} = V + 4\omega_r L_C I_{O1} \sin \omega_r t - 2V \cos \omega_r t \quad \text{V} \tag{7.51}$$

The windings of the energy-recovery transformer cannot conduct unless

$$|v_{LE2}| \geqslant V \quad \text{V} \tag{7.52}$$

If $v_{LE2} \geqslant V$, that is, if $v_{LE1} \geqslant nV$, then diodes D_1 and D_{22} conduct. If $v_{LE2} \leqslant V$, that is, $v_{LE1} \leqslant nV$, then diodes D_2 and D_{11} conduct. When either pair of diodes begin to conduct, then period I ends at $t = t_1$.

In the circuit mesh of Fig. 7.5a comprising L_2 and L_{E1}, v_{D2} becomes zero, and D_2 begins to conduct, when

$$v_{LE2} = -V \quad \text{V}: \qquad t = t_1 \quad \text{s} \tag{7.53}$$

that is, when

$$v_{LE1} = v_{L2} = v_{L1} = -nV \quad \text{V}: \qquad t = t_1 \quad \text{s} \tag{7.54}$$

Substitution from equation 7.44 in 7.54 yields

$$v_{L2} = -2\omega_r L_C I_{O1} \sin \omega_r t_1 + V \cos \omega_r t_1 = -nV \quad \text{V} \tag{7.55}$$

This equation may be solved for t_1, yielding

$$t_1 = \frac{1}{\omega_r}\left[\sin^{-1}\frac{nV}{\left[(2\omega_r L_C I_{O1})^2 + V^2\right]^{1/2}} + \tan^{-1}\frac{V}{2\omega_r L_C I_{O1}}\right] \quad \text{s} \quad (7.56)$$

At $t = t_1$ equation 7.54 shows that $v_{L2} < 0$, and period I ends some time after i_{Q2} has passed its peak value. This condition is shown in Fig. 7.7, where the time variations of the principal circuit currents and voltages during the commutation interval are illustrated. At $t = t_1$, from equation 7.42

$$i_{Q2} = 2I_{O1}\cos\omega_r t_1 + \frac{V}{\omega_r L_C}\sin\omega_r t_1 - I_{O1} = I_{Q2} \quad \text{A} \quad (7.57)$$

Also, at $t = t_1$, from equations 7.50 and 7.55

$$v_{AK1} = V(1+2n) \quad \text{V}: \qquad t = t_1 \quad \text{s} \quad (7.58)$$

Since $v_{AK2} = 0$,

$$v_{C2} = v_{L2} = -nV \quad \text{V}: \qquad t = t_1 \quad \text{s} \quad (7.59)$$

and

$$v_{C1} = v_{s1} + v_{s2} - v_{C2} = V(1+n) \quad \text{V}: \qquad t = t_1 \quad \text{s} \quad (7.60)$$

For the circuit mesh comprising C_2, source v_{s2}, and the load circuit,

$$v_{C2} - v_{s2} - v_O = 0 \quad \text{V}: \qquad t = t_1 \quad \text{s} \quad (7.61)$$

from which, by substitution from equation 7.59,

$$v_O = -\frac{V}{2}(1+2n) \quad \text{V}: \qquad t = t_1 \quad \text{s} \quad (7.62)$$

Period II   Throughout period II, while diodes D_2 and D_{11} are conducting, all circuit voltages remain clamped at the values which they reached at $t = t_1$. This period lasts for the interval $0 \leqslant t' \leqslant t'_1$, where

$$t' = t - t_1 \quad \text{s} \quad (7.63)$$

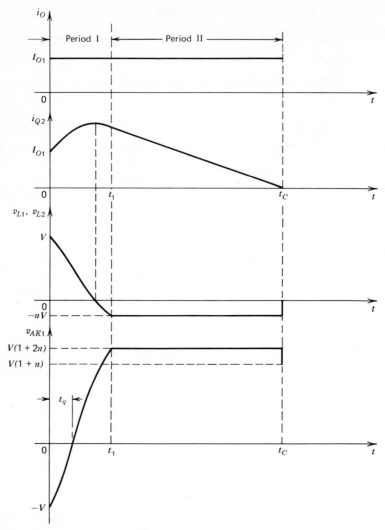

Fig. 7.7 Time variations of currents and voltages in the circuit of Fig. 7.5a during the commutation interval.

Thus

$$i_{C1} = i_{C2} = 0 \quad \text{A:} \qquad 0 \leqslant t' \leqslant t'_1 \quad \text{s} \tag{7.64}$$

The initial conditions for period II are shown in Table 7.3.

Table 7.3 Conditions at $t' = 0^+$ for Circuit of Fig. 7.5a

$i_O = I_{O1}$ A	: $\quad v_O = -\dfrac{V}{2}(1+2n)$ V
$v_{C1} = V(1+n)$ V	: $\quad v_{C2} = -nV$ V
$i_{C1} = 0$ A	: $\quad i_{C2} = 0$ A
$v_{L1} = v_{L2} = -nV$ V	
$i_{Q1} = 0$ A	: $\quad i_{Q2} = I_{Q2}$ A
$v_{AK1} = V(1+2n)$ V	: $\quad v_{AK2} = 0$ V
$i_{LE1} = 0$ A	: $\quad i_{LE2} = 0$ A
$v_{LE1} = -nV$ V	: $\quad v_{LE2} = -V$ V

During period II, diodes D_1 and D_{22}, as well as thyristor Q_1, are not conducting. The two capacitor currents are also zero. If the branches containing these five components are removed from the circuit of Fig. 7.5a, and the remaining circuit is slightly rearranged, then the circuit shown in Fig. 7.6b is obtained. The circuit loop comprising D_{11}, the two voltage sources, and L_{E2}, the secondary winding of the ideal transformer, may then be referred to the primary side of that transformer, giving the equivalent circuit shown in Fig. 7.6c. In this equivalent circuit, diode D'_{11} is D_{11} referred to the primary circuit, and source nV represents sources v_{s1} and v_{s2} in series, also referred to the primary circuit. It will be observed that source nV in this circuit can only absorb energy, owing to the presence of the diodes.

Voltage v_{L2} is clamped at the value given in equation 7.59 throughout period II while diodes D_2' and D_{11}' in the circuit of Fig. 7.6c are conducting. Thus

$$L_C \frac{di_{Q2}}{dt'} = -nV \quad \text{V} \tag{7.65}$$

Employing the initial conditions of equation 7.57, the solution of equation 7.65 is

$$i_{Q2} = I_{Q2} - \frac{nV}{L_C} t' \quad \text{A} \tag{7.66}$$

From the circuit of Fig. 7.6b, for node a,

$$i_{LE1} + i_{Q2} + I_{O1} = 0 \quad \text{A} \tag{7.67}$$

so that substitution from equation 7.66 yields

$$i_{LE1} = \frac{nV}{L_C} t' - I_{Q2} - I_{O1} \quad \text{A} \tag{7.68}$$

Also

$$i_{D11} = i_{LE2} = -ni_{LE1} \quad \text{A} \tag{7.69}$$

and

$$i_{D2} = i_{LE2} - i_{LE1} = -(n+1)i_{LE1} \quad \text{A} \tag{7.70}$$

Expressions for the diode currents may be obtained from these last three equations.

Period II ceases at $t' = t_1'$, when $i_{Q2} = 0$. Thus from equation 7.66

$$t_1' = \frac{L_C}{nV} I_{Q2} \quad \text{s} \tag{7.71}$$

and the length of the commutation interval is

$$t_C = t_1 + t_1' \quad \text{s} \tag{7.72}$$

The energy returned to source nV from inductor L_2 during period II is

$$W = \frac{1}{2} L_C I_{Q2}^2 \quad \text{J} \tag{7.73}$$

and this amount of energy will be recovered each half cycle, so that the average power saved is

$$P_E = \frac{2}{T}W = \frac{L_C I_{Q2}^2}{T} \quad \text{W} \tag{7.74}$$

The constant load current I_{O1} is assumed to continue to flow during period II; that is, in the manner of a normal reactive ac circuit, the load circuit is returning energy to the source during this part of the cycle.

The time t_q available for turn-off of thyristor Q_1 is indicated on the curve of v_{AK1} in Fig. 7.7. From this it may be seen that, at $t = t_q$, $v_{AK1} = 0$; that is, from equation 7.51

$$V \cos \omega_r t_q - 2\omega_r L_C I_{O1} \sin \omega_r t_q - \frac{V}{2} = 0 \quad \text{V} \tag{7.75}$$

Equation 7.75 shows that t_q is a function of the load current and has its maximum value on no load, when $I_{O1} = 0$. Thus from equation 7.75,

$$t_{q\,max} = \frac{\pi}{3\omega_r} = \frac{\pi}{3}\sqrt{2L_C C} \quad \text{s} \tag{7.76}$$

Equation 7.75 may be solved for t_q, giving

$$t_q = \frac{1}{\omega_r}[\cos^{-1}C - \tan^{-1}D] \quad \text{s} \tag{7.77}$$

where

$$C = \frac{V}{2\left[V^2 + (2\omega_r L_C I_{O1})^2\right]^{1/2}} \tag{7.78}$$

$$D = \frac{2\omega_r L_C I_{O1}}{V} \tag{7.79}$$

Alternatively the value of t_q may be obtained graphically from a diagram such as is shown in Fig. 7.8.

At the end of period II, it may be assumed that i_O begins to change, first falling to zero, and then increasing in a negative direction and achieving the value $-I_{O1}$ at the beginning of the commutation interval for thyristor Q_2. The fact that i_O falls to zero at some instant after Q_2 has been turned on, commutating Q_1, demonstrates the need for the continuous gating of each thyristor throughout alternate half-cycles of the output voltage.

The postcommutation interval of the half-cycle may be divided into periods III and IV. Both of these are very much greater than periods I and

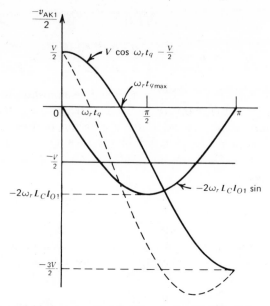

Fig. 7.8 Time t_q available for turn-off of thyristors in circuit of Fig. 7.5.

II of the commutation interval, as illustrated in Fig. 7.9. These two periods may be defined as follows:

$$\text{period III:} \qquad I_{O1} \geqslant i_O > 0 \quad \text{A:} \qquad t_C < t < t_2 \quad \text{s} \qquad (7.80)$$

$$\text{period IV:} \qquad 0 > i_O \geqslant -I_{O1} \quad \text{A:} \qquad t_2 < t < \frac{T}{2} \quad \text{s} \qquad (7.81)$$

Period III   Throughout period II, the voltages in the circuit of Fig. 7.5*a* have been clamped, and at the end of that period, when $t = t_C^-$ s, they still have the values given in Table 7.3. Of the circuit currents, $i_O = I_{O1}$ A, and $i_{Q1} = 0$ A, but by definition of period II,

$$i_{Q2} = 0 \quad \text{A:} \qquad t = t_C \quad \text{s} \qquad (7.82)$$

The diode currents may be determined from equations 7.69 and 7.70, giving

$$i_{D2} = (n+1)I_{O1} \quad \text{A:} \qquad t = t_C \quad \text{s} \qquad (7.83)$$

$$i_{D11} = n I_{O1} \quad \text{A:} \qquad t = t_C \quad \text{s} \qquad (7.84)$$

Since D_2 and D_{11} are still conducting in period III, all *branch* voltages remain clamped at their period II values. However in the two thyristor branches

$$v_{L1} = 0 \quad \text{V}: \qquad v_{L2} = 0 \quad \text{V} \tag{7.85}$$

thus

$$v_{AK2} = v_{LE1} = -nV \quad \text{V} \tag{7.86}$$

and for the circuit loop comprising Q_1, L_1, L_{E1}, L_{E2}, and D_{11}.

$$v_{AK1} = -v_{LE1} - v_{LE2} = nV + V = V(1+n) \quad \text{V} \tag{7.87}$$

and the voltages in these last three equations are clamped during period III.

When the energy stored in inductor L_2 is exhausted at the end of period II, and i_{Q2} has fallen to zero, the load circuit releases the energy stored in its own inductance to sources v_{s1} and v_{s2}. By a process exactly the same as that employed to obtain the equivalent circuit of Fig. 7.6c for period II, the equivalent circuit of Fig. 7.10a for period III is obtained. In this equivalent circuit, where an RL load is shown,

$$\frac{di_O}{dt''} + \frac{R}{L} i_O = -\frac{V}{2L}(1+2n) \quad \text{V} \tag{7.88}$$

where

$$t'' = t - t_C \quad \text{s} \tag{7.89}$$

Employing the initial conditions of equation 7.83, the solution of equation 7.88 is

$$i_O = -\frac{V}{2R}(1+2n) + \left[I_{D2} + \frac{V}{2R}(1+2n) \right] \epsilon^{-(R/L)t''} \quad \text{A} \tag{7.90}$$

At $t'' = t_1''$, i_O falls to zero, that is, at

$$t = t_C + t_1'' = t_2 \quad \text{s} \tag{7.91}$$

Setting i_O to zero in equation 7.90 would permit solution for t_1'' and hence the determination of t_2.

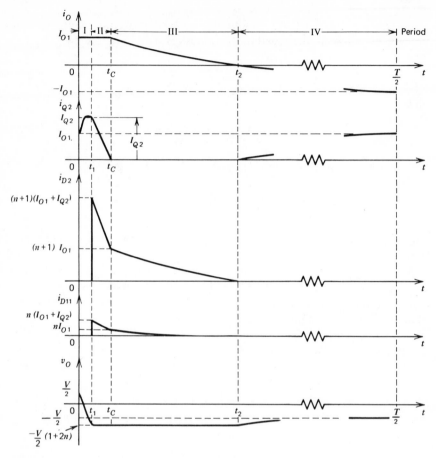

Fig. 7.9 Time variations of currents and voltages in the circuit of Fig. 7.5a.

During period III, the diode currents are

$$i_{D2} = -i_{LE1} + i_{LE2} = (n+1)i_O \quad \text{A} \tag{7.92}$$

$$i_{D11} = i_{LE2} = -ni_{LE1} = ni_O \quad \text{A} \tag{7.93}$$

The time variations of the circuit variables during period III are therefore as shown in Fig. 7.9.

Period IV At $t = t_2$, all circuit currents are momentarily zero. During period IV, no diodes conduct, nor yet does thyristor Q_1. The circuit to be considered during this interval is therefore that shown in Fig. 7.10b, which is simply Fig. 7.6c with an RL load circuit. Detailed analysis of this circuit

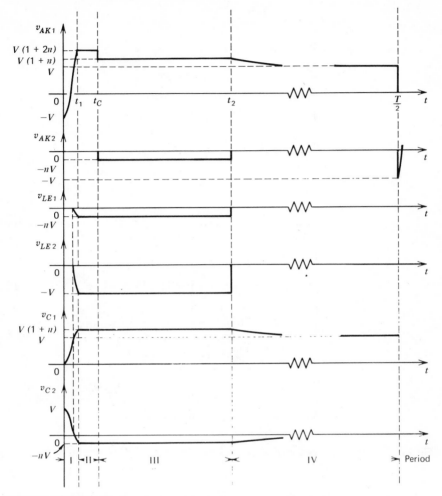

Fig. 7.9 (*Continued*)

is not necessary. It is sufficient to appreciate that, at $t = t_2^+$, when Q_2 begins to conduct, v_{AK2} becomes instantaneously zero, and v_{L2} acquires the value $-nV$ V. This inductor voltage decays exponentially as i_O approaches the value $-I_{O1}$, also exponentially, and v_O approaches the value $-V$ V. This occurs at $t = T/2$ s. During this interval $t_2 < t < T/2$ the capacitor voltages decay toward the values

$$v_{C1} = V \quad v: \qquad v_{C2} = 0 \quad V: \qquad t = \frac{T}{2} \quad s \qquad (7.94)$$

Thus voltage v_{AK1} also approaches the value V V.

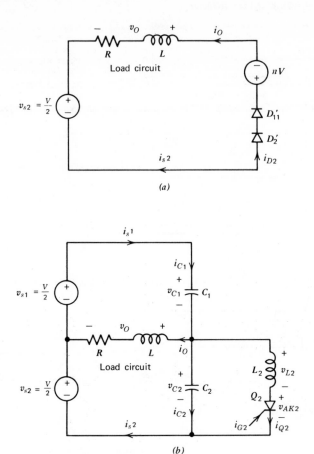

Fig. 7.10 Circuits applying during the post-commutation interval of thyristor Q_1 in Fig. 7.5a.

The time variations of the circuit variables during period IV are therefore as illustrated in Fig. 7.9.

7.2.4 Component Rating and Circuit Design—Voltage Commutation From the curves of Fig. 7.9 and the equations describing them in Section 7.2.3, the various components of the circuit of Fig. 7.5a may be accurately rated. It is doubtful, however, whether the engineering and computer time involved in such an operation is economically justifiable. An approximate but conservative rating procedure is therefore described here.

Voltage ratings of all components may be read off directly from Fig. 7.9. Those curves are drawn for a value of $n = 0.25$, but this is undesirably large and is employed there solely for the sake of clear illustration. The curve of

v_O in Fig. 7.9 shows that while the load voltage should ideally be $\pm V/2$ V, for an appreciable part of the time it is $\pm(1+2n)V/2$ V. In addition, the peak forward voltage applied to the thyristors is $V(1+2n)$ V, and this may become undesirably high if n is not given a small value; typically $n \approx 0.1$.

In inverters, it is the rms current rating of the thyristors that is of importance, and this may be obtained by Fourier analysis of the load current employing equations 7.8 to 7.10. Two or at most three terms of the current series will be adequate, even for load circuits of small inductance. This value may usually be employed to determine the thyristor current rating from the relationship

$$I_{QR} = \frac{I_R}{\sqrt{2}} \quad \text{A} \tag{7.95}$$

However if the output frequency of the inverter is high, then the pulse of commutation current, which in this circuit is carried by the main thyristors, will appreciably increase the value given in equation 7.95. In such a case, the rms current rating may be taken to be

$$I_{QR} = \left[\frac{I_R^2}{2} + I_{QC}^2 \right]^{1/2} \quad \text{A} \tag{7.96}$$

where

$$I_{QC} = \left[\frac{1}{T} \int_0^{t_C} i_Q^2 \, dt \right]^{1/2} \quad \text{A} \tag{7.97}$$

In equation 7.97 the expression for i_{Q2} in equation 7.42 must be employed.

The turn-off time for the thyristors may be specified from Fig. 7.8 where, if the load is variable, the maximum value of I_{O1} must be employed, since this gives the minimum value of t_q, the time available for turn-off.

The rating of diodes D_1 and D_2 may be taken to be the same as that of the thyristors. The rms current ratings of diodes D_{11} and D_{22} may then be taken as

$$I_{DR} = \frac{n}{n+1} I_{QR} \quad \text{A} \tag{7.98}$$

The current ratings of the windings of the energy-recovery circuit transformers may be determined from that of diodes D_{11} and D_{22}. Thus from equation 7.98,

$$I_{LE2} = \sqrt{2}\, I_{DR} = \frac{n\sqrt{2}}{n+1} I_{QR} \quad \text{A} \tag{7.99}$$

Consequently

$$I_{LE1} = \frac{I_{LE2}}{n} = \frac{\sqrt{2}}{n+1} I_{QR} \quad A \tag{7.100}$$

The rms current rating of the windings of the commutating inductances is obviously the same as that of the thyristors. The current in the commutating capacitors consists virtually of a single peak occurring during the interval $0 < t < t_1$, as the capacitor voltage changes by an amount $V(1+n)$ V. A good approximation to this current may be obtained from the average rate of change of voltage, and is therefore

$$I_C \cong \frac{CV(1+n)}{t_1} \quad A \tag{7.101}$$

and the rms capacitor current is then

$$I_{CR} = I_C \sqrt{\frac{t_1}{T}} \quad A \tag{7.102}$$

The design of the commutation circuit may again be based on the dimensionless quantity

$$x = \frac{V}{\omega_r L_C I_{O1}} \tag{7.103}$$

where

$$\omega_r = \frac{1}{[2L_C C]^{1/2}} \quad \text{rad/s} \tag{7.104}$$

and design experience has shown that $1.0 < x < 3.0$. From equation 7.75

$$G(x) = \omega_r t_q = \cos^{-1} C - \tan^{-1} D \quad \text{rad} \tag{7.105}$$

where

$$C = \frac{x}{2(x^2+4)^{1/2}} \tag{7.106}$$

$$D = \frac{2}{x} \tag{7.107}$$

Let t_q in equation 7.109 be

$$t_q = t_{\text{off}} + \Delta t \quad s \tag{7.108}$$

where Δt is a margin to allow for design approximations and tolerances in the building of the converter. From equations 7.105 and 7.108,

$$\omega_r = \frac{G(x)}{t_{\text{off}} + \Delta t} \quad \text{rad/s} \tag{7.109}$$

so that from equations 7.103 and 7.109,

$$L_C = \frac{V(t_{\text{off}} + \Delta t)}{xG(x)I_{O1}} \quad \text{H} \tag{7.110}$$

$$C = \frac{1}{2\omega_r^2 L_C} = \frac{xI_{O1}(t_{\text{off}} + \Delta t)}{2G(x)V} \quad \text{F} \tag{7.111}$$

It will be observed that this design procedure is identical with that employed in Section 6.2.4 for the voltage commutation circuit of the Type A chopper.

Example 7.4   For the circuit in Fig. 7.5a, $V = 220$ V and the current at the instant of commutation $I_{O1} = 250$ A. If $L_C = 50\,\mu$H, $C = 75\,\mu$F, and $n = 0.1$, calculate:

(a)   The time t_q available for turn-off of the thyristors.
(b)   The commutation interval.
(c)   The peak thyristor current and voltage.
(d)   The time $t_{q\,\text{max}}$ available for turn-off when I_{O1} is zero and compare this with the value obtained in a.

Solution

(a)   From equation 7.103,

$$x = \frac{V}{\omega_r L_C I_{O1}} = \frac{V}{I_{O1}}\left[\frac{2C}{L_C}\right]^{1/2} = \frac{220}{250}\left[\frac{2 \times 75}{50}\right]^{1/2} = 1.524$$

From equations 7.105 to 7.107

$$t_q = \frac{1}{\omega_r}[\cos^{-1} C - \tan^{-1} D]$$

and

$$C = \frac{x}{2[x^2+4]^{1/2}} = \frac{1.524}{2[1.524^2+4]^{1/2}} = 0.3030$$

$$D = \frac{2}{x} = \frac{2}{1.524} = 1.312$$

$$\therefore \quad t_q = [2L_CC]^{1/2}[\cos^{-1}0.3030 - \tan^{-1}1.312]$$

$$= [2\times50\times75]^{1/2}[1.263 - 0.920]10^{-6} = 29.70\times10^{-6} \quad \text{s}$$

(b)

$$t_C = t_1 + t_1' \quad \text{s} \tag{7.72}$$

From equations 7.56 and 7.103

$$t_1 = \frac{1}{\omega_r}\left[\sin^{-1}\frac{n}{\left[(2/x)^2+1\right]^{1/2}} + \tan^{-1}\frac{x}{2} \right]$$

$$= [2L_CC]^{1/2}\left[\sin^{-1}\frac{nx}{[x^2+4]^{1/2}} + \tan^{-1}\frac{x}{2} \right]$$

$$= [2\times50\times75]^{1/2}\left[\sin^{-1}\frac{0.1\times1.524}{[1.524^2+4]^{1/2}} + \tan^{-1}\frac{1.524}{2} \right]10^{-6}$$

$$= \sqrt{7500}\ [0.0606+0.6511]10^{-6} = 61.64\times10^{-6} \quad \text{s}$$

From equations 7.57 and 7.103,

$$I_{Q2} = I_{O1}[2\cos\omega_r t_1 + x\sin\omega_r t_1 - 1] \quad \text{A}$$

$$\omega_r t_1 = \frac{t_1}{[2L_CC]^{1/2}} = \frac{61.64\times10^{-6}}{\sqrt{7500}\ \times10^{-6}} = 0.7118^c$$

$$\therefore \quad I_{Q2} = 250[2\cos0.7118^c + 1.524\sin0.7118^c - 1]$$

$$= 250[1.514+0.995-1] = 377.4 \quad \text{A}$$

From equation 7.71

$$t'_1 = \frac{L_C}{nV} I_{Q2} = \frac{50 \times 377.4}{0.1 \times 220} \times 10^{-6} = 857.7 \times 10^{-6} \quad s$$

$$\therefore \quad t_C = (61.64 + 857.7)10^{-6} = 919.4 \times 10^{-6} \quad s$$

(c)   At the peak thyristor current, $v_{L2} = 0$. If this occurs at instant $t = t_p$ s then from equations 7.44 and 7.103

$$\omega_r t_p = \tan^{-1} \frac{x}{2} = 0.6511^c$$

From equations 7.42 and 7.103, the peak current is

$$I_{Q\,\text{peak}} = I_{O1}[2 \cos \omega_r t_p + x \sin \omega_r t_p - 1] \quad A$$

$$= 250[2 \cos 0.6511^c + 1.524 \sin 0.6511^c - 1]$$

$$= 250[1.5908 + 0.9236 - 1] = 378.6 \quad A$$

The peak thyristor voltage from Fig. 7.9 is

$$v_{AK\,\text{peak}} = V(1 + 2n) = 220 \times 1.2 = 264 \quad V$$

(d)   From equation 7.76

$$t_{q\,\text{max}} = \frac{\pi}{3} \sqrt{2 L_C C} = \frac{\pi}{3} \sqrt{7500} \times 10^{-6} = 90.69 \times 10^{-6} \quad s$$

This is three times as great as t_q for the loaded inverter.

7.2.5 Current Commutation of the Half-Bridge Inverter   Figure 7.11 shows the basic circuit of a half-bridge inverter with current commutation. This circuit should be compared with that in Fig. 6.15, which shows a two-quadrant chopper circuit with "type 1" commutation. The principle of operation of the commutation circuit is exactly the same in the two cases, and the inverter circuit may be analyzed and designed by the methods described in Sections 6.2.5 to 6.2.7 and 6.3.2a.

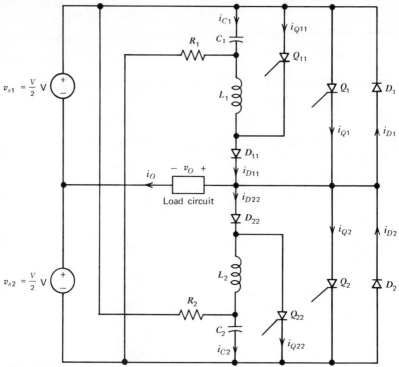

Fig. 7.11 Current commutation of the half-bridge inverter.

7.3 SINGLE-PHASE BRIDGE INVERTER

The power circuit of a "full-bridge," or more commonly "bridge," inverter is shown in Fig. 7.2a. If the load circuit branch of Fig. 7.2a is an RL series circuit, then this system may be represented by the equivalent circuit of Fig. 7.12a, where

$$v_s = V \quad \text{V}: \qquad 0 < t < \frac{T}{2} \quad \text{s}$$

$$v_s = -V \quad \text{V}: \qquad \frac{T}{2} < t < T \quad \text{s}$$

$$(7.112)$$

When several cycles of source voltage v_s have elapsed, the time variation of the current will have settled down to a periodic form such as is shown in Fig. 7.12b. The device currents of Fig. 7.2a that are identical with the load current during particular parts of the cycle are indicated on the current waveform. As in the half-bridge inverter, each pair of thyristors must be gated continuously throughout the half cycle.

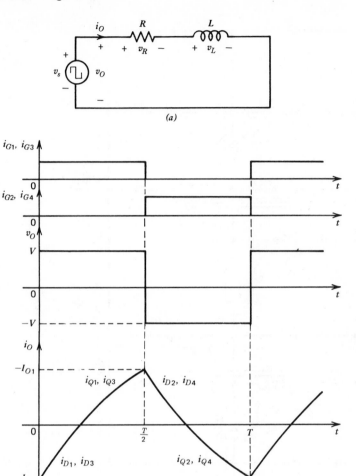

Fig. 7.12 Equivalent circuit and waveforms for the single-phase bridge inverter.

The load current may be determined by Fourier analysis as for the half-bridge rectifier. Owing to the increase in the amplitude of the rectangular voltage wave, however, the terms of the series expressed in equations 7.7 and 7.8 must be doubled when applied to the full-bridge inverter.

The bridge inverter will operate with load commutation, provided that the load branch is an underdamped RLC circuit and that inequality 7.12 is satisfied. The arrangements for complementary voltage commutation are shown in Fig. 7.13, and this circuit may be analyzed and designed

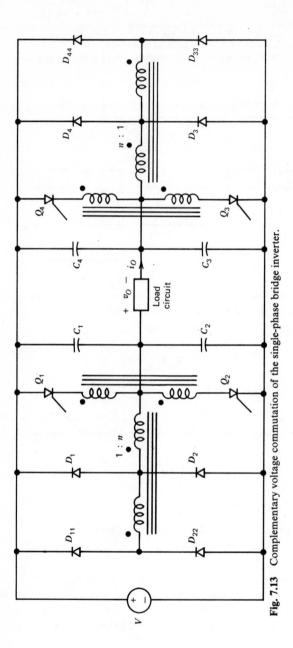

Fig. 7.13 Complementary voltage commutation of the single-phase bridge inverter.

employing equations 7.95 to 7.111. These equations are directly applicable to the bridge inverter by virtue of the source-voltage designations employed in the two cases. Current commutation may also be employed.

Example 7.5 In the bridge inverter of Fig. 7.2, $V = 300\,\text{V}$ and $T = 1/\omega = 10^{-3}\,\text{s}$. The load is an RLC series circuit in which $R = 0.5\,\Omega$, $\omega L = 10\,\Omega$, and $1/\omega C = 10.5\,\Omega$.

(a) Sketch to scale the waveform of v_O, the fundamental component of the waveform of i_O, and the waveforms of i_{Q1}, i_{Q2}, i_{D1}, i_{D2}, i_s, and v_{AK1} corresponding to that component of i_O (i.e., the higher harmonics of these variables are being neglected).

(b) Show that the power delivered by source V is equal to the power dissipated by the load circuit.

(c) Calculate the time available for turn-off of the thyristors with this load circuit.

Solution

(a) The fundamental component of the load voltage is

$$v_{O1} = \frac{4}{\pi} V \sin \omega t \quad \text{V}$$

The rms value of this component is

$$V_{1R} = \frac{2\sqrt{2}}{\pi} V = \frac{2\sqrt{2}}{\pi} \times 300 = 270 \quad \text{V}$$

The fundamental component of the load current is

$$i_{1O} = \frac{\sqrt{2}\,V_{1R}}{Z_1} \sin(\omega t - \phi_1) \quad \text{A}$$

where

$$Z_1 = \left[R^2 + \left(\omega L - \frac{1}{\omega C} \right)^2 \right]^{1/2}$$

$$= \left[0.5^2 + (-0.5)^2 \right]^{1/2} = 0.707\,\Omega$$

$$\phi_1 = \tan^{-1} \frac{\omega L - \dfrac{1}{\omega C}}{R} = \tan^{-1} \frac{-0.5}{0.5} = -45°$$

Thus

$$i_{1O} = \frac{270\sqrt{2}}{0.707} \sin(\omega t + 45°) = 540 \sin(\omega t + 45°) \quad A$$

The time variations of the circuit variables are shown in Fig. E7.5.
  (b)   The rms value of i_{O1} is

$$I_{1R} = \frac{540}{\sqrt{2}} \quad A$$

The power dissipated in the load circuit by this fundamental component is

$$P_O = RI_{1R}^2 = 0.5 \times \frac{540^2}{2} = 72.90 \times 10^3 \quad W$$

The power delivered by the source is

$$P_s = VI_s \quad W$$

where I_s is the average value of the fundamental component of the source current.

$$I_s = \frac{1}{\pi} \int_0^{\pi} \sqrt{2} \, I_{1R} \sin\left(\omega t + \frac{\pi}{4}\right) d(\omega t)$$

$$= \frac{\sqrt{2}}{\pi} I_{1R} \left[2\cos\frac{\pi}{4}\right] = \frac{\sqrt{2}}{\pi} \times \frac{540}{\sqrt{2}} \times \frac{2}{\sqrt{2}} = 243.1 \quad A$$

$$P_s = 300 \times 243.1 = 72.93 \times 10^3 = P_O \quad W$$

  (c)   As may be seen from Fig. E7.5, the thyristor current falls to zero before the end of each half cycle, so that at the end of each half cycle the diodes are conducting, and no forced commutation is required.
  The interval ωt_q is indicated on the waveform of v_{AK1}. From Fig. E7.5,

$$\omega t_q = \frac{\pi}{4} \quad rad$$

so that

$$t_q = \frac{T}{8} = 125 \times 10^{-6} \quad s$$

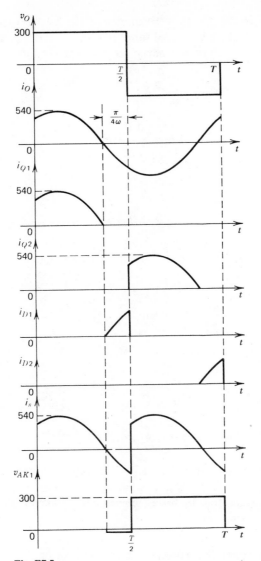

Fig. E7.5

Example 7.6 In the system of Example 7.5, the load circuit capacitance is increased until $1/\omega C = 9.5\,\Omega$. All other parameters are unchanged.

Sketch the waveforms and carry out the calculations specified in Example 7.5.

(a)

$$Z_1 = [0.5^2 + 0.5^2]^{1/2} = 0.707\ \Omega$$

$$\phi_1 = \tan^{-1}\frac{0.5}{0.5} = 45°$$

Thus

$$i_{O1} = 540\sin(\omega t - 45°)\quad\text{A}$$

The time variations of the circuit variables are shown in Fig. E7.6.

(b) As in Example 7.5,

$$I_{1R} = \frac{540}{\sqrt{2}}\quad\text{A}\qquad\text{and}\qquad P_O = 72.90\times10^3\quad\text{W}$$

The average value of the fundamental component of the source current is

$$I_s = \frac{1}{\pi}\int_0^\pi \sqrt{2}\,I_{1R}\sin\left(\omega t - \frac{\pi}{4}\right)d(\omega t) = 243.1\quad\text{A}$$

$$P_s = 300\times243.1 = 72.93\times10^3 = P_O\quad\text{W}$$

(c) As may be seen from Fig. E7.6, the thyristor current is still positive at the end of the half cycle, so that v_{AK1} is zero at this instant. The thyristors must therefore be forced commutated, since t_q in the uncommutated circuit does not exist.

7.4 VOLTAGE CONTROL OF SINGLE-PHASE INVERTERS

In most inverter applications, stepless control of the ratio of source voltage to ac voltage supplied to the load is necessary. Typical examples of such a requirement are:

1. An inverter system employing a battery as a dc source and required to deliver constant ac voltage to the load circuit must be able to eliminate the effect of battery voltage variation.

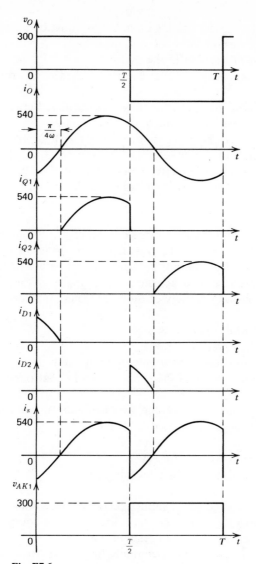

Fig. E7.6

2.   An inverter system driving an ac motor must be able to maintain an approximately constant ratio of output voltage to output frequency to avoid saturation of the iron in the motor.

The methods of voltage control commonly employed are:

a.   Introduction of an ac voltage controller between the output terminals of the inverter and the load.

b.   Introduction of equipment controlling the dc input voltage between the input terminals of the inverter and the source.

c.   Elaboration of the inverter control circuitry to permit variation of the ratio between the dc input voltage and the ac output voltage of the inverter itself.

The introduction of an ac voltage controller at the output terminals of the inverter is generally to be avoided, since it results in a very high harmonic content of the output voltage at lower values of that voltage.

If the available source is ac, then the dc input voltage to the inverter may be controlled by a variety of methods. These include:

1.   An ac-dc motor-generator set.
2.   A controlled rectifier.
3.   An induction regulator supplying an uncontrolled rectifier.
4.   A saturable reactor in the supply to an uncontrolled rectifier.

If the available source is dc, then two methods of controlling the input voltage to the inverter are available. These are:

1.   A dc-dc motor-generator set.
2.   A dc-to-dc voltage controller (chopper).

The main advantage of dc input voltage control is that, provided that the arrangement constitutes an ideal direct voltage source, the harmonic content of the ac output voltage remains constant at all voltages. The main disadvantage is that the current-commutating capability of an inverter with forced commutation is reduced as the dc input voltage is decreased. This point will be appreciated if Fig. 7.8 is examined, where it will be observed that a reduction in voltage V without a corresponding reduction in current I_{O1} will result in a reduction in t_q, the time available for turn-off of the thyristors. This effect may of course be guarded against to some extent by modification of the commutation-circuit parameters. But if this is carried too far, the commutating capacitor becomes prohibitively large. This problem is sometimes solved by using an auxiliary fixed direct voltage source for the commutation circuit, but the overall circuit of the inverter then becomes undesirably complicated.

A system of dc input voltage control that provides an ideal direct voltage source demands either a dc generator or a filter circuit before the inverter input terminals. Either of these alternatives is undesirable if a system with fast response is required. Furthermore, the introduction of another converter, either rotating or static, between the source and the load is undesirable, since it increases the losses in the system. The best method of voltage control is therefore that permitting variation of the ratio between the dc input voltage and the ac output voltage of the inverter itself, and this method is discussed in some detail in the immediately following sections. The techniques available differ in the harmonic content that they produce in the inverter output voltage, thus the acceptable harmonic content is the factor that determines the choice of technique. The three most commonly used techniques all employ pulse-width modulation and may be listed as:

1. Single-pulse modulation.
2. Multiple-pulse modulation.
3. Sinusoidal-pulse modulation.

For pulse-width modulation, forced commutation is essential. In the immediately following sections, the required voltage output waveforms are first illustrated, and the circumstances in which they can be achieved are then considered.

7.4.1 Single-Pulse Modulation The output voltage waveform for single-pulse modulation is illustrated in Fig. 7.14. This waveform should be compared with that in Fig. 7.2b. For the purpose of analysis it will be assumed that the start of each voltage pulse is retarded and the end of each pulse advanced by equal angular intervals, resulting in a variation of the pulse-width δ over the range $0 \leqslant \delta \leqslant \pi$ rad.

The waveform of v_O in Fig. 7.14 may be described by the series

$$v_O = \sum_{n=1,3,5\ldots}^{\infty} a_n \sin n\omega t + \sum_{n=1,3,5\ldots}^{\infty} b_n \cos n\omega t \quad \text{V} \qquad (7.113)$$

where

$$a_n = \frac{2}{\pi} \int_0^{\pi} V \sin n\omega t \, d(\omega t)$$

$$= \frac{2V}{\pi} \int_{(\pi-\delta)/2}^{(\pi+\delta)/2} \sin n\omega t \, d(\omega t) = \frac{4V}{n\pi} \sin \frac{n\delta}{2} \quad \text{V} \qquad (7.114)$$

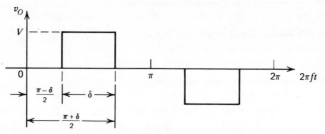

Fig. 7.14 Output voltage with single-pulse modulation.

and

$$b_n = \frac{2V}{\pi} \int_{(\pi-\delta)/2}^{(\pi+\delta)/2} \cos n\omega t \, d(\omega t) = 0 \quad \text{V} \qquad (7.115)$$

Thus

$$v_O = \sum_{n=1,3,5\ldots}^{\infty} a_n \sin n\omega t \quad \text{V} \qquad (7.116)$$

In Fig. 7.15, curves of the ratio $a_n/a_{1\,max}$ versus δ are shown for $n = 1, 3, 5$, and 7, where $a_{1\,max}$ is the amplitude of the fundamental component of the rectangular waveform obtained when $\delta = \pi$. From these curves it may be seen that, as δ is decreased, the harmonic content of the output voltage waveform increases until, when the amplitude of the fundamental component has been reduced to 20% of its maximum value, the amplitudes of the three harmonics illustrated are very nearly equal to that of the fundamental.

7.4.2 Multiple-Pulse Modulation The harmonic content at lower output voltages can be significantly reduced by using several pulses in each half cycle, thus obtaining an output voltage waveform such as is illustrated in Fig. 7.16a. The number of pulses per half cycle is

$$N = \frac{f_p}{2f} = \text{integer} \qquad (7.117)$$

where f_p is the frequency of pulses per second, and $f = 1/T$ is the output voltage frequency. For variation of the output voltage from zero to its maximum value V, the pulse-width δ^* must vary over the range $0 \leqslant \delta^* \leqslant \pi/N$.

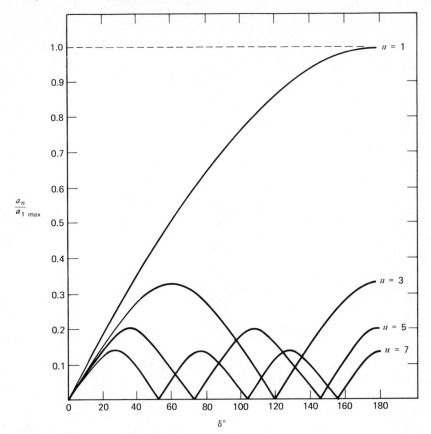

Fig. 7.15 Harmonic content of waveform in Fig. 7.14.

Expressions for the amplitudes a_n of the various output voltage harmonics are complicated but not difficult to obtain. They are most easily determined by deriving an expression for a general pair of pulses situated at $\omega t = \alpha$ and $\omega t = \pi + \alpha$, and then combining the effects of all such pairs of pulses in the cycle. Indeed, this is the method to be employed when programming the calculation for a digital computer. It is illustrated in Fig. 7.16b, where

$$v_O = \sum_{n=1,3,5\ldots}^{\infty} a_n \sin n\omega t + \sum_{n=1,3,5\ldots}^{\infty} b_n \cos n\omega t \quad \text{V} \qquad (7.118)$$

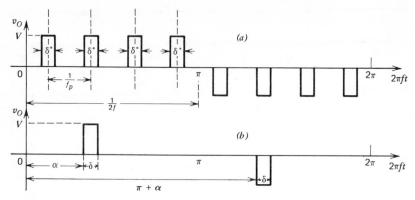

Fig. 7.16 Output voltage with multiple-pulse modulation.

in which

$$a_n = \frac{2V}{\pi} \int_{\alpha}^{\alpha + \delta} \sin n\omega t \, d(\omega t)$$

$$= \frac{2V}{n\pi} [\cos \alpha - \cos n(\alpha + \delta)] \quad \text{V} \tag{7.119}$$

$$b_n = \frac{2V}{\pi} \int_{\alpha}^{\alpha + \delta} \cos n\omega t \, d(\omega t)$$

$$= \frac{2V}{n\pi} [\sin n(\alpha + \delta) - \sin n\alpha] \quad \text{V} \tag{7.120}$$

Figure 7.17 shows curves of the ratios $a_n/a_{1\,\text{max}}$ for $n = 3, 5, 7$ as functions of δ^* for the frequency ratios $N = 3$ and $N = 10$. It can be seen that for $N = 10$, the third and fifth harmonic amplitudes approach the values existing in an unmodulated rectangular wave. For $N = 3$, the variation of harmonic amplitude with δ^* is much greater than for $N = 10$ for the harmonics illustrated in Fig. 7.17. Of course the amplitudes of some higher harmonics will be significantly greater for the larger value of N, but such harmonics produce negligible current in a load circuit with appreciable inductance and are readily filtered out for load circuits with little inductance.

7.4.3 Sinusoidal-Pulse Modulation The output voltage waveform for sinusoidal-pulse modulation is illustrated in Fig. 7.18a. In this waveform, pulse-width is a sinusoidal function of the angular position in the cycle of

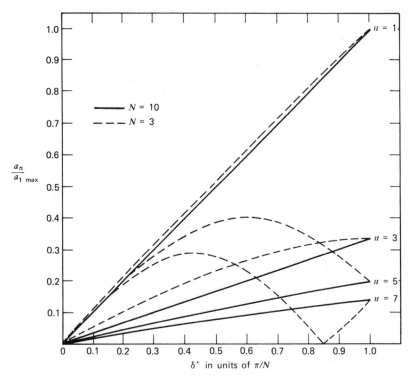

Fig. 7.17 Harmonic content of waveform in Fig. 7.16.

each pulse. A means of determining the positions and widths of the pulses is shown in Fig. 7.18b and corresponds closely to the actual techniques employed in the circuitry controlling the turn-on and the commutation of the thyristors.

The control functions consist of a sinusoidal wave of variable amplitude A and frequency $f = 1/T$ as well as a triangular wave of fixed amplitude A_p and frequency f_p with a direct component of magnitude A_p. This biassed triangular wave is reversed in polarity at the end of each half cycle of the output voltage. If N is the number of voltage pulses per half cycle, then

$$N = \frac{f_p}{2f} = \text{integer} \tag{7.121}$$

The angles for turn-on and commutation of the thyristors are determined by the intersections of these two waveforms. The output voltage is controlled by varying amplitude A over a range $0 \leqslant A \leqslant A_{\max}$, where $A_{\max}$

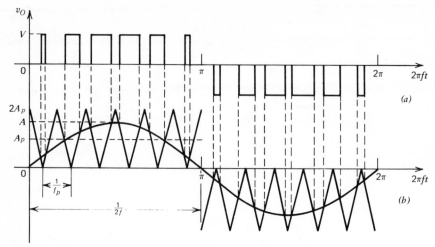

Fig. 7.18   Output voltage with sinusoidal-pulse modulation ($N = 6$).

$> 2A_p$. If $A_{\max}$ is made very large, then in the limit, the time variation of v_O approaches the rectangular voltage waveform of Fig. 7.2b.

Once again, the determination of the harmonic amplitudes is relatively complicated, necessitating integration over N angular intervals. If these amplitudes are computed, however, it is found that for $0 \leqslant A/A_p \leqslant 2$ all harmonics of order $n < 2N$ are eliminated. For $A/A_p > 2$, low-order harmonics appear, since pulse-width is no longer a sinusoidal function of the angular position of the pulse. In Fig. 7.19 are shown curves of the ratio $a_n / a_{1\max}$ as a function of A/A_p for $n = 3$, 5 and 7 with $N = 10$.

7.4.4 Pulse-Width Modulation in Bridge and Half-Bridge Inverters The waveforms of Figs. 7.14, 7.16, and 7.18 all show that intervals occur in the voltage cycle during which the voltage applied to the load circuit must be zero. This is not difficult to arrange with a full-bridge inverter. It simply means that during such intervals either thyristors Q_1 and Q_4 or thyristors Q_2 and Q_3 in Fig. 7.2a must be turned on simultaneously. In either case, two thyristor-diode pairs in series then constitute a short circuit between the terminals of the load circuit. In employing this method of producing the required waveform it must then be ensured that the number of commutations is as small as possible and that the thyristors are symmetrically utilized.

Production of the output voltage waveform shown in Fig. 7.16 demands a large number of commutations, since a thyristor must be commutated at the beginning and end of each pulse if the load is to be short-circuited and the load voltage reduced to zero between pulses. The number of commutations can be reduced if, at the end of each pulse, one of the two conducting

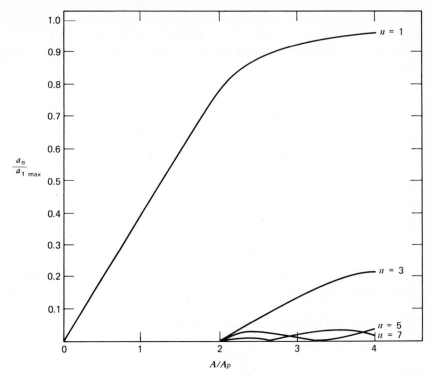

Fig. 7.19 Harmonic content of waveform in Fig. 7.18 ($N = 10$).

thyristors is commutated, but no other thyristor is turned on to provide the short-circuit across the load. At the beginning of the next pulse, it is then only necessary to turn on once more the thyristor commutated at the end of the preceding pulse. In this way the commutating losses in the circuit are reduced and the control circuitry somewhat simplified. The result of this procedure, however, is that the load voltage is not always defined between pulses, since it will depend on the nature of the load circuit. Analysis of the circuit under these conditions then becomes very complicated.

Inspection of the circuit of the half-bridge inverter in Fig. 7.1a shows that there is no way in which the load voltage v_O can be reduced to zero. At all times $v_O = \pm V/2$V. A method of pulse-width modulation that may be employed with a half-bridge inverter is illustrated in Fig. 7.20. Instead of being reduced to zero, the voltage is reversed for short intervals in each half cycle. When $\delta = 0$, the output voltage wave is that shown in Fig. 7.1b, in which the fundamental component has its maximum amplitude. If δ were increased to $\pi/3$ rad, the output voltage waveform would contain no fundamental component, since a triple-frequency rectangular voltage wave would be achieved. This type of voltage control is rarely used because of the high ratios of harmonic amplitudes to that of the fundamental component of the voltage wave.

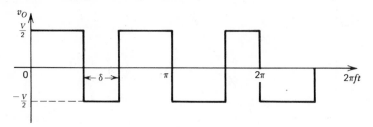

Fig. 7.20   Output voltage for half-bridge inverter.

Example 7.7   The single-phase full-bridge inverter of Fig. 7.2a is required to have a waveform such as is shown in Fig. 7.14, in which $\delta = 2\pi/3$ rad.

(a)   Draw one cycle of the required gate current waveforms for the four thyristors if they are to be symmetrically utilized.

(b)   If the load circuit is such that i_O passes through the zero value at $2\pi ft$ equal to:

(i)   0 and π rad
(ii) $\pi/3$ and $\pi + \pi/3$ rad
(iii) $-\pi/3$ and $\pi - \pi/3$ rad

draw in each case one cycle of the current waveform and indicate on it which devices are conducting in the various intervals throughout the cycle.

Solution

(a)   For the positive half-cycle of the voltage wave, thyristors Q_1 and Q_3 must conduct.

For the interval $5\pi/6 < 2\pi ft < 7\pi/6$ the load branch must be short-circuited by one pair of thyristors. Let these be Q_1 and Q_4.

For the negative half cycle of the voltage wave, thyristors Q_2 and Q_4 must conduct.

For the interval $11\pi/6 < 2\pi ft < 13\pi/6$ the load branch must be short-circuited by one pair of thyristors. To obtain symmetrical utilization of the thyristors, let these be Q_2 and Q_3. The necessary gating signals are shown in Fig. E7.7a.

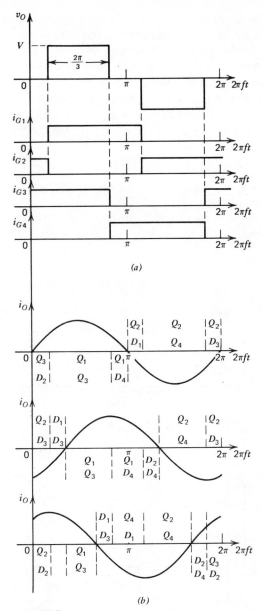

(a)

(b)

Fig. E7.7

411

Each thyristor must be commutated as the gating signal is removed. There will therefore be four commutations required per cycle.

(b) The exact current waveform will in each case depend on the load-circuit parameters, which are not given. For simplicity therefore, sinusoidal current waves are shown in Fig. E7.7b.

Which thyristors or diodes are conducting is determined by the direction of the load current at the instant considered and the thyristors which are turned on. The conducting devices are indicated on the waveforms shown in Fig. E7.7b.

Example 7.8 The single-phase full-bridge inverter of Fig. 7.2a has the gating signals shown in Fig. E7.8. At the end of each gating signal, the two thyristors are commutated.

Draw one cycle of the load-circuit current and voltage waveforms if the load circuit is such that i_O passes through the zero value at $2\pi ft$ equal to 0

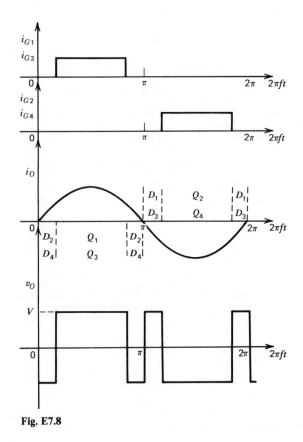

Fig. E7.8

and π rad. Indicate on the current waveform which devices are conducting in the various intervals throughout the cycle.

Solution   As in Example 7.7, the load-circuit parameters are not known, and the current is therefore represented by a sinusoidal waveform in Fig. E7.8.

When no thyristor is conducting, the load current necessarily flows through two diodes and the source, in one direction or the other. The load-circuit voltage is therefore never zero. Its waveform is shown in Fig. E7.8. This should be compared with the voltage waveform of Example 7.7.

Which devices are conducting is determined by the direction of the load current at the instant considered and by whether any thyristors are turned on. The conducting devices are indicated on the current waveform of Fig. E7.8.

7.5 REDUCTION OF OUTPUT-VOLTAGE HARMONICS

A quantity employed as a measure of the harmonic content of the output current of an inverter is the harmonic distortion K_I, defined as

$$K_I = \frac{I_{RI}}{I_1} \qquad (7.122)$$

where I_1 is the rms value of the fundamental component of the output current, and I_{RI} is the rms value of the harmonic components. Then I_{RI} may in turn be defined as

$$I_{RI} = \left[\sum I_{nR}^2 \right]^{1/2} \quad \text{A}: \qquad n \neq 1 \qquad (7.123)$$

where I_{nR} is the rms value of the nth harmonic.

In some applications (e.g., fast response power supplies) K_I must be less than 0.05, and this requires a low pass filter at the inverter output to remove the higher harmonics. The lower the frequencies of the harmonics that must be attenuated, the larger become the filter circuit elements and consequently the weight and bulk of the entire filter. Thus if lower order harmonics in the output voltage can be reduced or eliminated by other means, the size of the filter can be reduced. This also improves the speed of response to load change of the entire inverter-filter system.

7.5.1 Harmonic Reduction by Pulse-Width Modulation Techniques that may be employed to reduce the harmonic content of the unfiltered inverter output voltage depend in the first instance on the type of inverter circuit.

An output-voltage waveform that can be readily obtained from a half-bridge inverter is illustrated in Fig. 7.21. The method of modulating a waveform in this way was discussed in Section 7.4.4 in connection with voltage control. This waveform has quarter-wave symmetry and can therefore be represented by the series

$$v_O = \sum_{n=1,3,5\ldots}^{\infty} a_n \sin n\omega t \quad \text{V} \tag{7.124}$$

where

$$a_n = \frac{4}{\pi} \frac{V}{2} \left[\int_0^{\alpha_1} \sin n\omega t \, d(\omega t) - \int_{\alpha_1}^{\alpha_2} \sin n\omega t \, d(\omega t) + \int_{\alpha_2}^{\pi/2} \sin n\omega t \, d(\omega t) \right]$$

$$= \frac{4}{\pi} \frac{V}{2} \frac{[1 - 2\cos n\alpha_1 + 2\cos n\alpha_2]}{n} \quad \text{V} \tag{7.125}$$

If the third and fifth harmonics are to be eliminated, then it is necessary that

$$a_3 = 0: \qquad a_5 = 0 \tag{7.126}$$

Thus from equations 7.125 and 7.126 two simultaneous equations may be obtained and solved for α_1 and α_2, giving

$$\alpha_1 = 23.62°: \qquad \alpha_2 = 33.30° \tag{7.127}$$

This elimination of the third and fifth harmonics is achieved at the cost of reducing the amplitude of the fundamental component of the voltage

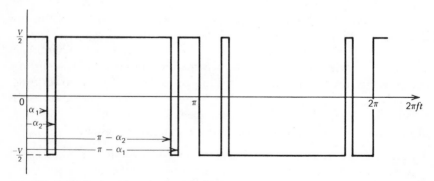

Fig. 7.21 Harmonic reduction in the half-bridge inverter.

wave to 0.839 of that of $a_{1\,\text{max}}$, the fundamental component of the un-modulated rectangular wave. This reduction in output voltage thus derates the inverter by some 16%. In addition, ten commutations per cycle are now needed, instead of only two for the unmodulated square wave. It must further be noted that, if control of the output voltage is also required, this can only be obtained by means external to the inverter.

The technique described in the preceding two paragraphs can also be applied to the full-bridge inverter. An alternative, and in some ways superior, method is to employ sinusoidal modulation of the output voltage, as described in Section 7.4.3 and illustrated in Figs. 7.18 and 7.19. Provided that the ratio $A/A_p \leqslant 2$, the lower-order harmonics are eliminated. In addition, voltage control is available over the range $0 \leqslant A/A_p \leqslant 2$. The main disadvantages of this system are:

1. The large number of commutations, which is a function of frequency f_p.
2. The reduction in the fundamental component, even at $A/A_p = 2$, as compared with that of the unmodulated rectangular output-voltage wave. From Fig. 7.19 it may be seen that this reduction exceeds 17%.

7.5.2 Harmonic Reduction by Transformer Connection The output voltages of two or more inverters having similar waveforms shifted in phase from one another may be combined by means of transformers to produce a combined voltage waveform with harmonic content less than that of the individual inverter waveforms. The transformer arrangement for combining two inverter outputs is shown in Fig. 7.22.

Fig. 7.23a shows two similar inverter output voltage waves with a phase displacement of $\pi/3$ rad and the resulting waveform when they have been

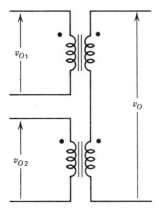

Fig. 7.22 Transformer connection for harmonic reduction.

combined by means of two $1:1$ ratio transformers connected as shown in Fig. 7.22. Figure 7.15 shows that the third-harmonic component of this waveform is zero, and this may be confirmed by adding the two series describing the two similar waveforms of Fig. 7.23 as follows:

$$v_{O1} = a_1 \sin \omega t + a_3 \sin 3\omega t + a_5 \sin 5\omega t + \cdots \quad \text{V} \tag{7.128}$$

$$v_{O2} = a_1 \sin\left(\omega t - \frac{\pi}{3}\right) + a_3 \sin 3\left(\omega t - \frac{\pi}{3}\right) + a_5 \sin 5\left(\omega t - \frac{\pi}{3}\right) + \cdots \quad \text{V} \tag{7.129}$$

$$v_O = v_{O1} + v_{O2} = \sqrt{3}\, a_1 \sin\left(\omega t - \frac{\pi}{6}\right) + \sqrt{3}\, a_5 \sin 5\left(\omega t + \frac{\pi}{6}\right) + \cdots \quad \text{V} \tag{7.130}$$

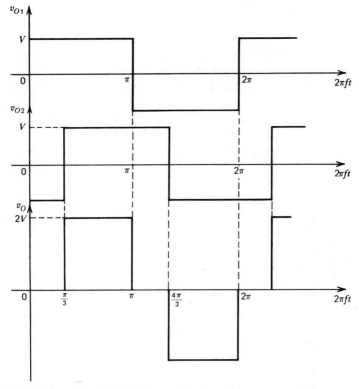

Fig. 7.23 Elimination of third harmonic by transformer connection.

Thus with a phase shift of $\pi/3$ rad, the third harmonic is eliminated, as indeed are all triplen harmonics, while the fifth harmonic amplitude is unchanged in relation to that of the fundamental component. On the other hand, the fundamental component is $\sqrt{3}/2$ of that in the individual waveforms, so that in effect the inverters have been derated by some 14%.

A further stage in harmonic reduction may be achieved by taking two such output voltages v_O, as are illustrated in Fig. 7.23, shifting them in phase by $\pi/5$ rad relative to one another, and combining them by means of the transformer arrangement of Fig. 7.22. For this purpose, the original voltage waveforms might be obtained by single-pulse-width modulation. The resultant waveform, which is without either third or fifth harmonic, is shown as v_O in Fig. 7.24. The inverters may be further derated as a consequence of such an additional harmonic reduction.

With the transformer method of harmonic reduction employed in two

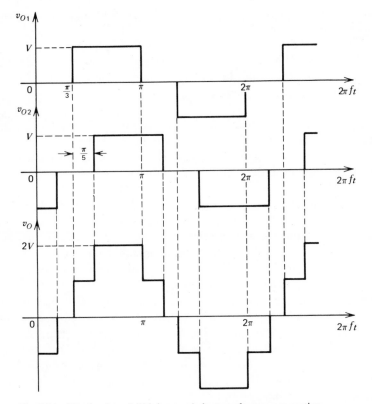

Fig. 7.24 Elimination of fifth harmonic by transformer connection.

cascaded stages of transformation, control of an output voltage without any third or fifth harmonics can be achieved by means of symmetrical pulse width modulation on all four inverters, provided only that the required phase displacements are maintained.

7.6 THREE-PHASE INVERTERS

A three-phase source may be obtained from a direct voltage source by simply employing three single-phase inverters and arranging that the gating signals for the three inverters shall be shifted in phase by 120° of the output frequency interval. The three outputs may then be brought to a three-phase transformer. The transformer primary windings must be isolated from one another, while the secondary windings may be connected in wye or delta to supply the load circuit. Commutation, voltage control, and harmonic reduction may be obtained by the methods already discussed in connection with single-phase inverters, so that component ratings may be determined and design carried out exactly as described in the earlier sections of this chapter.

As indicated in Section 7.1 and Fig. 7.3, a three-phase source may also be obtained from a three-phase bridge inverter. This inverter is discussed in some detail in this and succeeding sections. There are two possible patterns of gating signals. These are:

1.   Three thyristors turned on at any instant; this results in output-voltage waves that are defined under all conditions of load.
2.   Two thyristors turned on at any instant; this results in undefined output-voltage waves under some load conditions.

Whichever pattern of gating signals is employed, it is necessary that signals be applied and removed at 60° intervals of the output voltage waveforms. Thus there are six distinct periods of operation in one cycle. The thyristors in Fig. 7.3a are numbered in the sequence in which the gating signals are applied to them to give positive-sequence voltages v_{AB}, v_{BC}, v_{CA} at the output terminals A, B, C.

When any thyristor is turned on, that thyristor and the diode connected in antiparallel with it constitute a short circuit. Thus when, for example, thyristor Q_1 in Fig. 7.3a is turned on, output terminal A is brought to the potential of the positive source terminal. Conversely, when Q_4 is turned on (at which time Q_1 must be turned off), terminal A is brought to the potential of the negative source terminal. Similar considerations determine the potentials of terminals B and C.

7.6.1 Three Thyristors Turned On at Any Instant The pattern of gating signals shown in Fig. 7.3b results in three thyristors being turned on simultaneously. The potentials of the output terminals during the periods for which these gating signals are applied may be determined, and from them the line-to-line output voltages may be obtained. The resulting voltage waveforms are shown in Fig. 7.3b. These represent a balanced set of three-phase alternating voltages. These voltages are unaffected by the nature of the load circuit, which may have any combination of resistance, inductance, and capacitance, may be balanced or unbalanced, linear or nonlinear. The line-to-line voltage waveforms are reproduced in Fig. 7.25a on a horizontal scale calibrated in radians. On this diagram also are indicated the six periods and the thyristors that are turned on during these periods.

Voltage waveform v_{AB} in Fig. 7.25a may be described by the Fourier series

$$v_{AB} = \sum_{n=1,3,5\ldots}^{\infty} \frac{4V}{n\pi} \cos \frac{n\pi}{6} \sin n\left(\omega t + \frac{\pi}{6}\right) \quad V \qquad (7.131)$$

(This series may be obtained by the procedure described in Section 2.6.1, but may also be obtained with considerably less labor by shifting the waveform for v_{AB} in Fig. 7.25a by $-\pi/6$ rad. Since it now has quarter-wave symmetry, the waveform is readily analyzed giving the expression of equation 7.131 without the phase angle of $-\pi/6$ rad. If the phase angle is then included, the Fourier series is referred to the origin of the waveforms in Fig. 7.25). The expressions for the other two line-to-line voltages are then

$$v_{BC} = \sum_{n=1,3,5\ldots}^{\infty} \frac{4V}{n\pi} \cos \frac{n\pi}{6} \sin n\left(\omega t - \frac{\pi}{2}\right) \quad V \qquad (7.132)$$

$$v_{CA} = \sum_{n=1,3,5\ldots}^{\infty} \frac{4V}{n\pi} \cos \frac{n\pi}{6} \sin n\left(\omega t - \frac{7\pi}{6}\right) \quad V \qquad (7.133)$$

It should be noted that the triplen harmonics ($n = 3, 6, 9, \ldots$, etc.) in the series of equations 7.131 to 7.133 are all zero.

If the load circuit is linear and delta connected, the load-branch currents may be directly obtained from the voltage series of equations 7.131 to 7.133. If the load circuit is linear and wye connected, then superposition

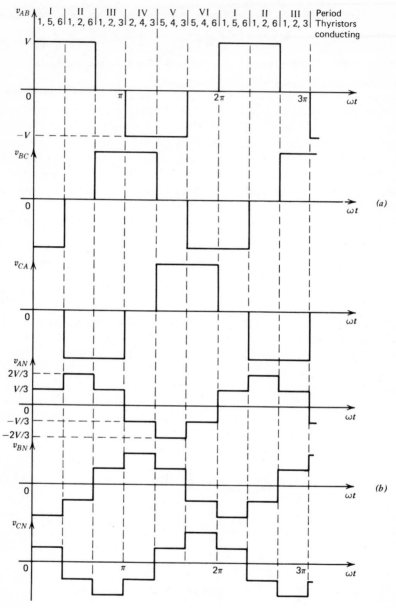

Fig. 7.25 Output voltage waveforms for gating signals of Fig. 7.3*b* with balanced resistive load circuit.

may be employed to determine the load-branch currents and line-to-neutral voltages.

At the end of each of the periods shown in Fig. 7.25a, the gating signal is removed from one thyristor, which, under most load conditions, will require to be commutated. Six commutations per cycle are therefore necessary. The thyristor current at the instant of commutation may be calculated to any desired degree of accuracy by employing the necessary number of terms of equations 7.131 to 7.133.

A special case of particular interest is the performance of this inverter when supplying a balanced, wye-connected resistive load. Figure 7.26a shows such a load circuit, and Fig. 7.26b shows the equivalent circuit of the system during three consecutive periods of the line-to-line voltage cycle shown in Fig. 7.25. By applying voltage division to the circuits of Fig. 7.26b, the line-to-neutral voltages for the load circuit may be determined, and their waveforms are shown in Fig. 7.25b.

With a resistive load, only the thyristors conduct, and in theory the diodes could be removed from the circuit of Fig. 7.3a without affecting its operation. For this ideal condition of operation, it is possible to determine a factor for the inverter-load combination that may be employed as the base for a figure of merit applying to any other inverter-load combination. This quantity is called the derating factor D_R, and is defined as

$$D_R = \frac{P_Q}{P_O} \tag{7.134}$$

where P_O is the output power of the inverter, and

$$P_Q = N_Q V_{FB} I_{\text{RMS}} \quad \text{W} \tag{7.135}$$

in which N_Q is the number of load-current carrying thyristors in the inverter circuit, while V_{FB} and I_{RMS} are defined in Fig. 3.24a, being respectively the rated Repetitive Peak Forward Voltage of the thyristor and the rated RMS Forward Current.

Thus for the three-phase bridge inverter, in which three thyristors are turned on at any instant to supply a balanced resistive load of R Ω per phase,

$$P_O = \frac{3}{\pi} \int_0^\pi \frac{v_{AN}^2}{R} \, d(\omega t) \quad \text{W} \tag{7.136}$$

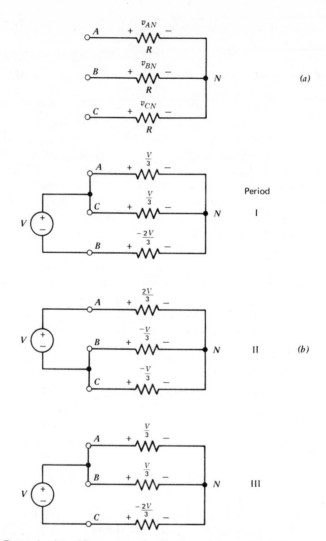

Fig. 7.26   Determination of line-to-neutral voltage waveforms of Fig. 7.25*b*.

and since, from Fig. 7.26*b*,

$$v_{AN} = \frac{V}{3} \quad \text{V}: \qquad 0 < \omega t < \frac{\pi}{3} \quad \text{rad}$$

$$v_{AN} = \frac{2V}{3} \quad \text{V}: \qquad \frac{\pi}{3} < \omega t < \frac{2\pi}{3} \quad \text{rad} \qquad (7.137)$$

$$v_{AN} = \frac{V}{3} \quad \text{V}: \qquad \frac{2\pi}{3} < \omega t < \pi \quad \text{rad}$$

substitution in equation 7.136 yields

$$P_O = \frac{2V^2}{3R} \quad \text{W} \tag{7.138}$$

The output power may also be expressed as

$$P_O = 3RI_R^2 \quad \text{W} \tag{7.139}$$

where I_R is the rms value of the per phase load current. Thus

$$I_R = \left(\frac{P_O}{3R}\right)^{1/2} = \frac{\sqrt{2}}{3}\frac{V}{R} \quad \text{A} \tag{7.140}$$

On the assumption that thyristors are available for which $V_{FB} = V$ and $I_{RMS} = I_R / \sqrt{2}$, then from equation 7.135,

$$P_Q = 6V\frac{1}{\sqrt{2}}\frac{\sqrt{2}\,V}{3R} = \frac{2V^2}{R} \quad \text{W} \tag{7.141}$$

and the derating factor is

$$D_R = \frac{P_Q}{P_O} = 3 \tag{7.142}$$

that is, the inverter is able to deliver to the load a power equal to one third of the combined maximum power rating of the six main thyristors.

7.6.2 Two Thyristors Turned On at Any Instant Inspection of the gating signal waveforms shown in Fig. 7.3b shows that at $t = T/2$ or $\omega t = \pi$, i_{G1} is cut off, and simultaneously i_{G4} is applied. In practice, a commutation interval must elapse between the removal of i_{G1} and the application of i_{G4}, since if thyristor Q_1 were not allowed time to turn off, then application of i_{G4} would result in short-circuit of the source through thyristors Q_1 and Q_4. Indeed it is a disadvantage of the pattern of gating signals shown in Fig. 7.3b that, even if a normally sufficient interval is provided between the end of one gating signal and the beginning of the next, any failure of commutation, for whatever reason, will result in destructive short circuit of the source through two thyristors.

The danger of short-circuit may be considerably reduced by adopting the pattern of gating signals shown in Fig. 7.27a. It will there be seen that an angular interval of $\pi/3$ rad elapses between the end of the gating signal on one thyristor and the beginning of the gating signal on the thyristor connected in series with it. This provides ample time for the first thyristor to turn off. In addition, delayed commutation of, say, thyristor Q_1 due to some malfunction only results in an additional path for current to pass through the load circuit. While this may cause momentary unbalance of the load currents, it will not result in a destructive short-circuit current, unless of course commutation fails completely, and Q_1 is still conducting at $\omega t = \pi$, when Q_4 is turned on.

Six forced commutations per cycle are necessary under most load conditions for this pattern of gating signals also, so that the cycle may again be divided into the six periods shown in Fig. 7.27. Since each thyristor is commutated as its gating signal is removed, the potentials of only two of the output terminals will be defined at any instant of the cycle if the load circuit is other than resistive. Thus the analysis of the inverter performance with this type of control becomes very complicated for a "general" load circuit. The load terminal that is not at a particular instant connected to a source terminal by a short-circuit composed of a turned-on-thyristor and anti-parallel-connected diode will find a potential that depends on the nature of the load circuit. Moreover this potential will not be constant during the $\pi/3$ angular interval during which the terminal is "free."

For a balanced, wye-connected, resistive load the output voltage waveforms are defined, and the line-to-neutral voltage waveforms may be determined from the series of equivalent circuits shown in Fig. 7.28. These waveforms are shown in Fig. 7.27b, and from these the line-to-line output voltage waveforms are readily obtained. That of v_{AB} is shown in Fig. 7.27c. For this inverter-load system, it is possible to determine a derating factor. Thus from equation 7.136 and the waveform of v_{AN} in Fig. 7.27b

$$P_O = \frac{3}{\pi} \int_0^{2\pi/3} \frac{1}{R} \left(\frac{V}{2} \right)^2 d(\omega t) = \frac{V^2}{2R} \quad \text{W} \tag{7.143}$$

and from equation 7.140,

$$I_R = \left(\frac{P_O}{3R} \right)^{1/2} = \frac{1}{\sqrt{6}} \frac{V}{R} \quad \text{A} \tag{7.144}$$

On the assumption that thyristors are available for which $V_{FB} = V$, and $I_{RMS} = I_R/\sqrt{2}$, then from equation 7.135,

$$P_Q = 6V \frac{1}{2\sqrt{3}} \frac{V}{R} = \sqrt{3} \frac{V^2}{R} \quad \text{W} \tag{7.145}$$

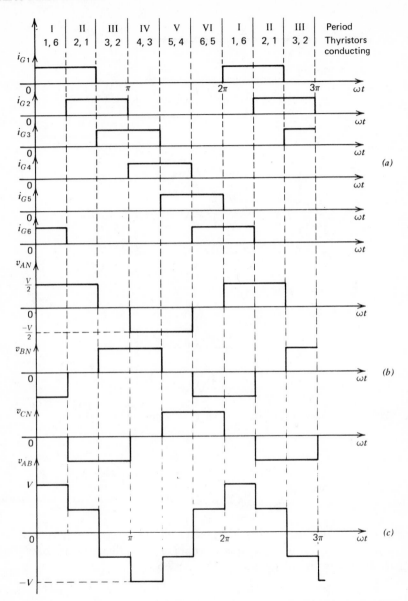

Fig. 7.27 Output voltage waveforms for circuit of Fig. 7.3a with balanced resistive load circuit.

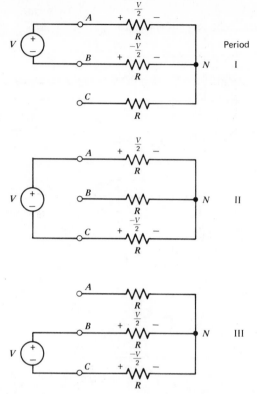

Fig. 7.28 Determination of line-to-neutral voltage waveforms of Fig. 7.27*b*.

and

$$D_R = \frac{P_Q}{P_O} = 2\sqrt{3} \tag{7.146}$$

Comparison of this result with the value given in equation 7.142 shows that the thyristors are utilized to less advantage with this pattern of gating signals than with that discussed in Section 7.6.1.

7.7 SERIES INVERTERS

Series inverters are inverter systems in which reactive elements are placed in series with the load circuit to provide load commutation. They are commonly employed when a high output frequency of some 200 Hz to 100 kHz is required. Such frequencies may be necessary in induction heating, fluorescent lighting, or other relatively fixed-load applications. Since load commutation is employed, then if the existing parameters of the load circuit are not such as to form an underdamped system of the necessary frequency of oscillation, additional inductance and/or capacitance must be added.

In this section, one simple form of the series inverter is analyzed in detail. Two other series inverter configurations are also illustrated.

7.7.1 The Basic Series Inverter Circuit The circuit of Fig. 7.29a illustrates the basic principle of the series inverter. The thyristors are turned on alternately by short pulses of gating current applied at the intervals

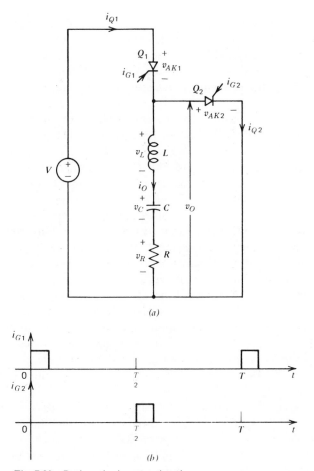

(a)

(b)

Fig. 7.29 Basic series inverter circuti.

required to give the necessary load current angular frequency of $\omega = 2\pi f$ rad/sec. These signals are illustrated in Fig. 7.29b. The load circuit, whatever its original nature, is provided with the reactance necessary to produce an underdamped oscillatory current of ringing frequency ω_r. The load current must go to zero, and thyristor Q_1 must turn off before thyristor Q_2 is turned on, since if this is not the case, the source will be short-circuited through the two thyristors. A limiting condition is therefore

$$\frac{\pi}{\omega} - \frac{\pi}{\omega_r} = t_{q\min} \quad \text{s} \tag{7.147}$$

where $t_{q\min}$ is the minimum permissible time available for turn-off of either thyristor.

During the interval $0 < t < (T/2) - t_{q\min}$, $i_O > 0$, and the capacitor builds up a positive voltage, When i_O, which is oscillatory, falls to zero, Q_1 turns off, since the gating signal has by then been removed. After time $t_{q\min}$ has elapsed, Q_2 is turned on. During the interval $T/2 < t < T - t_{q\min}$, $i_O < 0$, energy stored in the capacitor is released, and a negative half cycle of i_O takes place.

This qualitative description of the circuit operation immediately reveals that the circuit of Fig. 7.29a is not practical. Toward the end of the first half-wave of current i_O, when $di_O/dt < 0$, and hence $v_L < 0$, v_R approaches zero. Thus when i_O reaches zero and Q_1 turns off, v_L disappears instantaneously, and the capacitor voltage v_C is instantaneously applied to thyristor Q_2. Due to the high value of dv/dt, Q_2 turns on, and since Q_1 has not had time to turn off, short-circuit of the source results.

It is thus clearly necessary, as it has been in the converters discussed earlier, to add to the circuit elements that protect the thyristors against excessive dv/dt, and also against excessive di/dt. The necessary arrangements of inductors and snubber circuits are shown in Fig. 7.30. Hitherto, it has been permissible to disregard the effect of these protective elements on the operation of the main power circuit, because their parameters were such as to allow them to be approximated by short circuits or open circuits. In the high-frequency applications for which series inverters are employed, this is no longer the case, and the protective elements of the circuit of Fig. 7.30 must be taken into account when analyzing the circuit performance.

An analysis of the complete circuit of Fig. 7.30 should not be beyond the capabilities of a reader who has followed this text thus far, but it is more than is needed to illustrate the principle of operation, which is what is of interest here. For this reason, a simple case is discussed in which the actual load circuit is considered to be purely resistive.

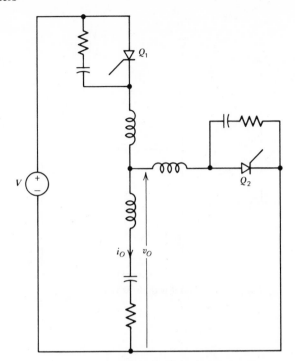

Fig. 7.30 Series inverter with protective circuit elements.

7.7.2 The Series Inverter with Resistive Load A circuit arrangement that may be employed when the load circuit is purely resistive is shown in Fig. 7.31. In this circuit R represents the actual load; C is a series capacitor introduced to give the necessary ringing frequency for load commutation in conjunction with either of the two inductors, for which

$$L_1 = L_2 = L \quad \text{H} \tag{7.148}$$

The gating-current signals required for this circuit are identical with those prescribed for the basic circuit of Fig. 7.29a, and these are shown again in Fig. 7.32. In this circuit, when the current i_O falls to zero at the end of the positive half cycle, and Q_1 turns off, the instantaneous disappearance of voltage v_{L1} has no effect on thyristor Q_2, which has been brought to the forward voltage $v_{AK2} = v_C$ at low rate of dv/dt. Thyristor Q_2 does not conduct, therefore, until the gating signal i_{G2} is applied. Again, near the end of the second half cycle, when $i_{Q2} = -i_O$ becomes zero, and thyristor Q_2 turns off, thyristor Q_1 has been brought at a low rate of dv/dt to a

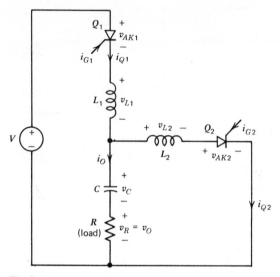

Fig. 7.31   Series inverter with resistive load.

potential

$$v_{AK1} = V - v_C \quad \text{V} \tag{7.149}$$

Thyristor Q_1 does not conduct, therefore, until the gating signal i_{G1} is again applied. A detailed analysis may now be undertaken.

When the circuit of Fig. 7.31 has been in operation for a few cycles, all of the variables will have settled down to a cyclical pattern. In each half cycle, the load current i_O increases positively or negatively from zero as Q_1 or Q_2 is turned on and then falls again to zero, when Q_1 or Q_2 turn off. An interval $t_{q\min}$ must then elapse before another gating signal is applied to initiate the next half cycle. The cycle may therefore be divided into four intervals, during two of which the currents and voltages are varying. These two intervals are separated by two intervals, each $t_{q\min}$ seconds in length, during which all currents are zero, and all voltages are constant. These intervals are shown in Fig. 7.32.

Since $L_1 = L_2$, the ringing frequency of the RLC circuit will be the same for each half cycle; that is, during each half cycle the load current will flow for the same interval of π/ω_r, s. Moreover, since the capacitor must release as much charge during one half cycle as it stores during the other, then the area enclosed between the current wave and the horizontal axis must be the same for each half cycle. The current wave therefore has alternating symmetry.

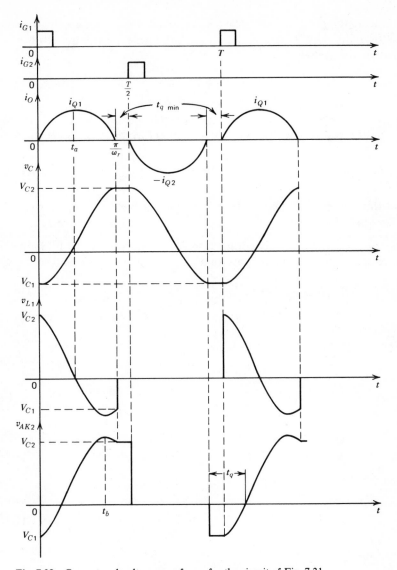

Fig. 7.32 Current and voltage waveforms for the circuit of Fig. 7.31.

During the positive half cycle of i_O, capacitor C is charged by the source via thyristor Q_1. At the end of that half cycle it will therefore have a positive voltage

$$v_C = V_{C2} \quad \text{V:} \qquad t = \frac{\pi}{\omega_r} \quad \text{s} \qquad\qquad (7.150)$$

During the negative half cycle of i_O, capacitor C is discharged via thyristor Q_2. At the end of that half cycle it will therefore have a voltage

$$v_C = V_{C1} < V_{C2} \quad \text{V:} \qquad t = \frac{T}{2} + \frac{\pi}{\omega_r} \quad \text{s} \qquad\qquad (7.151)$$

and this will be the voltage of the capacitor at instant $t = 0$, shown in Fig. 7.32, when the positive half wave of i_O is about to commence.

Interval I The initial conditions for this interval, in addition to that expressed in equation 7.151, are that all currents are zero, and

$$v_{AK1} = V - V_{C1} \quad \text{V:} \qquad t = 0^- \quad \text{s} \qquad\qquad (7.152)$$

$$v_{AK2} = V_{C1} \quad \text{V:} \qquad t = 0^- \quad \text{s} \qquad\qquad (7.153)$$

When thyristor Q_1 is turned on,

$$v_{AK1} = 0 \quad \text{V:} \qquad t = 0^+ \quad \text{s} \qquad\qquad (7.154)$$

$$i_O = 0 \quad \text{A:} \qquad t = 0^+ \quad \text{s} \qquad\qquad (7.155)$$

and since

$$v_R = R i_O = 0 \quad \text{V:} \qquad t = 0^+ \quad \text{s} \qquad\qquad (7.156)$$

then for the left-hand mesh of the circuit in Fig. 7.31,

$$V = v_{L1} + v_C = \frac{L di_O}{dt} + V_{C1} \quad \text{V:} \qquad t = 0^+ \quad \text{s} \qquad\qquad (7.157)$$

so that

$$\frac{di_O}{dt} = \frac{V - V_{C1}}{L} \quad \text{A/s:} \qquad t = 0^+ \quad \text{s} \qquad\qquad (7.158)$$

During interval I, for which $0 < t < \pi/\omega_r$ s,

$$V = v_{L1} + v_C + v_R$$

$$= \frac{L di_O}{dt} + \frac{1}{C} \int_0^t i_O \, dt + V_{C1} + R i_O \quad \text{V} \qquad\qquad (7.159)$$

and differentiation of equation 7.159 yields

$$\frac{d^2 i_O}{dt^2} + \frac{R}{L} \frac{di_O}{dt} + \frac{1}{LC} = 0 \quad A/s^2 \qquad (7.160)$$

The solution of equation 7.160, employing the initial conditions of equations 7.155 and 7.158 is

$$i_O = i_{Q1} = \frac{(V - V_{C1})}{\omega_r L} \epsilon^{-\zeta t} \sin \omega_r t \quad A \qquad (7.161)$$

where

$$\zeta = \frac{R}{2L} \quad s^{-1} \qquad (7.162)$$

$$\omega_r = [\omega_0^2 - \zeta^2]^{1/2} \quad rad/s \qquad (7.163)$$

$$\omega_0 = \frac{1}{\sqrt{LC}} \quad rad/s \qquad (7.164)$$

The remaining circuit variables during this interval may now be determined. These are:

$$v_O = v_R = R i_O = \frac{R(V - V_{C1})}{\omega_r L} \epsilon^{-\zeta t} \sin \omega_r t \quad V \qquad (7.165)$$

$$v_{L1} = \frac{L di_O}{dt} = \frac{\omega_0}{\omega_r} (V - V_{C1}) \epsilon^{-\zeta t} \cos (\omega_r t + \psi) \quad V \qquad (7.166)$$

where

$$\psi = \tan^{-1} \frac{\zeta}{\omega_r} \quad rad \qquad (7.167)$$

$$v_C = V - v_{L1} - v_R = V - \frac{\omega_0}{\omega_r} (V - V_{C1}) \epsilon^{-\zeta t} \cos (\omega_r t - \psi) \quad V \quad (7.168)$$

$$v_{AK2} = V - v_{L1} = V - \frac{\omega_0}{\omega_r} (V - V_{C1}) \epsilon^{-\zeta t} \cos (\omega_r t + \psi) \quad V \quad (7.169)$$

Interval I ends when i_O falls to zero at

$$t = \frac{\pi}{\omega_r} = t_1 \quad s \qquad (7.170)$$

By substitution from equation 7.170 in equations 7.165, 7.166, 7.168, and 7.169,

$$v_O = v_R = 0 \quad \text{V}: \qquad t = t_1^- \quad \text{s} \tag{7.171}$$

$$v_C = V + (V - V_{C1})\epsilon^{-\zeta\pi/\omega_r} = V_{C2} \quad \text{V}: \qquad t = t_1^- \quad \text{s} \tag{7.172}$$

$$v_{L1} = (V - V_{C1})\epsilon^{-\zeta\pi/\omega_r} \quad \text{V}: \qquad t = t_1^- \quad \text{s} \tag{7.173}$$

$$v_{AK2} = v_C \quad \text{V}: \qquad t = t_1^- \quad \text{s} \tag{7.174}$$

Interval II   At $t = t_1^+$ s, when i_O and di_O/dt are both zero, all currents in the circuit are zero, and from equation 7.166,

$$v_{L1} = 0 \quad \text{V}: \qquad t = t_1^+ \quad \text{s} \tag{7.175}$$

For the right-hand mesh of the circuit in Fig. 7.31,

$$v_{AK2} = v_C = V_{C2} \quad v: \qquad t = t_1^+ \quad \text{s} \tag{7.176}$$

while for the left-hand mesh of that circuit,

$$v_{AK1} = V - v_C = V - V_{C2} \quad \text{V}: \qquad t = t_1^+ \quad \text{s} \tag{7.177}$$

The circuit remains in this static condition throughout interval II, that is, for $t_1 < t < T/2$ s.

Interval III   The initial conditions for this interval are those existing in the static condition of interval II. Thus when thyristor Q_2 is turned on,

$$v_{AK2} = 0 \quad \text{V}: \qquad t' = 0^+ \quad \text{s} \tag{7.178}$$

where

$$t' = t - \frac{T}{2} \quad \text{s} \tag{7.179}$$

Also

$$i_O = 0 \quad \text{A}: \qquad t' = 0^+ \quad \text{s} \tag{7.180}$$

$$v_R = Ri_O = 0 \quad \text{V}: \qquad t' = 0^+ \quad \text{s} \tag{7.181}$$

For the right-hand mesh of the circuit in Fig. 7.31,

$$v_C = V_{C2} = V_{L2} = \frac{L\,di_{Q2}}{dt} = -\frac{L\,di_O}{dt} \quad \text{V}: \qquad t' = 0^+ \quad \text{s} \tag{7.182}$$

so that

$$\frac{di_O}{dt} = -\frac{V_{C2}}{L} \quad \text{A/s:} \quad t' = 0^+ \quad \text{s} \tag{7.183}$$

During interval III, for which $0 < t' < \pi/\omega_r$ s,

$$v_C + v_R = v_{L2} \quad \text{V} \tag{7.184}$$

or

$$\frac{1}{C} \int_0^{t'} i_O \, dt' + V_{C2} + R i_O = -\frac{L \, di_O}{dt'} \quad \text{V} \tag{7.185}$$

and differentiation of equation 7.185 yields

$$\frac{d^2 i_O}{(dt')^2} + \frac{R}{L} \frac{di_O}{dt'} + \frac{i_O}{LC} = 0 \quad \text{A/s}^2 \tag{7.186}$$

The solution of equation 7.186, employing the initial conditions of equations 7.180 and 7.183, is

$$i_O = -i_{Q2} = -\frac{V_{C2}}{\omega_r L} \epsilon^{-\zeta t'} \sin \omega_r t' \quad \text{A} \tag{7.187}$$

where ζ and ω_r have been defined in equations 7.162 to 7.164.

Since the current waveform has alternating symmetry, it follows from equations 7.161 and 7.187 that

$$V - V_{C1} = V_{C2} \quad \text{V} \tag{7.188}$$

The remaining circuit variables during this interval may now be determined, while substituting for V_{C2} from equation 7.188. These are

$$v_O = v_R = -\frac{R(V - V_{C1})}{\omega_r L} \epsilon^{-\zeta t'} \sin \omega_r t' \quad \text{V} \tag{7.189}$$

$$v_{L2} = -\frac{L \, di_O}{dt'} = \frac{\omega_0}{\omega_r} (V - V_{C1}) \epsilon^{-\zeta t'} \cos (\omega_r t' + \psi) \quad \text{V} \tag{7.190}$$

$$V_C = v_{L2} - v_R = \frac{\omega_0}{\omega_r} (V - V_{C1}) \epsilon^{-\zeta t'} \cos (\omega_r t' - \psi) \quad \text{V} \tag{7.191}$$

$$v_{AK1} = V - v_{L2} = V - \frac{\omega_0}{\omega_r} (V - V_{C1}) \epsilon^{-\zeta t'} \cos (\omega_r t' + \psi) \quad \text{V} \tag{7.192}$$

Interval III ends when i_O becomes zero at

$$t' = \frac{\pi}{\omega_r} = t_1' \quad \text{s}$$

(7.193)

Thus by substitution in equations 7.189 to 7.192,

$$v_O = v_R = 0 \quad \text{V}: \qquad t' = t_1'^{-} \quad \text{s}$$

(7.194)

$$v_{L2} = (V - V_{C1})\epsilon^{-\zeta\pi/\omega_r} \quad \text{V}: \qquad t' = t_1'^{-} \quad \text{s}$$

(7.195)

$$v_C = -(V - V_{C1})\epsilon^{-\zeta\pi/\omega_r} = V_{C1} \quad \text{V}: \qquad t' = t_1'^{-} \quad \text{s}$$

(7.196)

$$v_{AK1} = V - (V - V_{C1})\epsilon^{-\zeta\pi/\omega_r} \quad \text{V}: \qquad t' = t_1'^{-} \quad \text{s}$$

(7.197)

Interval IV At $t' = t_1'^{+}$ s, when i_O and di_O/dt' are both zero, all currents in the circuit are zero, and

$$v_{L2} = -\frac{L\,di_O}{dt'} = 0 \quad \text{V}: \qquad t' = t_1'^{+} \quad \text{s}$$

(7.198)

For the left-hand mesh of the circuit in Fig. 7.31,

$$v_{AK1} = V - v_C = V - V_{C1} \quad \text{V}: \qquad t' = t_1'^{+} \quad \text{s}$$

(7.199)

while for the right-hand mesh of the circuit

$$v_{AK2} = v_C = V_{C1} \quad \text{V}: \qquad t' = t_1'^{+} \quad \text{s}$$

(7.200)

The circuit remains in this static condition throughout interval IV, that is, for $(T/2) + \pi/\omega_r < t < T$ s. From equation 7.196

$$V_{C1} = \frac{-V}{\epsilon^{\zeta\pi/\omega_r} - 1} \quad \text{V}$$

(7.201)

so that substitution for V_{C1} in equation 7.188 yields

$$V_{C2} = \frac{V\epsilon^{\zeta\pi/\omega_r}}{\epsilon^{\zeta\pi/\omega_r} - 1} \quad \text{V}$$

(7.202)

The waveforms of the circuit variables throughout the cycle are shown in Fig. 7.32. The waveform of voltage v_{L2} is identical with that of v_{L1}, but displaced $T/2$ s from it. Similarly the waveform of voltage v_{AK1} is identical with that of v_{AK2}, but displaced $T/2$ from it.

7.7.3 Component Ratings Since the current waveform of i_{Q2} is simply the reflection of that of i_{Q1}, it follows that the two thyristors and the two inductors have identical current ratings. In addition, the waveform of voltage v_{AK1} is similar to that of v_{AK2}, so that the two thyristors have identical voltage ratings.

For the ideal circuit of Fig. 7.31, the average output power P_O dissipated in the resistive load R must be equal to the average input power P_s provided by the source. Thus

$$P_s = \frac{1}{T} \int_0^T v i_{Q1} \, dt \quad \text{W} \tag{7.203}$$

Current i_{Q1} flows only during the interval $0 < t < \pi/\omega_r$ s, so that substitution from equation 7.161 in 7.203 yields

$$P_s = \frac{\omega}{2\pi} \int_0^{\pi/\omega_r} \frac{V(V - V_{C1})}{\omega_r L} \epsilon^{-\zeta t} \sin \omega_r t \, dt$$

$$= \frac{\omega V(V - V_{C1})}{2\pi \omega_0^2 L} (1 + \epsilon^{-\zeta \pi/\omega_r}) \quad \text{W} \tag{7.204}$$

and this expression could be further simplified. Also since

$$P_s = P_O = R I_R^2 \quad \text{W} \tag{7.205}$$

the rms value of the load current is

$$I_R = \left(\frac{P_s}{R} \right)^{1/2} \quad \text{A} \tag{7.206}$$

and this is also the rms current rating of the capacitor C. The rms value of the thyristor and inductor currents is then

$$I_{QR} = \frac{I_R}{\sqrt{2}} \quad \text{A} \tag{7.207}$$

and the average value of the thyristor currents may be seen from equation 7.203 to be

$$I_Q = \frac{1}{T} \int_0^T i_{Q1} \, dt = \frac{P_s}{V} \quad \text{A} \tag{7.208}$$

The inductors must not saturate at the peak values of i_{Q1} and i_{Q2}. If the expression for i_{Q1} in equation 7.161 is differentiated and the derivative equated to zero, it is found that the maximum value of i_{Q1} occurs when $t = t_a$, that is, when

$$\omega_r t_a = \tan^{-1} \frac{\omega_r}{\zeta} = \frac{\pi}{2} - \psi \quad \text{rad} \tag{7.209}$$

where ψ is defined in equation 7.167, so that from equation 7.161,

$$I_{Q\text{peak}} = \frac{V - V_{C1}}{\omega_0 L} \epsilon^{-(\zeta/\omega_r)(\pi/2 - \psi)} \quad \text{A} \tag{7.210}$$

From Fig. 7.32, the peak voltage applied to the capacitor is seen to be

$$v_{C\text{peak}} = V_{C2} \quad \text{V} \tag{7.211}$$

while the peak forward voltage applied to each thyristor is somewhat greater than V_{C2}. If the expression for v_{AK2} in equation 7.169 is differentiated and the derivative equated to zero, it is found that the maximum value of v_{AK2} occurs when $t = t_b$, that is, when

$$\omega_r t_b = \pi - 2\psi \quad \text{rad} \tag{7.212}$$

so that from equation 7.169

$$v_{AK\text{peak}} = V + (V - V_{C1})\epsilon^{-(\zeta/\omega_r)(\pi - 2\psi)} \quad \text{V} \tag{7.213}$$

The expression in equations 7.210 and 7.213 may be further simplified by substitution from equations 7.188 and 7.202.

The actual time t_q available for turn-off is shown on the curve of v_{AK2} in Fig. 7.32. This is greater than the value $t_{q\text{min}}$ obtained from equation 7.147. This difference may simply be accepted as a factor of safety. On the other hand, t_q may be accurately determined by substituting t_q for t in equation (7.169), setting v_{AK2} to zero, and solving the resulting transcendental equation numerically.

For other than purely resistive loads, an analysis basically similar to the foregoing can be carried out. The complexity of the load circuit, however, will result in high-order differential equations, and these call for numerical solution.

7.7.4 Series Bridge Inverters Two bridge configurations of series inverter circuits are shown in Fig. 7.33a and b. For each of these, load commutation is employed so that the parameters of the RLC circuits and the operating frequency must be such that sufficient time t_q is provided in which the thyristors can turn off.

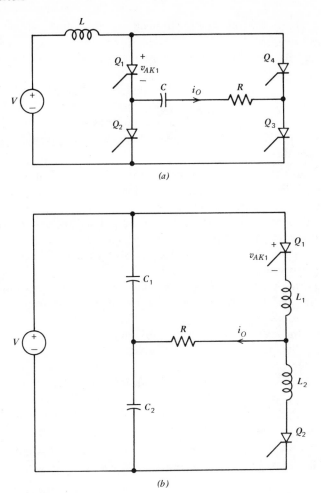

Fig. 7.33 Series bridge inverters.

In the full-bridge circuit of Fig. 7.33a, thyristors Q_1 and Q_3 are turned on simultaneously and conduct during the interval $0 < t < \pi/\omega_r$ s, after which a short interval elapses before thyristors Q_2 and Q_4 are turned on. The inverter is shown with a resistive load and capacitance added for commutation. If the load-circuit branch also possessed inductance, danger of short-circuit of the source similar to that discussed in connection with the basic circuit of Fig. 7.29 would arise. This would necessitate additional circuit elements to reduce the rate of change of forward voltage applied to one pair of thyristors when the current in the other pair fell to zero. It should be noted that energy is supplied by the source during both the positive and negative half-cycle of the load current. This has the advantage of requiring a lower rms source current for a given power output than would be the case with the inverter of Fig. 7.31.

The half-bridge circuit of Fig. 7.33b again shows a resistive load branch. In this circuit also, the source supplies energy during both half-cycles of the output current. As indicated earlier, the presence of inductance in the load branch would necessitate protective elements to prevent short-circuit of the source through the two thyristors.

7.7.5 General Comments on Series Inverters The foregoing brief and very basic discussion of series inverters permits some broad conclusions to be drawn about their usefulness. In general, the series inverter must be designed to provide a constant power output at a constant frequency. No-load operation is not possible, and protection against short-circuit of the output demands special control features to detect excessive output current or capacitor voltage.

However, the power and basic control circuitry of the series inverter are very simple, and full control of variation of output power can be permitted, provided that a corresponding variation in output frequency is acceptable. In addition, since series inverters operate at high frequency, only physically small inductors and capacitors are required, so that compact units can be built.

The series inverter may be considered as an alternative ac source to a high-frequency motor generator set, and since efficient high-frequency generators are difficult to design and expensive to build, the inverter is often the more economical alternative.

Example 7.9 In the series inverter circuit of Fig. 7.31, $V = 300$ V, $R = 3\ \Omega$, $L = 40\ \mu$H, $C = 5\ \mu$F, and the output frequency f_O is to be 8000 Hz. Determine:

(a) The minimum time provided for turn-off $t_{q\,\text{min}}$.
(b) The peak thyristor voltage.
(c) The peak thyristor current.
(d) The power output.
(e) The rms thyristor current.
(f) The derating factor for the thyristors.
(g) Also sketch to scale the waveforms of i_O and v_C.

Solution

(a) From equation 7.162 to 7.164,

$$\omega_r = \left[\frac{1}{LC} - \frac{R^2}{4L^2} \right]^{1/2} = \left[\frac{10^{12}}{40 \times 5} - \frac{3^2 \times 10^{12}}{4 \times 40^2} \right]^{1/2} = 59950 \quad \text{rad/s}$$

From equation 7.147

$$t_{q\min} = \frac{\pi}{2\pi \times 8000} - \frac{\pi}{59950} = 10.09 \times 10^{-6} \quad \text{s}$$

(b)

$$\zeta = \frac{R}{2L} = \frac{3 \times 10^6}{2 \times 40} = 37.5 \times 10^3 \quad \text{s}^{-1}$$

$$\psi = \tan^{-1}\frac{\zeta}{\omega_r} = \tan^{-1}\frac{37.5}{59.95} = 0.5589 \quad \text{rad}$$

From equation 7.202

$$V_{C2} = \frac{300\epsilon^{37.5\pi/59.95}}{\epsilon^{37.5\pi/59.95} - 1} = 348.9 \quad \text{V}$$

From equations 7.188 and 7.213,

$$v_{AK\,\text{peak}} = V + V_{C2}\epsilon^{-(\zeta/\omega_r)(\pi - 2\psi)}$$

$$\frac{\zeta}{\omega_r}(\pi - 2\psi) = \frac{37.5}{59.95}(\pi - 2 \times 0.5589) = 1.266$$

Thus

$$v_{AK\,\text{peak}} = 300 + 348.9\epsilon^{-1.266} = 398.4 \quad \text{V}$$

(c) The peak thyristor current occurs at $t = t_a$, where, from equation 7.209

$$t_a = \frac{1}{\omega_r}\left(\frac{\pi}{2} - \psi\right) = \frac{10^{-6}}{0.05995}\left(\frac{\pi}{2} - 0.5589\right)$$

$$= 16.88 \times 10^{-6} \quad \text{s}$$

From equation 7.120

$$I_{Q\,\text{peak}} = \frac{V_{C2}}{\omega_0 L}\epsilon^{-\zeta t_a}$$

$$\omega_0 = [\zeta^2 + \omega_r^2]^{1/2} = [37.5^2 + 59.95^2]^{1/2} \times 10^3$$

$$= 70.71 \times 10^3\,\text{rad/s}$$

Thus

$$I_{Q\,peak} = \frac{348.9\epsilon^{-37.5\times16.88\times10^{-3}}}{70.71\times40\times10^{-3}} = 65.50 \quad A$$

(d)   From equations 7.188 and 7.204

$$P_s = \frac{\omega V V_{C2}}{2\pi\omega_0^2 L}[1+\epsilon^{-(\zeta\pi/\omega_r)}]$$

$$\frac{\zeta\pi}{\omega_r} = \frac{37.5}{59.95}\pi = 1.965$$

Thus

$$P_s = \frac{16\pi\times10^3\times300\times348.9}{2\pi\times70.71^2\times40}(1+\epsilon^{-1.965})$$

$$= 4774 \quad W$$

(e)   From equations 7.206 and 7.207

$$I_R = \left(\frac{P_s}{R}\right)^{1/2} = \left(\frac{4774}{3}\right)^{1/2} = 39.89 \quad A$$

and

$$I_{QR} = \frac{I_R}{\sqrt{2}} = \frac{39.89}{\sqrt{2}} = 28.21 \quad A$$

(f)   On the assumption that thyristors are available for which V_{FB} $= v_{AK\,peak}$ and $I_{RMS} = I_{QR}$, then from equation 7.135,

$$P_Q = 2\times398.4\times28.21 = 22480 \quad W$$

and

$$D_R = \frac{P_Q}{P_O} = \frac{P_Q}{P_s} = \frac{22480}{4774} = 4.708$$

(g)   The waveforms of i_O and v_C are shown in Fig. E7.9.

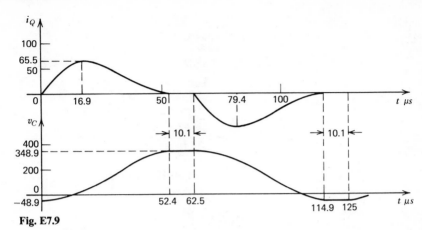

Fig. E7.9

Example 7.10 In the series inverter of Example 7.9, $R = 1$ Ω; all other parameters are unchanged. Determine the quantities and draw the waveforms specified in Example 7.9 for this modified load circuit.

Solution

(a) As in the previous example,

$$\omega_0 = \frac{1}{\sqrt{LC}} = 70.71 \times 10^3 \quad \text{rad/s}$$

Now

$$\zeta = \frac{R}{2L} = \frac{10^6}{80} = 12500 \quad \text{s}^{-1}$$

and

$$\omega_r = [\omega_0{}^2 - \zeta^2]^{1/2} = [70.71^2 - 12.5^2]^{1/2} \times 10^3$$

$$= 69600 \quad \text{rad/s}$$

$$t_{q\,min} = \frac{\pi}{2\pi \times 8000} - \frac{\pi}{69600} = 17.36 \times 10^{-6} \quad \text{s}$$

(b)
$$\psi = \tan^{-1} \frac{\zeta}{\omega_r} = \tan^{-1} \frac{12.5}{69.6} = 0.1777 \quad \text{rad}$$

$$\frac{\zeta \pi}{\omega_r} = 0.5642$$

$$V_{C2} = \frac{300\epsilon^{0.5642}}{\epsilon^{0.5642} - 1} = 695.8 \quad \text{V}$$

$$\frac{\zeta}{\omega_r}(\pi - 2\psi) = \frac{12.5}{69.6}(\pi - 2 \times 0.1777) = 0.5004$$

$$v_{AK\,\text{peak}} = 300 + 695.8\epsilon^{-0.5004} = 721.9 \quad \text{V}$$

(c)
$$t_a = \frac{1}{69600}\left(\frac{\pi}{2} - 0.1777\right) = 20.02 \times 10^{-6} \quad \text{s}$$

$$I_{Q\,\text{peak}} = \frac{695.8 \times 10^3}{70.71 \times 40}\epsilon^{-12.5 \times 20.02 \times 10^{-3}} = 191.5 \quad \text{A}$$

(d)
$$P_s = \frac{16\pi \times 10^3 \times 300 \times 695.8(1 + \epsilon^{-0.5642})}{2\pi \times 70.71^2 \times 40}$$

$$= 13100 \quad \text{W}$$

(e)
$$I_{QR} = \frac{1}{\sqrt{2}}\left(\frac{13100}{1}\right)^{1/2} = 80.93 \quad \text{A}$$

(f)
$$D_R = \frac{2 \times 721.9 \times 80.93}{13100} = 8.92$$

(g) The waveforms of i_O and v_C are shown in Fig. E7.10.

7.8 AC MOTOR DRIVES

The ability to construct a static alternating-voltage source of controllable frequency immediately presents the possibility of variable-speed drives employing synchronous motors or standard Design A, B, or C squirrel-

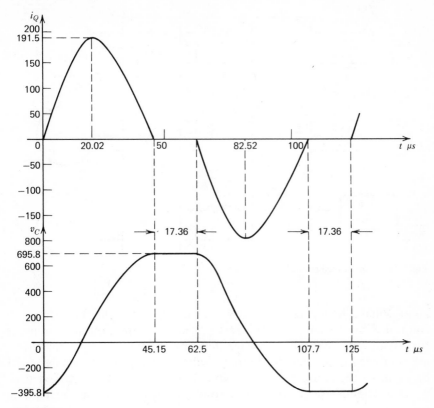

Fig. E7.10

cage induction motors. The only qualification that must be made to this statement is that not only the frequency, but also the voltage, of the source must be controllable. This is due to the fact that if the excitation frequency of an ac motor is reduced without a reduction of applied voltage, the magnetic circuit of the machine will saturate. This can be avoided if the ratio of voltage to frequency is maintained constant over the speed range to be employed. Only at low frequencies, when resistance as compared with reactance of the motor windings becomes appreciable, is it necessary to increase this ratio.

Three possible combinations of power-semiconductor converters that will give the desired variable-frequency variable-voltage source are illustrated in Fig. 7.34. The variable-frequency output in each case is provided by the inverter. In *a*, the inverter itself provides the required voltage control by methods such as are described in Section 7.4. In *b*, the voltage control is provided by the controlled rectifier, which varies the

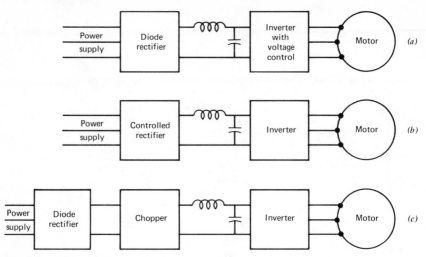

Fig. 7.34 Motor speed control by frequency variation.

input voltage to the inverter. In c, the input voltage to the inverter is controlled by the chopper. This last system may appear cumbersome at first sight, but it has the advantage that three relatively simple converters are combined to provide the desired result. In each case, smoothing of the dc input to the converter is necessary. It should also be remarked that, while a three-phase motor is employed in each system, a single-phase source will provide the necessary power input to the controlled or uncontrolled rectifier.

7.8.1 Synchronous Motor Drive An extremely accurate speed-control system may be obtained employing a synchronous motor excited by one of the converter combinations shown in Fig. 7.34. The angular speed of rotation of a synchronous motor is linked to the angular frequency of the stator supply by the relationship.

$$\omega_m = \omega_{\text{syn}} = \frac{2}{p} \omega_s \quad \text{rad/s} \tag{7.214}$$

Since the inverter frequency ω_s can be controlled by means of a crystal oscillator to an accuracy of up to 0.01% of any desired value, this degree of accuracy is exactly reflected in the accuracy of the speed control achieved.

A model commonly employed to represent a synchronous motor is the per-phase equivalent circuit shown in Fig. 7.35. In that circuit, V_s is the line-to-neutral voltage of the stator supply, I_s is the stator line current, and X_s is the per-phase "synchronous reactance" of the motor. Voltage V_g is

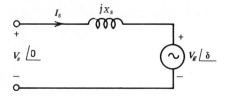

Fig. 7.35 Per-phase equivalent circuit of a synchronous motor.

the "excitation voltage" or per-phase electromotive force induced in the stator winding and is a function of the rotor or "field" current I_f of the motor. If stator losses and the effect of saliency of the rotor poles are neglected, as they have been in the evolution of the model of Fig. 7.35, then it may be shown that the air-gap torque developed by the motor is

$$T = -\frac{p}{2}\frac{3}{\omega_s}\frac{V_g V_s}{X_s}\sin\delta \quad \text{n-m} \tag{7.215}$$

where δ is the phase angle of $\overline{V}_g$ relative to $\overline{V}_s$. The useful torque at the motor coupling will be somewhat less than the air-gap torque of equation 7.215, due to friction and windage losses. For constant rotor excitation and fixed stator supply voltage and frequency

$$T = -T_{po}\sin\delta \quad \text{n-m} \tag{7.216}$$

This relationship between T and δ is illustrated in Fig. 7.36, where the maximum torque, occurring at $\delta = \pm\pi/2$ rad is $\pm T_{po}$, the pull-out torque at which the machine pulls out of synchronism. The excitation voltage V_g

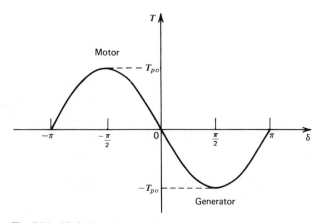

Fig. 7.36 Variation of torque of a synchronous motor.

is, as in a dc machine, directly proportional to the rotor speed, so that for constant field current I_f,

$$V_g \propto \omega_s \qquad (7.217)$$

Also

$$X_s \propto \omega_s \qquad (7.218)$$

Thus if the frequency of the stator supply ω_s is varied, and the stator terminal voltage V_s is varied also to maintain a constant ratio V_s/ω_s, then equation 7.215 shows that the pull-out torque will be constant provided that the rotor excitation is constant. The speed-torque characteristics for the range of frequency over which the model of Fig. 7.35 may be applied are as illustrated in Fig. 7.37. Operation in the third and fourth quadrants of the speed-torque diagram may be obtained by reversing the phase sequence of the stator supply, and this may be achieved by reversing the sequence of the gating signals applied to the inverter. This change must not be made abruptly, as is done when plugging an induction motor, since the synchronous machine would pull out of synchronism. Instead, the speed of the motor must be brought to zero before the phase sequence can be reversed.

As may be seen from Figs. 7.36 and 7.37, the load driven by the motor may be decelerated by regenerative braking. If such braking is to be employed, then the dc source to the inverter must be able to accept negative current. Of the three arrangements shown in Fig. 7.34, only b is capable of doing so, and the controlled rectifier shown in that diagram would need to be a dual converter.

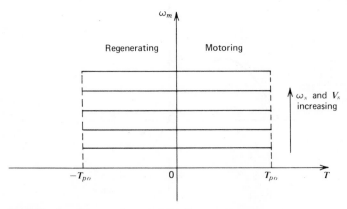

Fig. 7.37 Speed torque characteristics for a synchronous motor.

A block diagram of a possible synchronous motor drive is shown in Fig. 7.38. This is an open-loop system in which the current limit element consists of an overload protection device. For four quadrant operation, the inverter logic must be capable of reversing the inverter output phase sequence if the polarity of the command signal Ω_{REF} is reversed. The converter logic must be capable of detecting the need for regenerative braking and applying gating signals to the appropriate controlled bridge rectifier.

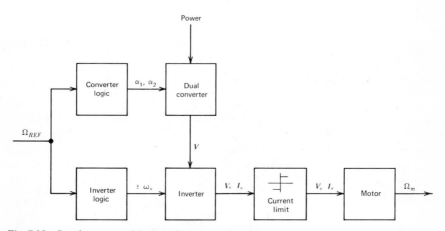

Fig. 7.38   Steady-state model of synchronous motor drive.

When a large number of low power drives are required to rotate in exact synchronism, synchronous reluctance motors are employed. These are machines with polyphase stators and unwound rotors carrying salient poles. They operate in essentially the same way as do the larger machines, but they unavoidably operate at a lagging and somewhat low power factor.

7.8.2 Induction Motor Drives The system of Fig. 7.38 may also be employed to drive an induction motor. The speed control in such a system is not as accurate as that given by the synchronous motor, since there is no feedback loop that corrects for the slip of the induction motor. The control is sufficiently accurate for many purposes, however.

For the analysis of induction motor operation with variable frequency, the equivalent circuit of Fig. 4.26 must be modified by replacing the fixed reactances with the frequency-dependent parameters shown in Fig. 7.39. As before, the synchronous speed of the motor at any stator frequency ω_s

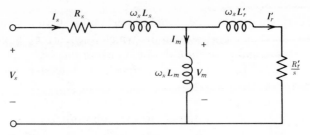

Fig. 7.39 Per-phase equivalent circuit of an induction motor.

may be defined as

$$\omega_{\text{syn}} = \frac{2}{p}\omega_s \quad \text{rad/s} \qquad (7.219)$$

where p is the number of stator poles. The per-unit slip may be defined as

$$s = \frac{\omega_{\text{syn}} - \omega_m}{\omega_{\text{syn}}} \qquad (7.220)$$

where ω_m is the speed of the motor.

Equation 4.51 gives an expression for the motor output torque if rotational losses are neglected, that is, for the internal torque developed by the motor. This is

$$T_O = \frac{3}{\omega_m}(1-s)\frac{R_r'}{s}(I_r')^2 \quad \text{n-m} \qquad (7.221)$$

In an integral horsepower induction motor, the stator leakage impedance is small, so that the equivalent circuit of Fig. 7.39 may be replaced by the approximate circuit of Fig. 7.40 with only negligible effect on the torque calculated from equation 7.221. From Fig. 7.40,

$$I_r' = \frac{V_s}{\left\{[R_s + (R_r'/s)]^2 + \omega_s^2(L_s + L_r')^2\right\}^{1/2}} \quad \text{A} \qquad (7.222)$$

Also, from equations 7.219 and 7.220,

$$\frac{\omega_m}{1-s} = \frac{2}{p}\omega_s \quad \text{rad/s} \qquad (7.223)$$

so that from equations 7.221 to 7.223,

$$T_O = \frac{3}{\omega_s}\frac{p}{2}\frac{R_r'}{s}\frac{V_s^2}{[R_s + (R_r'/s)]^2 + \omega_s^2(L_s + L_r')^2} \quad \text{n-m} \qquad (7.224)$$

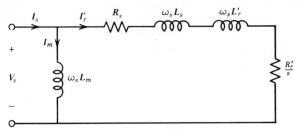

Fig. 7.40 Approximate equivalent circuit of an induction motor.

At any given value of ω_s, torque developed is seen from equation 7.224 to be a function of slip only. Maximum positive and negative values of torque and the slips at which they occur may therefore be determined by equating dT_O/ds to zero and solving for s. If this is done, the solutions are found to be

$$s = \pm \frac{R_r'}{\left[R_s^2 + \omega_s^2 (L_s + L_r')^2 \right]^{1/2}} \qquad (7.225)$$

Substitution of the positive value of s from equations 7.225 in equation 7.224 gives the maximum positive or motoring torque as

$$T_{m\,max} = \frac{3}{\omega_s} \frac{p}{4} \frac{V_s^2}{\left[R_s^2 + \omega_s^2 (L_s + L_r')^2 \right]^{1/2} + R_s} \quad \text{n-m} \qquad (7.226)$$

and this is the breakdown torque of the motor. Substitution of the negative value of s gives the maximum negative or regenerative braking torque as

$$T_{g\,max} = - \frac{3}{\omega_s} \frac{p}{4} \frac{V_s^2}{\left[R_s^2 + \omega_s^2 (L_s + L_r')^2 \right]^{1/2} - R_s} \quad \text{n-m} \qquad (7.227)$$

Under normal operating conditions

$$\omega_s (L_s + L_r') \gg R_s \quad \Omega \qquad (7.228)$$

so that

$$|T_{m\,max}| \cong |T_{g\,max}| \cong T_{O\,max} = \frac{3p}{4} \frac{V_s^2}{\omega_s^2 (L_s + L_r')} \quad \text{n-m} \qquad (7.229)$$

Equation 7.229 shows that, provided that the ratio V_s/ω_s is held constant, the breakdown torque developed by the motor will not change appreciably with speed.

The standstill or starting torque developed by the motor occurs when $s = 1$, so that from equation 7.224 at any given value of ω_s

$$T_{START} = \frac{3}{\omega_s} \frac{p}{2} \frac{R_r' V_s^2}{(R_s + R_r')^2 + \omega_s^2 (L_s + L_r')^2} \quad \text{n-m} \qquad (7.230)$$

When ω_s is large,

$$\omega_s (L_s + L_r') \gg (R_s + R_r') \quad \Omega \qquad (7.231)$$

so that from equation 7.230,

$$T_{START} \cong \frac{3p}{2} \frac{R_r' V_s^2}{\omega_s^3 (L_s + L_r')^2} \quad \text{n-m} \qquad (7.232)$$

Thus over the upper range of ω_s in which V_s/ω_s is held constant, the standstill torque will decrease with increase in stator frequency. When ω_s is small,

$$\omega_s (L_s + L_r') \ll (R_s + R_r') \quad \Omega \qquad (7.233)$$

and

$$T_{START} \cong \frac{3p}{2} \frac{R_r' V_s^2}{\omega_s (R_s + R_r')^2} \quad \text{n-m} \qquad (7.234)$$

Thus over the lower range of ω_s in which V_s/ω_s is held constant, the standstill torque will increase with stator frequency. It follows that somewhere in the total speed range for which V_s/ω_s is held constant, a maximum value of standstill torque will occur, and this will determine the lowest value to which ω_s should be reduced.

Figure 7.41 shows a family of induction motor speed-torque characteristics for a wide range of stator frequencies. The speed ω_{BASE} is the synchronous speed of the motor when the maximum voltage available from the inverter of Fig. 7.38 is applied to the motor terminals. This would normally be the rated voltage of the motor and would be applied at rated frequency, thus establishing the ratio V_s/ω_s. Below speed ω_{BASE}, the characteristics represent operating conditions for which the ratio V_s/ω_s is held constant. The characteristics with synchronous speed greater than ω_{BASE} represent operating conditions in which V_s is fixed while ω_s in-

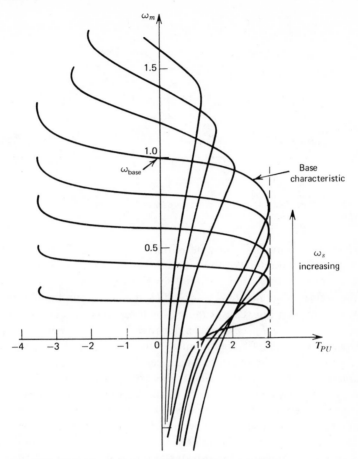

Fig. 7.41 Induction motor—variable frequency operation.

creases. Thus from equation 7.229,

$$T_{O\,\mathrm{max}} \propto \frac{1}{\omega_s^2} \quad \text{n-m}: \qquad \omega_{\mathrm{syn}} > \omega_{\mathrm{BASE}} \quad \text{rad/s} \qquad (7.235)$$

These curves show that at speeds greater than ω_{BASE}, the load torque must be reduced if there is not to be a danger of exceeding the breakdown torque of the motor.

The induction motor in a variable-frequency speed control system may be required to operate in any one of the three quadrants of the speed-torque characteristics shown in Fig. 7.41. Operation in the first two quadrants may readily be visualized from Fig. 7.42. The speed-torque

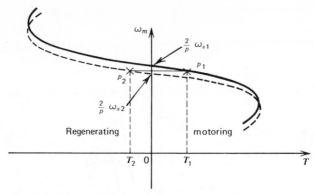

Fig. 7.42 Induction motor—regenerative braking.

characteristic shown in full line is that corresponding to stator frequency ω_{s1}. On this characteristic the motor is operating at point p_1, of which the torque coordinate T_1 represents a positive or motoring torque. Reduction of the stator frequency to ω_{s2}, for which the corresponding characteristic is shown in broken line, causes the motor to operate at point p_2, of which the torque coordinate T_2 represents a negative braking or regenerating torque.

Operation in the first and fourth quadrants is illustrated in Fig. 7.43, where reversal of the phase sequence results in rotation of the speed-torque characteristic through 180° about the origin. If the motor is operating at point p_1 on the positive-sequence characteristic, and the phase sequence is abruptly reversed giving operation at point p_2 on the negative-sequence characteristic, then the positive motoring torque T_1 is replaced by the negative braking torque T_2. Such a transition as this could only take place at a very low value of ω_s, that is, when the forward speed of the motor had already been brought to the lower limit of the controlled-speed range.

The angular frequency of the currents induced in the rotor circuits of an induction motor is

$$\omega_r = s\omega_s \quad \text{rad/s} \tag{7.236}$$

so that from equation 7.223,

$$\frac{\omega_m}{1-s} = \frac{2}{p}\frac{\omega_r}{s} \quad \text{rad/s} \tag{7.237}$$

Substitution from equation 7.237 in equation 7.221 then yields

$$T_O = \frac{3p}{2}\frac{R_r'(I_r')^2}{\omega_r} \quad \text{n-m} \tag{7.238}$$

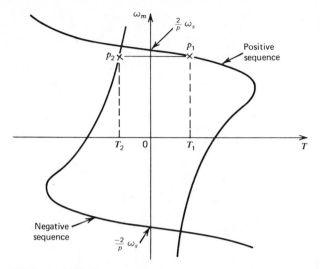

Fig. 7.43 Induction motor—plugging.

This expression for torque shows that if ω_r is held constant, then the output torque per rotor ampere is also constant and will be large when ω_r is small. It is thus advantageous to devise a control system by means of which ω_r may be held constant at a low value. Moreover, from equations 7.219, 7.220, and 7.236,

$$\omega_s = \frac{p}{2}\omega_m + \omega_r \quad \text{rad/s} \tag{7.239}$$

Thus if in a control system both ω_s and ω_r are generated, then an exact value of ω_m may be obtained. Such a system is illustrated in the block diagram of Fig. 7.44, where the reference signals are motor speed Ω_m and rotor frequency Ω_r.

When, in the system of Fig. 7.44, a positive error voltage $V_{REF} - V_T$ is applied to the speed control unit, indicating a demand for an increase of speed, the reference current I_{REF} is increased, and consequently the motor terminal voltage is raised so that the resulting increase of stator current and torque accelerates the motor. The positive slip occurring during this condition of operation results in a $+1$ output of the polarity sensor and the satisfaction of equation 7.239 at the input to the inverter logic unit. If a reduction in motor speed were called for by a reduction of Ω_m, then s would be negative due to the reduction in ω_s, and so from equation 7.236 would be ω_r. The negative error signal $V_{REF} - V_T$ would then result in a -1 output of the polarity sensor, again satisfying equation 7.239, where ω_r

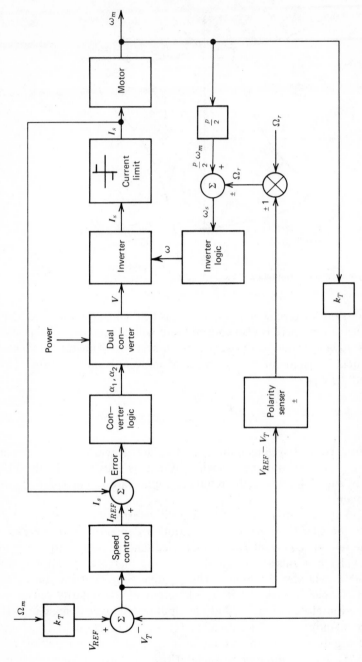

Fig. 7.44 Induction-motor drive with rotor slip frequency control.

has become negative. If a reverse drive were required, then the phase sequence of ω_s could be changed in the inverter logic. This change could be made to depend on the polarity of V_{REF}.

PROBLEMS

7.1   In the inverter of Fig. 7.1*a* $V = 500$ V and $T = 2000$ μs. The load is an RLC series circuit for which $R = 1.2$ Ω, $\omega L = 10$ Ω, $1/\omega C = 10$ Ω. The gating signals are as shown in Fig. 7.1*b*.

(a)   Sketch to scale the waveform of v_O, i_O, i_{Q1}, i_{D1} and v_{AK1}. Higher harmonics than the fundamental component may be neglected.
(b)   Calculate the rms and average thyristor and diode currents.
(c)   If the turn-off time of the thyristors is 50 μs, state whether forced commutation will be required and determine the thyristor current at the instant of commutation.

7.2   Repeat problem 7.1 if, in the load circuit, $R = 1.2$ Ω, $\omega L = 7.92$ Ω, $1/\omega C = 10$ Ω.

7.3   Repeat problem 7.1 if, in the load circuit, $R = 1.2$ Ω, $\omega L = 10$ Ω, $1/\omega C = 7.92$ Ω.

7.4   If the load circuit for the inverter of Fig. 7.1*a* is that shown in Fig. P7.1, $T = 2500$ μs and t_{off} for the thyristors is 50 μs, calculate the range of R over which load commutation would take place. Higher harmonics than the fundamental component may be neglected.

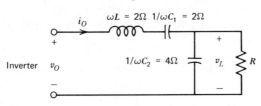

Fig. P7.1

7.5   The inverter of Fig. 7.5*a* is operating on open circuit.

(a)   Draw the circuits applying the commutation of thyristor Q_1.
(b)   Show from first principles that the following expressions are valid:

$$\omega_r t_1 = \frac{\pi}{2} + \sin^{-1} n \quad \text{rad} \tag{i}$$

$$t_q = \frac{\pi}{3} \sqrt{2 L_C C} \quad \text{s} \tag{ii}$$

$$I_{Q2} = \frac{V}{\omega_r L_C} \sin \omega_r t_1 \quad \text{A} \tag{iii}$$

7.6 The inverter of Fig. 7.5a is operating on open circuit. At $t = 0$, thyristor Q_2 is turned on, commutating Q_1. If at this instant v_{C1} is zero, sketch to scale the waveforms during the commutation interval of i_{Q1}, i_{Q2}, v_{AK1}, v_{AK2}, v_{L1}, and v_{L2}.

7.7 For the inverter of Fig. 7.5a, $V = 220$ V, the current at the instant of commutation I_{O1} is 150 A, and the minimum acceptable time to be available for commutation t_q is 20 μs. If the dimensionless design factor x is chosen at 1.5:

(a) Calculate the required values of L_C and C.

(b) With the values of L_C and C determined in a, calculate the time available for turn-off t_q if I_{O1} is (i) zero and (ii) 300 A.

7.8 For the inverter of Fig. 7.5a, $V = 220$ V, $L_C = 50$ μH, $C = 75$ μF, and the minimum acceptable time to be available for commutation t_q is 15 μs. Determine the maximum current I_{O1} that can be commutated.

7.9 For the inverter of Fig. 7.11, $V = 220$ V, $L_1 = 8$ μH, and $C = 20$ μF.

(a) If the current at the instant of commutation I_{O1} is 150 A, calculate the time available for turn off t_q and the commutation interval.

(b) Explain the importance of the commutation interval in this circuit.

7.10 For the half-bridge inverter of Fig. 7.1a:

(a) Draw a diagram including the commutation circuit elements and employing the "Type 2" current commutation illustrated in Fig. 6.16.

(b) Without performing a complete analysis, give a verbal explanation of the operation of the circuit.

7.11 For the bridge inverter of Fig. 7.2 repeat the calculations of Example 7.5 if the load-circuit parameters are $R = 0.5$ Ω, $\omega L = 10$ Ω, and $1/\omega C = 10$ Ω.

7.12 For the bridge inverter of Fig. 7.2:

(a) Draw a diagram including the commutation circuit elements and employing the "Type 1" current commutation illustrated in Fig. 6.15.

(b) Without performing a complete analysis, give a verbal explanation of the operation of the circuit.

7.13 Repeat Problem 7.12 employing "Type 2" current commutation illustrated in Fig. 6.16.

7.14 Repeat Problem 7.4 for the bridge inverter of Fig. 7.2.

7.15 The bridge inverter of Fig. 7.2 is supplying a series RL load circuit. By writing and solving the appropriate differential equations show that the current at the instant of commutation under steady-state operating conditions is

$$I_{O1} = \frac{V}{R}\left[\frac{1-\epsilon^{-T/2\tau}}{1+\epsilon^{-T/2\tau}}\right] \quad A$$

where $\tau = L/R$, and T is the period of the output voltage v_O.

7.16   The output voltage of the bridge inverter of Fig. 7.2 is controlled by single-pulse modulation:

(a)   Calculate the rms values of the fundamental, fifth, and seventh harmonic components of the output voltage for a pulse width of 90°.

(b)   Sketch the required gating waveforms if the output voltage v_O is to be defined under all load conditions for a pulse-width of 90°.

7.17   The output voltage of the bridge inverter of Fig. 7.2 is controlled by single-pulse modulation. If the load circuit is purely resistive, and the pulse-width is 90°, sketch to scale the waveforms of v_O, i_O, i_{Q1}, i_{Q2}, i_{D1}, i_{D2}, and i_s.

7.18   Repeat Problem 7.17 for a purely inductive load.

7.19   For the bridge inverter of Fig. 7.2, $V = 300$ V and the load is an RLC series circuit in which $R = 1.2\ \Omega$, $\omega L = 1.2\ \Omega$, and $1/\omega C = 0.8\ \Omega$. The output voltage is controlled by single pulse modulation and the pulse-width is 90°. Determine the current I_{O1} at the instant of commutation. Harmonics higher than the seventh may be neglected.

7.20   A load circuit for the bridge inverter of Fig. 7.2 is shown in Fig. P7.2. Voltage control by single pulse modulation is employed. Determine the harmonic distortion factor K_v for voltage v_L appearing across the 2-Ω resistance when the pulse width is (a) 90°; (b) 180°. Harmonics higher than the seventh may be neglected.

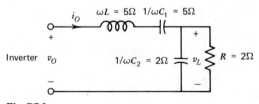

Fig. P7.2

7.21   The rms value of the fundamental component of the output in the bridge inverter of Fig. 7.2 must be 0.45 V, when V is the source voltage. Calculate the pulse-width required, sketch to scale the waveform of v_O, and determine the harmonic distortion factor K_v for voltage v_O if:

(a) Voltage control by single-pulse modulation is employed.
(b) Voltage control by multiple-pulse modulation is employed in which there are 10 pulses per half cycle of v_O.

Harmonics higher than the seventh may be neglected.

7.22 The transformer connection shown in Fig. 7.22 is employed to combine the output voltages v_{O1} and v_{O2} of two bridge inverters. These voltages have a relative phase displacement of $60°$ and the voltage control of the inverters is by means of single-pulse modulation. The pulse-width is $90°$.

(a) Sketch to scale the waveforms of v_{O1}, v_{O2}, and v_O.
(b) State whether v_O contains any third harmonic and explain your reasoning.

7.23 Using a number of single-phase bridge inverters, operating without pulse-width control, and transformers of unity turns ratio, show how an output voltage waveform can be generated that contains no third, fifth, or seventh harmonics.

7.24 The three-phase inverter of Fig. 7.3 is supplying a purely inductive star-connected load circuit. The star-point of the load is terminal N. Three-thyristor gating is employed, as illustrated in Fig. 7.3b. Sketch to scale the waveforms of v_{AB}, v_{AN}, i_A, i_B, i_C, i_{Q1}, i_{Q4}, i_{D1}, i_{D4}, and i_s, and show that the average power from the direct voltage source V is zero.

7.25 Repeat Problem 7.24 but with two-thyristor gating, as illustrated in Fig. 7.27a.

7.26 The three-phase inverter of Fig. 7.3 with three-thyristor gating is supplying a star-connected three-phase load circuit. Each branch of the load circuit comprises a resistance $R = 1.2\ \Omega$ in series with an inductance for which $\omega L = 2.4\ \Omega$. The direct source voltage is 300 V.

(a) Determine the Fourier series describing v_{AB}.
(b) From the series obtained in a, derive that describing v_{AN}, where N is the star point of the load circuit.
(c) Determine the current at the instant of commutation neglecting the effect of harmonics higher than the fifth.

7.27 In the three-phase inverter of Fig. 7.3, voltage control by pulse modulation and three-thyristor gating are employed. Control is by means of one $45°$ pulse for every $60°$ interval.
Sketch to scale the gating waveforms for the six thyristors such that the output voltages are defined under all load conditions.

7.28 For the series inverter of Fig. 7.31, $V = 300$ V, $R = 1\ \Omega$, the operating frequency is 10 kHz, and $t_{q\,min} = 10\ \mu s$.

(a) Calculate the values of L and C required if the peak thyristor voltage is to be 350 V.

(b) For the values of L and C obtained in a, determine the output power and the derating factor.

7.29 For the series inverter of Fig. 7.33a derive expressions for:

(a) The minimum time available for turn-off $t_{q\,min}$.
(b) The steady-state variation of i_O.
(c) The maximum value of thyristor voltage v_{AK1}.

7.30 Repeat Problem 7.29 for the series inverter of Fig. 7.33b.

EIGHT

AC-TO-AC CONVERTERS

The inverters discussed in Chapter 7 are able to provide ac-to-ac conversion by means of an intermediate dc link, as illustrated in Fig. 7.34. This chapter discusses direct ac-to-ac converters that convert energy from an m-phase ac source at a given frequency to an n-phase ac load circuit at some other desired frequency. These converters in general have the following features:

1. Due to the elimination of one or more converters needed to provide an intermediate dc link, these converters are more efficient than the systems of Fig. 7.34.
2. In most cases, output voltage control is inherent in the converter.
3. Line or load commutation is normally employed. Forced commutation is possible but complicated.
4. Input power factor correction and harmonic reduction are usually necessary.

The simplest form of ac-to-ac converter, described in Section 8.1, is the dual converter discussed in Section 5.5 and illustrated with a dc load circuit in Fig. 5.31. In the simplified diagram of a dual converter shown in Fig. 8.1a it may be seen that, if the voltage and current in the load branch are varied cyclically, an alternating output is obtained in which the two controlled rectifiers function during parts of each cycle either as rectifiers or as dc to fixed-frequency ac inverters. When the dual converter is employed in this way, it is called a cycloconverter.

If the controlled rectifiers in Fig. 8.1a were single-phase rectifiers of the types shown in Figs. 5.2 and 5.3, then the system would constitute a

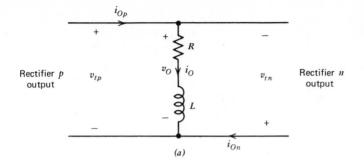

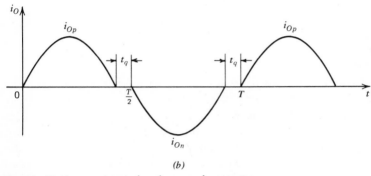

(b)

Fig. 8.1   Dual converter employed as a cycloconverter.

single-phase to single-phase cycloconverter. On the other hand, if the controlled rectifiers were three-phase bridges of the type shown in Fig. 5.6, then a three-phase to single-phase cycloconverter would be formed. Three such converters could be employed with their single-phase outputs coupled by a three-phase transformer to provide a three-phase to three-phase cycloconverter.

A somewhat idealized output-current waveform is shown in Fig. 8.1b. Line commutation is employed, and the zero-current intervals of duration t_q become necessary under certain conditions of operation to ensure that the sources of the two rectifiers are not short-circuited.

Single-phase to single-phase cycloconverters are not widely used for reasons that will become apparent when they are discussed in more detail. However, they form a useful example by means of which basic operating principles may be established, and these principles may then be applied to more complicated systems.

The operating characteristics of all cycloconverters are such that it is necessary to restrict the maximum output frequency to a value that is only a fraction of the source frequency. They are particularly suitable for high-power (up to 5000 hp) variable-speed ac motor drives in which low speed is required.

Section 8.2 considers converters that provide an output frequency that is an integral multiple of the source frequency. The multiplication factor is related to the number of source phases. These employ line commutation. The circuit of a frequency tripler with a resistive load branch is shown in Fig. 8.2a. It may be readily understood by considering it as a combination of three single-phase voltage controllers supplying a common load circuit and excited by the line-to-neutral voltages of a three-phase source. The angle of retard for this load may be varied over the range

$$\frac{2\pi}{3} + \omega_s t_q \leqslant \alpha \leqslant \pi \quad \text{rad} \tag{8.1}$$

where ω_s is the source frequency, and t_q is the time available for turn-off of the thyristors.

In Fig. 8.2b are shown the waveforms of the variables of the frequency tripler circuit with the angle of retard reduced to the lower limit, resulting in maximum output. The waveforms of the output voltage and current show that the output frequency is three times that of the source. Only one thyristor conducts at any instant, and the zero-current intervals of duration t_q prohibit short-circuit of two phases of the source through two thyristors.

The tripler of Fig. 8.2 is line-commutated; however, if an oscillatory series RLC load circuit is provided, load commutation may be employed

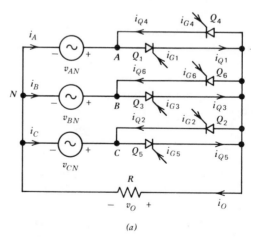

(a)

Fig. 8.2   Frequency tripler with line commutation.

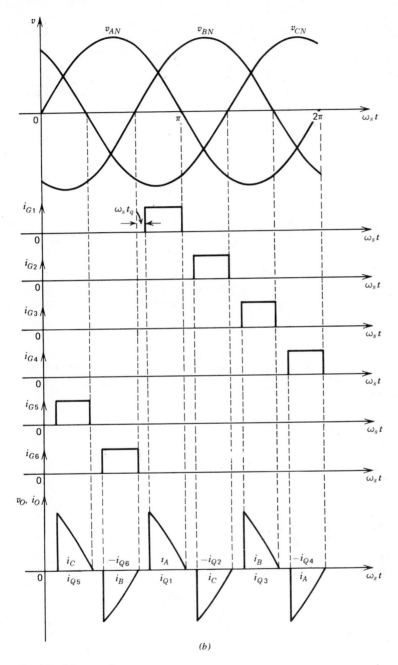

(b)

Fig. 8.2 (*Continued*)

and the converter can provide an output frequency that may be a nonintegral multiple of the source frequency. This load-commutated multiplier—sometimes called a cycloinverter—is shown in Fig. 8.3a, where two equal inductances, one for the positive and one for the negative half cycle of the output current, are introduced in series with the RC load circuit. The positioning of the inductors in the circuit limits the value of dv/dt applied to the thyristors, and this circuit may be compared in this regard with that of the series inverter of Fig. 7.31.

Figure 8.3b shows the waveforms of the load-commutated multiplier. The ringing frequency of the circuit must be such that the zero-current interval $t_q > t_{off}$ is present between successive half cycles of the current wave.

In the circuit of Fig. 8.3a, the three sources would normally be the emf's in the secondary windings of a three-phase transformer. Owing to the connection of the load-circuit branch to the star point of these windings, they would carry the output current. This may be avoided by employing the circuit shown in Fig. 8.4, where the capacitance of the ringing circuit is provided by three capacitors connected to the three-phase input lines. The output current waveform of this converter is identical with that of Fig. 8.3b.

This type of converter finds wide application in induction heating (180- to 3000-Hz output from a 60-Hz source) where the resistance shown in the load circuit of Fig. 8.4 would represent a parallel-compensated induction heating load.

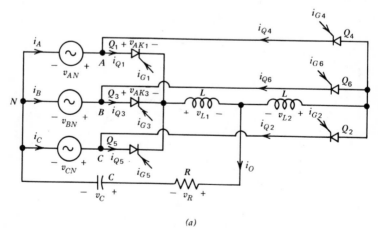

(a)

Fig. 8.3 Frequency multiplier with load commutation.

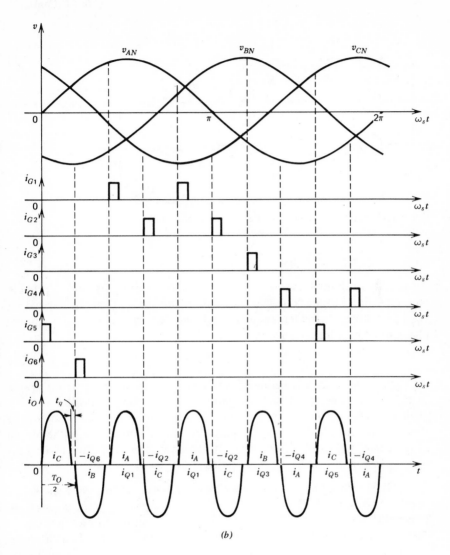

(b)

Fig. 8.3 (*Continued*)

467

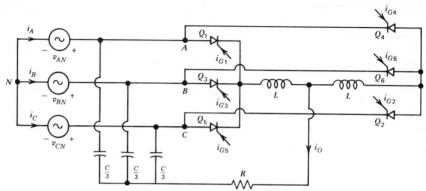

Fig. 8.4 Modified frequency multiplier with load commutation.

8.1 CYCLOCONVERTERS

As already explained, cycloconverters employ line commutation to provide an alternating output voltage of angular frequency ω_O from an alternating voltage source of greater angular frequency ω_s. In the case of a single-phase to single-phase cycloconverter,

$$\omega_s \geqslant 9\omega_O \quad \text{rad/s} \tag{8.2}$$

As in the case of the dual converter, for the basic cycloconverter illustrated in Fig. 8.1 it is essential that the gating signals to one controlled rectifier shall be blanked out when the other is supplying current. When this is done, each controlled rectifier presents an infinite impedance to the output of the other. If this blanking of signals were not arranged, the currents i_{Op} and i_{On} in Fig. 8.1 could be unbounded.

8.1.1 Single-Phase to Single-Phase Cycloconverter   Figure 8.5*a* shows the main power circuit of a single-phase to single-phase cycloconverter. In effect this is a dual converter in which the two controlled rectifiers are of the type shown in Fig. 5.3 and employ a common center-tapped transformer as their source. The transformer may be replaced in the circuit by the two voltage sources of Fig. 8.5*b*, and this may be rearranged to give the equivalent circuit of Fig. 8.5*c*. This is in the form of the circuit of Fig. 5.8 and permits the interpretation of the modes of operation of the cycloconverter in terms of the controlled rectifier diagram of Fig. 5.10. This shows that the cycloconverter can operate with continuous or discontinuous output current.

Operation with Discontinuous Output Current   In the ac load circuit of Fig. 8.5*c*, there is no source of emf, so that in Fig. 5.10, $m = 0$, and discon

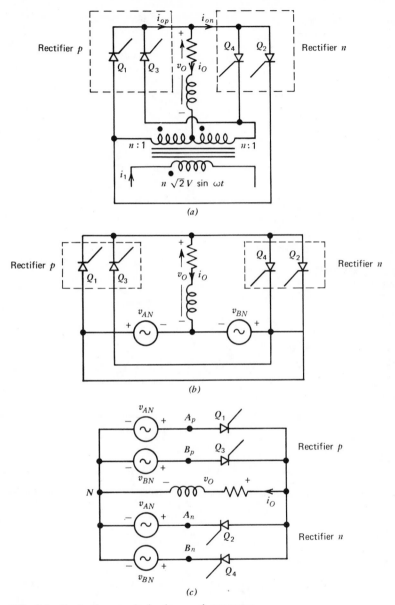

Fig. 8.5 Single-phase to single-phase cycloconverter.

tinuous-current operation will ensue if the initial current is zero, and

$$\phi \leqslant \alpha < \pi \quad \text{rad} \tag{8.3}$$

where

$$\tan \phi = \frac{\omega_s L}{R} \quad \text{rad} \tag{8.4}$$

Figure 8.6 shows the waveforms of the variables for the operation of the cycloconverter in which inequality 8.3 is satisfied, but the output frequency is greatly in excess of the maximum desirable value. In addition to the gating signals i_{G1} to i_{G4}, Fig. 8.6 shows two blanking signals P and N. Only if signal P is applied do the gating signals to rectifier p appear. Conversely, only if signal N is applied do the gating signals to rectifier n appear. Omitted gating signals are shown shaded in Fig. 8.6. The output frequency of the cycloconverter is thus established by the frequency of the blanking signals of periodic time T_O and duration $T_O/2$ s.

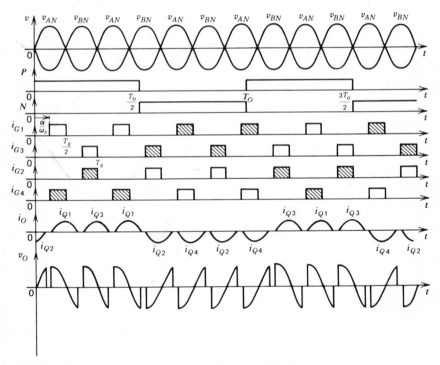

Fig. 8.6 Discontinuous-current operation of circuit of Fig. 8.5.

The waveforms of i_O and v_O shown in Fig. 8.6 are very far from sinusoidal; however, it can be seen that the periodic time of the fundamental component of each will be T_O. It will also be noted—most easily in the case of the current wave—that the two "half waves" of the current and voltage are not identical. In the negative half wave there are four pulses of current, whereas in each of the positive half waves illustrated, there are only three. Clearly a waveform such as this has a negative direct component of current. However, if the time axis in Fig. 8.6 were sufficiently extended, it would be found that after the completion of a few cycles of the output variables, this situation would be reversed, with four positive pulses of output current alternating with only three negative pulses. Such a waveform would have a positive direct component of current. This "alternating direct component" in fact forms a subharmonic component of the output variables, and it is to restrict the amplitude of this subharmonic that the constraint of inequality 8.2 is imposed.

Operation with discontinuous output current in RL load circuits necessarily implies operation at low amplitudes of fundamental components of current and voltage; that is, the converter is operating at low output power.

Operation with Continuous Output Current   Figure 8.7 shows the waveforms for the operation of the cycloconverter with continuous output current. From Fig. 5.10 it may be seen that this condition arises when the initial current is zero, and

$$0 \leqslant \alpha < \phi \quad \text{rad} \tag{8.5}$$

Continuous output current implies operation at high amplitudes of fundamental components of current and voltage waveforms; that is, the converter is operating at high output power. The maximum power output is achieved when $\alpha = 0$.

From the point of view of output harmonic content, continuous-current output may appear from Figs. 8.6 and 8.7 to be something of an improvement over discontinuous current. This is illusory, however, since the maximum amplitude of the current waveform in Fig. 8.7 will be much greater than that in Fig. 8.6. It might also be thought that the replacement of the two single-phase rectifiers forming the dual converter by three-phase rectifiers would improve this state of affairs. This change, however, would simply tend to make the voltage and current waveforms more rectangular, so that the harmonic content would still be very great. The solution to this problem lies in varying the delay angle α throughout each half cycle of the output waveforms, and the manner in which this is done is the subject of the following section.

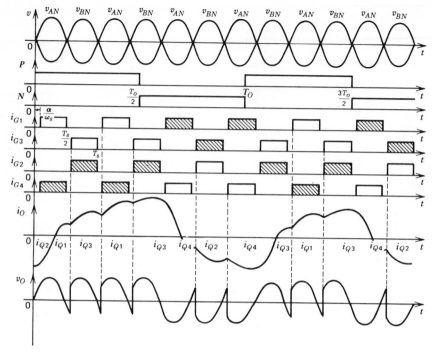

Fig. 8.7   Continuous current operation of circuit of Fig. 8.5.

A possibility that is not illustrated in Fig. 8.7, but that exists under conditions of continuous-current operation, is that of short-circuit of the source through two thyristors. In Fig. 8.7, whenever the output current i_O passes through zero it is transferred from one thyristor to another, but the pairs of thyristors involved are always Q_1 and Q_2 or Q_3 and Q_4. Inspection of Fig. 8.5 shows that Q_1 may safely be turned on while Q_2 is conducting, and conversely, since both of these thyristors are in series with the same voltage source v_{AN}. A similar statement may be made concerning thyristors Q_3 and Q_4.

Immunity from danger of short-circuit of the source does not exist with all load circuits, however, and the situation illustrated in Fig. 8.8 might well arise. This shows that the output current oscillation causes thyristor Q_3 to continue to conduct until the pulse of i_{G2} is due to commence. If Q_2 were in fact turned on, then thyristors Q_2 and Q_3 in series would short circuit the source. To guard against this possibility, protection equipment is fitted, and rules are embodied in the design of the control circuits that may

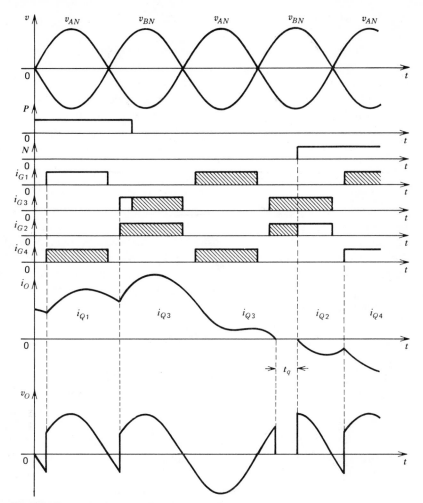

Fig. 8.8 Precautions against short-circuit of the source.

be formulated as follows:

1. A gating pulse is applied to rectifier p if, and only if, signal P is applied to the gating circuits.

2. Signal P is applied to the gating circuits after the current from rectifier n has fallen to zero.

3. The interval between the instant at which the current from rectifier n becomes zero and the instant of application of signal P must exceed the turn-off time of the thyristors in rectifier n.

A converse set of rules governs the application of the gating signals to rectifier n.

The effect of the gating rules on the waveforms of the cycloconverter variables is shown in Fig. 8.8. An interval of zero current of duration t_q now appears between the two half waves of the output current i_O. It may be noted that the pulse of i_{G3} which turns on thyristor Q_3 at the end of the positive half-wave of current is cut off short by the removal of signal P without affecting the conduction of Q_3. Also, at the end of the interval t_q, the pulse of i_{G2} begins later than it would normally have done, since it was delayed by the absence of signal N.

In Figs. 8.6 and 8.8, whenever the output current i_O is zero, the output voltage v_O is also shown to be zero. This represents the true state of affairs if the cycloconverter is supplying a passive load circuit. However if the converter were employed to drive a synchronous motor, then an alternating emf of frequency ω_O would exist in the load circuit branch. Whenever the motor was open-circuited therefore by cessation of the converter output current, the emf induced in the stator windings by the rotor field (i.e., the excitation voltage) would appear at the output terminals of the cycloconverter.

Reduction of Output Harmonics   In cycloconverters operating with continuously variable output frequency, it is often important to reduce the harmonics in the output current at low frequency (e.g., to reduce torque pulsations in ac drives). If the cycloconverter were able to provide a purely sinusoidal output voltage waveform, then a purely sinusoidal current would flow in any linear load circuit. However, the output voltage waveform is necessarily made up of segments of the source voltage waveforms. Nevertheless it is possible to arrange that these segments shall be such as to produce the minimum current harmonics, so giving a close approximation to a sinusoidal output current. The following discussion of how this is achieved deals with the continuous-current operating condition, since this is the important and common situation.

Since it is desired that the waveform of the actual output voltage shall be such as to produce as nearly as possible the effect of a sinusoidal output voltage, the delay angle at which each thyristor in the rectifiers is turned on is controlled with reference to an ideal output voltage wave that may be defined as

$$v_O^* = \sqrt{2}\, V_O^* \sin \omega_O t \quad \text{V} \tag{8.6}$$

The source voltages in Fig. 8.5 may be defined as

$$v_{AN} = \sqrt{2}\, V \sin \omega_s t \quad \text{V} \tag{8.7}$$

$$v_{BN} = -\sqrt{2}\, V \sin \omega_s t \quad \text{V} \tag{8.8}$$

In Fig. 8.9 is shown one cycle of v_O^*. A second curve is also shown in broken line representing the ideal output current i_O^* produced by v_O^* in a load circuit with a leading power factor. For the first part of the cycle both v_O^* and i_O^* are positive, and rectifier p is operating as a rectifier.

For any instant during the cycle of v_O^* there is a corresponding delay angle α which, if applied continuously to the thyristors of the rectifier delivering current, would give an average output voltage equal to the value of v_O^* at that instant. Thus from equation 5.11 the necessary angle α is given by

$$v_O^* = \frac{2\sqrt{2}\,V}{\pi}\cos\alpha \quad \text{V} \tag{8.9}$$

where

$$\alpha = \omega_s t - 2n\pi \quad \text{rad} \tag{8.10}$$

In equation 8.10, n is an integer giving the number of complete cycles of source voltage that have elapsed since $t = 0$. To achieve an output voltage that will have the desired effect, it is therefore necessary to arrange that during the first cycle of source voltage thyristor Q_1 in Fig. 8.5 is turned on at instant t_1, when

$$v_{O1}^* = \frac{2\sqrt{2}\,V}{\pi}\cos\omega_s t_1 \quad \text{V} \tag{8.11}$$

A curve of the function $(2\sqrt{2}\,V/\pi)\cos\omega_s t$ is shown in Fig. 8.9. The first intersection of this curve with that of v_O^* defines the instant t_1 which satisfies equation 8.11 and determines the value of α_1, the delay angle at which gating signal i_{G1} commences. This gating signal is shown in Fig. 8.9. Thyristor Q_1 is turned on at $\omega_s t_1 = \alpha_1$, so that voltage v_{AN} appears across the load circuit, and the succeeding segment of the wave of output voltage v_O is described by equation 8.7. This output voltage wave also is shown in Fig. 8.9. Necessarily

$$0 < \omega_s t_1 < \pi \quad \text{rad} \tag{8.12}$$

At instant t_3, thyristor Q_3 must be turned on, and the desired output voltage is

$$v_{O3}^* = \frac{2\sqrt{2}\,V}{\pi}\cos\alpha_3 \quad \text{V} \tag{8.13}$$

If α_3 is the delay angle of thyristor Q_3 measured from the instant of commencement of the positive half cycle of voltage v_{BN}, then

$$\omega_s t_3 = \alpha_3 + \pi \quad \text{rad} \tag{8.14}$$

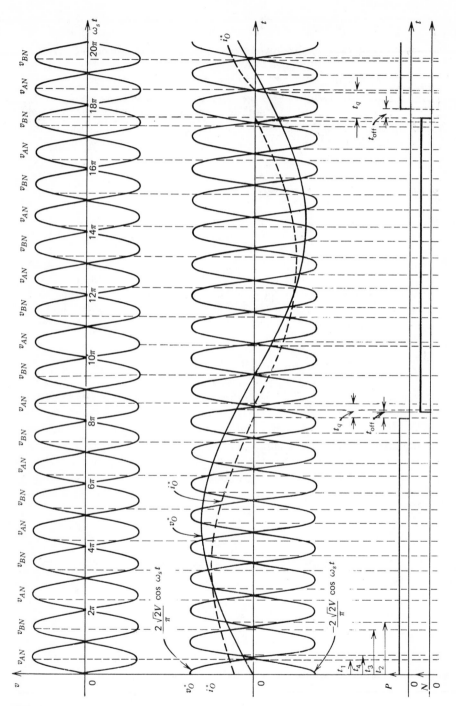

476

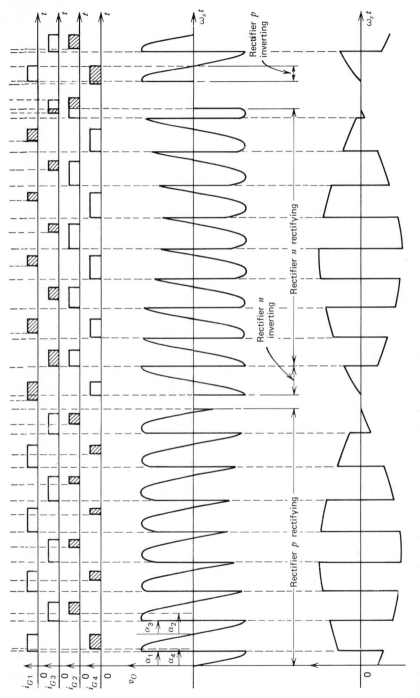

Fig. 8.9 Reduction of output harmonics—determination of α. ($\omega_s = 9.5\omega_O$).

and

$$v_{O3}^* = \frac{2\sqrt{2}\,V}{\pi} \cos(\omega_s t_3 - \pi) = -\frac{2\sqrt{2}\,V}{\pi} \cos \omega_s t_3 \quad \text{V} \qquad (8.15)$$

A curve of $-(2\sqrt{2}\,V/\pi)\cos\omega_s t$ is also shown in Fig. 8.9. From equation 8.14, necessarily

$$\pi < \omega_s t_3 < 2\pi \quad \text{rad} \qquad (8.16)$$

Thus the first intersection of the curve v_O^* with that of $-(2\sqrt{2}\,V/\pi)\cos\omega_s t$ within the range specified in inequality 8.16 defines the instant t_3 that satisfies equation 8.15 and determines the value of α_3, the delay angle at which gating signal i_{G3} commences. This gating signal also is shown in Fig. 8.9. Thyristor Q_3 is turned on at $\omega_s t_3 = \alpha_3$, so that voltage v_{BN} appears across the load circuit, and the succeeding segment of the wave of output voltage v_O is described by equation 8.8. The succeeding values of α_1 and α_3 may be determined from the corresponding intersections during each cycle of the source voltages.

During the first part of the cycle of v_O^* in Fig. 8.9, while $i_O^* > 0$ and thyristors Q_1 and Q_3 are being alternately turned on, the gating signals for thyristors Q_4 and Q_2 are also being generated, but do not reach the thyristors, owing to the absence of signal N. These gating signals must commence at delay angles α_4 and α_2 such that, if they were indeed applied to rectifier n, they would yield an average output voltage equal to that of v_O^* at the instant of their application. This would constitute a negative voltage for rectifier n, and would signify inverter operation. During the first cycle of the source voltages the gating signal for thyristor Q_4 will commence when

$$v_{O4}^* = -\frac{2\sqrt{2}\,V}{\pi} \cos \omega_s t_4 \quad \text{V} \qquad (8.17)$$

at an intersection for which

$$0 < \omega_s t_4 < \pi \quad \text{rad} \qquad (8.18)$$

Correspondingly, the gating signal for thyristor Q_2 will commence when

$$v_{O2}^* = \frac{2\sqrt{2}\,V}{\pi} \cos \omega_s t_2 \qquad (8.19)$$

at an intersection for which

$$\pi < \omega_s t_2 < 2\pi \quad \text{rad} \qquad (8.20)$$

These signals are shown shaded in Fig. 8.9.

When the positive half wave of current ceases in Fig. 8.9, the waveform of v_O is interrupted, and signal P is removed. After an interval that must be at least as long as t_{off} for the thyristors, signal N commences, and gating signals i_{G2} and i_{G4} are applied to the thyristors. After an interval $t_q > t_{\text{off}}$ negative current begins to flow from rectifier n acting as an inverter. This condition of operation persists until v_O^* goes negative, from which point on rectifier n acts as a rectifier. The modes of operation throughout the output cycle of the two controlled rectifiers that form the cycloconverter are indicated in Fig. 8.9. The maximum output voltage amplitude V_{Om}^* is achieved when, from equations 8.6 and 8.9,

$$V_{Om}^* = \frac{2V}{\pi} \quad \text{V} \tag{8.21}$$

While this method of control results in an output current waveform with a low harmonic content, it has the disadvantage of introducing a large input current harmonic at the beat frequency between the source frequency and the output frequency. The waveform of the input current i_1 is shown in Fig. 8.9, where it may be seen to have a very high harmonic content.

It remains to indicate how the delay angles for the cycloconverter may be generated and the rules governing the avoidance of short-circuit applied. This is illustrated in the block diagram of Fig. 8.10. If the system into which this control scheme was incorporated were a variable-speed drive of a synchronous motor, then the system command would be the reference speed Ω_R, and the system response would be the motor speed Ω_m. The effect of an increase of Ω_R, giving a positive error, would be that of initiating a ramp increase in ω_O and V_O^*, that would be coupled to avoid saturation of the magnetic circuit of the motor. This would cause an increase in the amplitude of v_O^* in Fig. 8.9 relative to the fixed amplitude $2\sqrt{2}\,V/\pi$. The limit to v_O^*, giving the maximum speed of the drive, is defined by equation 8.21. Comparison of v_O^* with the signal $\pm(2\sqrt{2}\,V/\pi)\cos\omega_s t$ determines the values of the delay angles for the four thyristors of Fig. 8.5.

Control of the P and N signals is also illustrated in Fig. 8.10. The negative half-wave of i_O is supplied to the P signal generator, and when i_O becomes zero, the P signal is emitted. A delay unit then delays its application to the rectifier p gating circuits until interval $t_q > t_{\text{off}}$ has elapsed, and the thyristor in rectifier n has turned off. The P signal is then applied, and gating signals i_{G1} and i_{G3} are emitted at the appropriate instants. An exactly similar procedure, initiated by means of the positive half-wave of i_O, controls gating signals i_{G2} and i_{G4} in rectifier n.

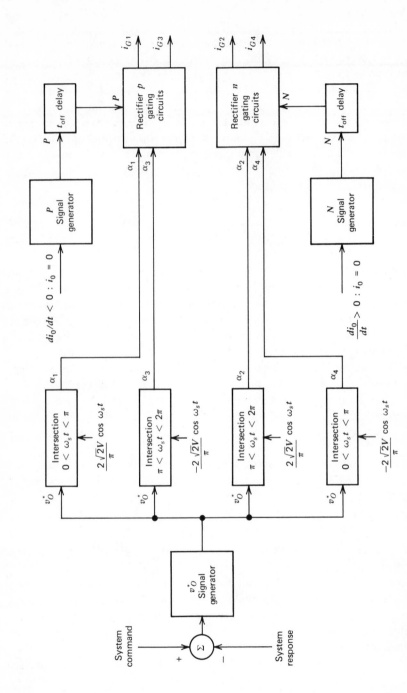

Fig. 8.10 Modulation of α for reduction of output harmonics.

Rating of Circuit Components As each rectifier forming the cycloconverter will operate continuously for a relatively long period (e.g., if the output frequency is 1 Hz, then each rectifier operates for 0.5 s), the component ratings are calculated as if for a continuously operating rectifier. The only exception to this rule is that of the heat sinks, which, because of their high thermal capacity, may be made smaller than those for an independent rectifier. Since the actual power loss in one of the rectifiers of a cyclocon-verter will be only one half of that for the independent rectifier, the heat-sink rating for the thyristors of the cycloconverter may be reduced to half of that for the thyristors of an independent rectifier.

The procedure of rating the rectifiers for continuous operation provides a margin of safety that makes it permissible to ignore the output current harmonics when sinusoidal voltage modulation is employed. Thus the rms rated output current may be taken as

$$I_R = \frac{V_{Om}^*}{Z_O} \quad \text{A} \tag{8.22}$$

where V_{Om}^* is the maximum rms value of the signal of desired voltage v_O^*, and Z_O is the load-circuit impedance at output frequency ω_O. The rms thyristor current is then

$$I_{QR} = \frac{I_R}{\sqrt{2}} \quad \text{A} \tag{8.23}$$

The maximum instantaneous thyristor current is

$$i_{Q\,\text{peak}} = \frac{\sqrt{2}\, V_{Om}^*}{Z_O} \quad \text{A} \tag{8.24}$$

and the average thyristor current is

$$I_Q = \frac{2}{\pi} I_R \quad \text{A} \tag{8.25}$$

From equation 5.31, the peak value of forward and reverse voltage applied to the thyristors of the circuit in Fig. 8.5a will be

$$v_{AK\,\text{peak}} = \pm 2\sqrt{2}\, V \quad \text{V} \tag{8.26}$$

The rms current in the secondary winding of the transformer is the same as that in one of the thyristors, that is,

$$I_2 = I_{QR} = \frac{I_R}{\sqrt{2}} \quad \text{A} \tag{8.27}$$

so that the rating of the whole secondary winding is

$$S_2 = 2VI_{QR} = \sqrt{2}\, VI_R \quad \text{VA} \tag{8.28}$$

Since current flows in the primary winding during both half cycles, but in each half of the secondary winding during alternate halfcycles, it follows that the rms primary current is

$$I_1 = \frac{\sqrt{2}\, I_2}{n} = \frac{I_R}{n} \quad \text{A} \tag{8.29}$$

where n is the turns ratio between the primary winding and one half of the secondary winding. The rating of the primary winding is therefore

$$S_1 = nV\frac{I_R}{n} = VI_R \quad \text{VA} \tag{8.30}$$

The maximum output power of the cycloconverter is

$$P_O = V_{Om}^* I_R \cos\phi_O \quad \text{W} \tag{8.31}$$

where ϕ_O is the angle of the load circuit impedance at frequency ω_O, and therefore $\cos\phi_O$ is the load-circuit power factor for sinusoidal output voltage and current. On the assumption that the internal losses of the cycloconverter are negligible, P_O is also the input power to the transformer primary winding. Thus the input power factor (PF) for maximum output power is

$$PF = \frac{P_O}{S_1} = \frac{V_{Om}^* I_R \cos\phi_O}{VI_R} \tag{8.32}$$

and since, from equation 8.21,

$$\frac{V_{Om}^*}{V} = \frac{2}{\pi} \tag{8.33}$$

then

$$PF = 0.636 \cos\phi_O \tag{8.34}$$

Equation 8.34 makes clear one of the biggest disadvantages of the single-phase to single-phase cycloconverter; namely, that the input power factor is only 0.636 of the load-circuit power factor. This low power factor is due to the high harmonic content of the input current.

8.1.2 Three-Phase to Single-Phase Cycloconverter The circuit diagram of a three-phase to single-phase cycloconverter is shown in Fig. 8.11a, where it may be seen that the two controlled rectifiers forming the dual converter are three-phase bridge rectifiers. The pairs of terminals marked a and a', b and b', c and c' are joined together and connected to the three-phase supply. An equivalent circuit for this cycloconverter may also be derived and is shown in Fig. 8.11b. If the discussion of single-phase to single-phase cycloconverters of Sections 8.1.1a and b is considered in conjunction with the discussion of the three-phase controlled rectifier in Section 5.3, then it may readily be seen how the three-phase to single-phase cycloconverter can operate to produce output current waveforms similar to those of Figs. 8.6 and 8.7.

The major advantage of the three-phase as opposed to the single-phase source of the same frequency lies in the increased rate of sampling that the three-phase source makes possible. In other words, during any given period, there are three times as many pulses of output current from the three-phase supplied converter. This means that, for a given level of subharmonic in the current wave, the output frequency for the three-phase excited converter may be three times as great as that for the single-phase excited converter. In the three-phase case therefore, typically

$$\omega_s \geqslant 3\omega_O \quad \text{rad/s} \tag{8.35}$$

The fact that an increased number of source phases does not necessarily reduce the output-current harmonics of a cycloconverter has already been remarked in Section 8.1.1b. As in the case of the single-phase excited converter, an output current wave with low harmonic content can be achieved by modulation of the delay angles of all the thyristors in the converter. The method of doing this and of rating the devices when it is done will therefore form the subject matter of the remainder of this section.

Reduction of Output Harmonics Output harmonics may be reduced in a three-phase to single-phase cycloconverter by modulation of α in exactly the same way as for the single-phase excited converter. However the existence of a three-phase source and the necessity for 12 distinct gating signals complicates the problem of determining the required delay angles.

In Fig. 8.11b is shown the equivalent circuit of the cycloconverter of Fig. 8.11a. The relationship between these two circuits has been fully discussed

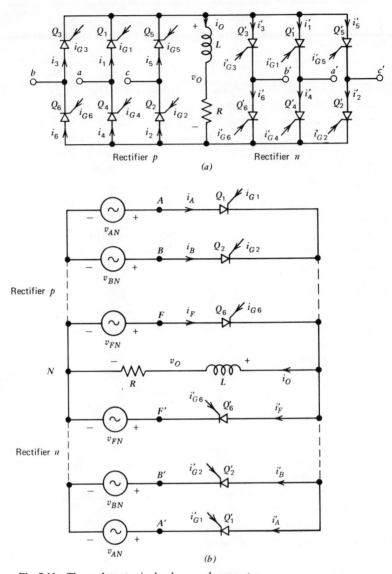

Fig. 8.11 Three-phase to single-phase cycloconverter.

in Section 5.3 and is summarized in Table 5.1 and Fig. 5.19. The source voltages of Fig. 8.11b are defined in equations 5.42 and 5.43 and are

$$v_{AN} = v_{ab} = \sqrt{2}\, V \sin \omega_s t \quad \text{V}$$

$$v_{BN} = -v_{ca} = \sqrt{2}\, V \sin \left(\omega_s t - \frac{\pi}{3} \right) \quad \text{V}$$

$$v_{CN} = v_{bc} = \sqrt{2}\, V \sin \left(\omega_s t - \frac{2\pi}{3} \right) \quad \text{V}$$

$$(8.36)$$

$$v_{DN} = -v_{ab} = \sqrt{2}\, V \sin \left(\omega_s t - \pi \right) \quad \text{V}$$

$$v_{EN} = v_{ca} = \sqrt{2}\, V \sin \left(\omega_s t - \frac{4\pi}{3} \right) \quad \text{V}$$

$$v_{FN} = -v_{bc} = \sqrt{2}\, V \sin \left(\omega_s t - \frac{5\pi}{3} \right) \quad \text{V}$$

As before, the ideal output voltage wave is defined as

$$v_O^* = \sqrt{2}\, V_O^* \sin \omega_O t \quad \text{V} \qquad (8.37)$$

In Fig. 8.12 is shown one cycle of v_O^* and the ideal output current i_O^* for a load circuit with leading power factor. From equation 5.50 the delay angle α which is required as each thyristor is turned on is given by

$$v_O^* = \frac{3\sqrt{2}\, V}{\pi} \cos \alpha \quad \text{V} \qquad (8.38)$$

where

$$\alpha = \omega_s t - \frac{\pi}{3} \quad \text{rad} \qquad (8.39)$$

and $t = 0$ at the commencement of the cycle of the voltage applied to the thyristor for which α is to be determined. The angle $\pi/3$ appears in equation 8.39 owing to the datum employed for measuring α, and this is illustrated in Fig. 5.20 and arises out of the definition of delay angle.

The dual converter of the equivalent circuit of Fig. 8.11b is made up of two six-phase half-wave rectifiers. The delay angles determined for the thyristors of this circuit may be related to those required by the thyristors of the real circuit of Fig. 8.11a by means of Table 5.1. To achieve an

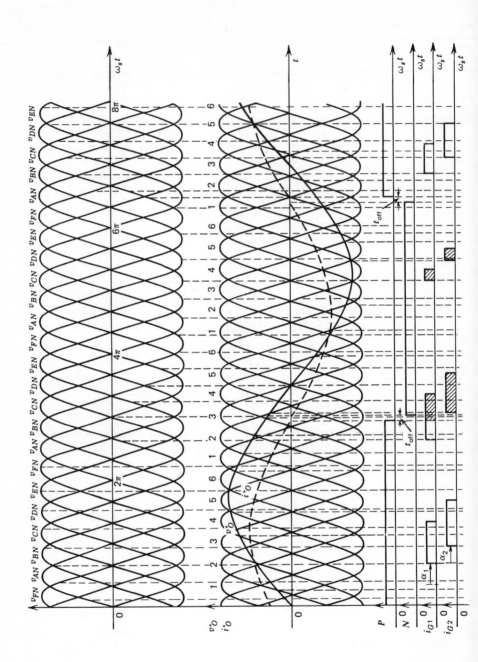

486

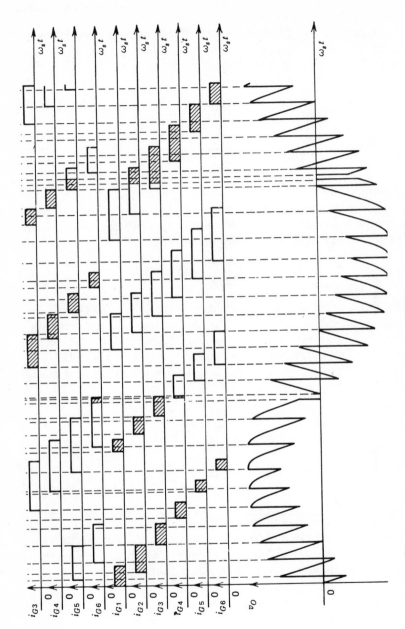

Fig. 8.12 Waveforms of the variables in the equivalent circuit of Fig. 8.11b ($\omega_s = 3.5\omega_0$).

output voltage that will produce an output current with low harmonic content, it is necessary to arrange that, during the first cycle of voltage v_{AN}, thyristor Q_1 in Fig. 8.11b is turned on at instant t_1, when

$$v_{O1}^* = \frac{3\sqrt{2}\ V}{\pi} \cos\left(\omega_s t_1 - \frac{\pi}{3}\right) \quad \text{V} \tag{8.40}$$

and

$$\frac{\pi}{3} < \omega_s t_1 < \frac{4\pi}{3} \quad \text{rad} \tag{8.41}$$

A curve (numbered 1) of the function $(3\sqrt{2}\ V/\pi)\cos(\omega_s t - \pi/3)$ is shown in Fig. 8.12. The intersection of this curve with that of v_O^* within the range defined by inequality 8.41 defines the instant t_1 which satisfies equation 8.40. The delay angle may then be determined from

$$\alpha_1 = \omega_s t_1 - \frac{\pi}{3} \quad \text{rad} \tag{8.42}$$

This is the delay angle at which i_{G1} commences and thyristor Q_1 is turned on. Voltage v_{AN} then appears across the load circuit, and the succeeding segment of the wave of output voltage v_O is described by the first of equations 8.36. This output voltage wave also is shown in Fig. 8.12.

At instant t_2, thyristor Q_2 must be turned on when the desired output voltage is

$$v_{O2}^* = \frac{3\sqrt{2}\ V}{\pi} \cos\left(\omega_s t_2 - \frac{2\pi}{3}\right) \quad \text{V} \tag{8.43}$$

and

$$\frac{2\pi}{3} < \omega_s t_2 < \frac{5\pi}{3} \quad \text{rad} \tag{8.44}$$

A curve (numbered 2) of the function $(3\sqrt{2}\ V/\pi)\cos(\omega_s t - 2\pi/3)$ is shown in Fig. 8.12. The intersection of this curve with that of v_O^* within the range defined by inequality 8.44 defines the instant t_2 which satisfies equation 8.43. The delay angle may then be determined from

$$\alpha_2 = \omega_s t_2 - \frac{2\pi}{3} \quad \text{rad} \tag{8.45}$$

This is the delay angle at which thyristor Q_2 is turned on. The succeeding segment of the wave of v_O is then described by the second of equations 8.36. Repetition of the foregoing procedure determines the remaining delay angles for rectifier p throughout the cycle of v_O.

During the first part of the cycle of v_O^* in Fig. 8.12, while $i_O^* > 0$, the gating signals for rectifier n are also being generated, but the gating currents are inhibited by the absence of signal N. The gating signal for thyristor Q_1' will commence when equation 8.40 is satisfied by an angle $\omega_s t_1'$ in the interval

$$\frac{4\pi}{3} < \omega_s t_1' < \frac{7\pi}{3} \quad \text{rad} \tag{8.46}$$

that is, by an intersection in the half cycle of the cosine function 1 succeeding that defined by inequality 8.41. The resulting pulse is shown shaded in the waveform of i_{G1}' to indicate that it is not applied to the thyristor. In similar fashion the remaining delay angles for rectifier n may be determined throughout the cycle of v_O^*, and when $i_O^* < 0$, and signal N is present, the relevant gating signals are applied to the thyristors of rectifier n.

The output voltage waveform of Fig. 8.12 is a great improvement over that of Fig. 8.9, and it may also be noted that the zero-current intervals in Fig. 8.12 are shorter. This is due to the fact that at the end of each half cycle of i_O^*, a gating signal has already commenced when signal N or P appears, so that a thyristor is immediately turned on. In this case, therefore, the time t_q available for turn off is no greater than the minimum value t_{off} introduced between the beginnings and ends of the N and P signals; this is not necessarily the case under other load conditions. The maximum rms output voltage V_{Om}^* is achieved when, from equations 8.37 and 8.38

$$V_{Om}^* = \frac{3V}{\pi} \quad \text{V} \tag{8.47}$$

If a waveform of line current were required, it would be necessary to prepare a set of waveforms for the circuit of Fig. 8.11a corresponding to that shown in Fig. 5.25. The control circuit for generating the required gating signals would employ the principles illustrated in the block diagram of Fig. 8.10.

Rating of Circuit Components As in the case of the single-phase to single-phase cycloconverter, it will be assumed that output frequencies may be so low as to make it advisable to rate all components excepting the heat sinks for continuous operation.

The rms rated output current may again be taken to be

$$I_R = \frac{V_{Om}^*}{Z_O} \quad \text{A} \tag{8.48}$$

and from equation 5.66, the rms thyristor current is

$$I_{QR} = \frac{I_R}{\sqrt{3}} \quad A \tag{8.49}$$

The maximum instantaneous thyristor current is

$$i_{Q\,\mathrm{peak}} = \frac{\sqrt{2}\, V_{Om}^*}{Z_O} \quad A \tag{8.50}$$

The half-cycle average output current is

$$I_O = \frac{2\sqrt{2}\, I_R}{\pi} \quad A \tag{8.51}$$

and since each thyristor of the circuit of Fig. 8.11a conducts two of the six current pulses occurring in each cycle, the average thyristor current is

$$I_Q = \frac{2\sqrt{2}}{3\pi} I_R \quad A \tag{8.52}$$

From equation 5.63, the peak value of the forward and reverse voltage applied to the thyristors will be

$$v_{AK\,\mathrm{peak}} = \pm\sqrt{2}\, V \quad V \tag{8.53}$$

If the line current to the converter is assumed to be the secondary current of a three-phase transformer, then from equation 5.67 the rms value will be

$$I_2 = \sqrt{2}\, I_{QR} = \frac{\sqrt{2}}{\sqrt{3}} I_R \quad A \tag{8.54}$$

Thus from equation 5.68, the transformer rating is

$$S_2 = \sqrt{3}\, VI_2 = \sqrt{2}\, VI_R = S_1 \quad VA \tag{8.55}$$

The maximum output power of the cycloconverter is

$$P_O = V_{Om}^* I_R \cos\phi_O \tag{8.56}$$

so that the input power factor at this power is

$$PF = \frac{P_O}{S_1} = \frac{V_{Om}^* I_R \cos\phi_O}{\sqrt{2}\, V I_R} \tag{8.57}$$

and since from equation 8.47

$$\frac{V_{Om}^*}{V} = \frac{3}{\pi} \tag{8.58}$$

then

$$PF = 0.675 \cos\phi_O \tag{8.59}$$

This is only marginally better than the input power factor of the single-phase to single-phase cycloconverter.

8.2 LINE-COMMUTATED FREQUENCY MULTIPLIERS

The operation of a line-commutated frequency tripler has already been briefly discussed at the beginning of this chapter and is illustrated in Fig. 8.2. The factor by which the input frequency of these converters is multiplied is related to the number of source phases m and is therefore necessarily an integer. The particular multiplier discussed in this section has a three-phase input, and the output frequency is three times the source frequency. Multipliers with other integral multiplication factors can indeed be built, and an analysis similar to what follows can be carried out. In general, the larger values of multiplication result in poorer input power factor and lower utilization of the semiconductor devices.

In Fig. 8.2, each phase of the source and the associated pair of thyristors may be considered as a single-phase ac controller supplying the load circuit over a limited part of the input cycle. The analysis that follows will refer to a simple series RL load circuit, since then it may be based on the design curves of Figs. 4.6, 4.7, and 4.8 relating to a single-phase voltage controller supplying such a load. If it should be required to design a multiplier with a more complicated load circuit, then it will be necessary to develop the equations for the operation of a half-wave rectifier with such a circuit in a similar manner to that employed in Section 3.3.1.

From Fig. 8.2 it may be seen that the maximum permissible conduction angle for the tripler is

$$\gamma_{\max} = \frac{\pi}{3} - \omega_s t_q \quad \text{rad} \tag{8.60}$$

where t_q is the time available for turnoff. If two phases of the source are not to be short circuited, a zero-current interval of duration $t_q \geqslant t_{\text{off}}$ must exist at each current reversal in the output waveform.

The permissible range of angle of retard α for the tripler supplying a series RL load circuit may be obtained from Fig. 4.6, where only that part of the diagram that lies below the line $\gamma = \gamma_{\text{max}}$ may be employed. The intersection of this line with the curve for the appropriate value of ϕ, where

$$\phi = \tan^{-1} \frac{\omega_s L}{R} \quad \text{rad} \tag{8.61}$$

and ω_s is the source frequency, gives the lower limit for α. As for a single-phase voltage controller, α is measured with respect to the zero of the line-to-neutral voltage wave for each phase. The upper limit of α for all load circuits is 180°.

For any value of α, I_N the normalized average thyristor current, and I_{RN} the normalized rms thyristor current, may be obtained from Figs. 4.7 and 4.8 respectively. The rms load current for the tripler will then be

$$I_R = \sqrt{3} \left(\sqrt{2} \, I_{RN} I_{\text{BASE}} \right) \quad \text{A} \tag{8.62}$$

and since

$$I_{\text{BASE}} = \frac{\sqrt{2} \, V}{Z} \quad \text{A} \tag{8.63}$$

where

$$Z = \left[R^2 + \omega_s^2 L^2 \right]^{1/2} \quad \Omega \tag{8.64}$$

and V is the rms line-to-neutral source voltage, then

$$I_R = \frac{2\sqrt{3} \, I_{RN} V}{Z} \quad \text{A} \tag{8.65}$$

The average thyristor current is

$$I_Q = I_N \times I_{\text{BASE}} \quad \text{A} \tag{8.66}$$

Since when neither element of an antiparallel connected pair of thyristors is conducting, one of the line-to-line voltages is applied to them, it follows that

$$v_{AK\text{peak}} = \pm \sqrt{3} \times \sqrt{2} \, V \quad \text{V} \tag{8.67}$$

The rms line current, which will be the secondary winding current of the three-phase transformer supplying the converter, is

$$I_2 = \sqrt{2}\, I_{RN} I_{BASE} = \frac{2V}{Z} I_{RN} \quad \text{A} \qquad (8.68)$$

so that the apparent power input to the converter, which is also the transformer secondary rating, is

$$S_2 = 3VI_2 = \frac{6V^2}{Z} I_{RN} \quad \text{VA} \qquad (8.69)$$

The load current i_O which is shared by the secondary windings of the three-phase transformer forms a zero sequence component which does not appear in the currents in the transformer primary windings. Thus if the turns ratio of primary to secondary windings is n, and the transformer is Y-Y connected, then the rms primary current will be

$$I_1 = \frac{1}{n} \left[I_2^2 - \left(\frac{I_R}{3} \right)^2 \right]^{1/2} \quad \text{A} \qquad (8.70)$$

and substitution from equations (8.65) and (8.68) in (8.70) yields

$$I_1 = \frac{I_2}{n} \left[1 - \frac{1}{3} \right]^{1/2} = \sqrt{\frac{2}{3}} \, \frac{I_2}{n} \quad \text{A} \qquad (8.71)$$

The transformer primary rating is thus

$$S_1 = 3nVI_1 = \sqrt{6}\, VI_2 = \sqrt{\frac{2}{3}} \, S_2 \quad \text{VA} \qquad (8.72)$$

The output power of the converter is

$$P_O = RI_R^2 = 12 \frac{V^2}{Z^2} I_{RN}^2 R \quad \text{W} \qquad (8.73)$$

so that the power factor on the primary side of the three-phase transformer is

$$\text{PF} = \frac{P_O}{S_1} \qquad (8.74)$$

Substitution from equations 8.69, 8.72, and 8.73 yields

$$PF = \sqrt{6}\, I_{RN} \cos\phi \qquad (8.75)$$

The derating factor is

$$D_R = \frac{6\sqrt{6}\, VI_{RN}I_{BASE}}{P_O} = \frac{\sqrt{3}\, Z}{RI_{RN}} = \frac{\sqrt{3}}{I_{RN}\cos\phi} \qquad (8.76)$$

Example 8.1 A line-commutated frequency tripler is required to supply a resistive load from a 60-Hz source. For the thyristors to be employed, $t_{off} = 100\,\mu s$. Calculate the transformer primary power factor and the derating factor.

Solution From equation 8.60,

$$\gamma_{max} = \frac{\pi}{3} - 120\pi \times 10^{-4} = 1.009^c = 57.8°$$

From Fig. 4.6, the corresponding value of α is 122°, and from Fig. 4.8

$$I_{RN} = 0.21$$

From equation 8.75

$$PF = \frac{2 \times 0.21}{[1 - \frac{1}{3}]^{1/2}} = 0.514$$

From equation 8.76

$$D_R = \frac{\sqrt{3}}{0.21} = 8.24$$

Example 8.2 A frequency tripler is supplied by a $Y\text{-}Y$ connected three-phase transformer of unity turns ratio from a 230-V line-to-line 60-Hz source. The load-circuit parameters are $L = 1\,mH$, $R \cong 0$. For the thyristors employed, t_{off} is 125 μs.

(a) Determine the value of the delay angle for maximum output current.

(b) Determine the maximum rms output current.

(c) Calculate the peak value of the thyristor voltage and the maximum rms thyristor current.

(d) Calculate the transformer secondary and primary winding currents.

(e) Sketch approximately to scale the waveforms of the transformer secondary and primary currents.

Solution

(a) From equation 8.60

$$\gamma_{max} = \frac{\pi}{3} - 120\pi \times 1.25 \times 10^{-4}$$

$$= 1.000^c = 57.3°$$

From Fig. 4.6, the corresponding value of $\alpha = 151°$
(b) From Fig. 4.8, $I_{RN} = 0.03$

$$Z \cong \omega_s L = 120\pi \times 10^{-3} = 0.377 \quad \Omega$$

$$I_{BASE} = \sqrt{2} \times \frac{230}{\sqrt{3}} \times \frac{1}{0.377} = 498 \quad A$$

From equation 8.62,

$$I_R = \sqrt{6} \times 0.03 \times 498 = 36.6 \quad A$$

(c) The peak value of thyristor voltage is

$$v_{AKpeak} \cong \sqrt{2} \times 230 = 325 \quad V$$

The rms thyristor current for $\alpha = 151°$ is

$$I_{QR} = I_{RN} I_{BASE} = 0.03 \times 498 = 14.9 \quad A$$

(d) From equation 8.68,

$$I_2 = \sqrt{2} \times 14.9 = 21.1 \quad A$$

From equation 8.71

$$I_1 = 21.1[1 - \tfrac{1}{3}]^{1/2} = 17.3 \quad A$$

(e) The waveforms of I_1 and I_2 are shown in Fig. E8.2. On the assumption that the output waveform of i_O is sinusoidal its amplitude will be approximately $\sqrt{2} I_R = 51.8 A$.

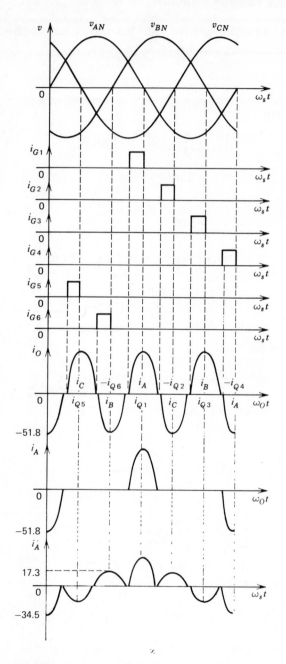

Fig. E8.2

From the circuit of Fig. 8.2a

$$i_A = i_{Q1} - i_{Q4} \quad A$$

and this waveform is shown in Fig. E8.2. The waveforms of i_B and i_C would be similar, but shifted by $2\pi/3\omega_s$ and $4\pi/3\omega_s$ s from it.

Since the transformer turns ratio is unity and the output current i_O forms a third harmonic at source frequency, it follows that the primary current in phase A of the transformer will be

$$i'_A = i_{Q1} - i_{Q4} - \frac{i_O}{3} \quad A$$

and this waveform also is shown in Fig. E8.2.

8.3 LOAD-COMMUTATED AC-TO-AC CONVERTERS (CYCLOINVERTERS)

Figure 8.3 has been employed at the beginning of this chapter to explain the principle of operation of a load-commutated frequency multiplier or ac-to-ac converter. In one respect this diagram differs significantly from all others of a similar nature so far discussed, and that is in the apparent irregularity of the sequence of gating signals applied to the thyristors. The means by which these signals are controlled is shown in more detail in Fig. 8.13.

Currents i_{Gp} and i_{Gn} in Fig. 8.13 constitute two series of pulses, the first of which may be applied to thyristors Q_1 Q_3 or Q_5, while the second may be applied to thyristors Q_2, Q_4, or Q_6. For each thyristor a permissive signal P_1 to P_6 is generated. If the permissive signal and the i_{Gp} or i_{Gn} pulse occur simultaneously, then a gating current is applied to a thyristor. Thus, when, for example, P_3 and i_{Gp} are applied simultaneously, thyristor Q_3 is turned on. Thyristors Q_1, Q_3, and Q_5 may be considered to form a "positive" three-phase half-wave rectifier supplying positive current to the load circuit, while thyristors Q_2, Q_4, and Q_6 form a "negative" rectifier. Since the P signals each extend only for one third of a cycle of the source voltage, it is not possible for more than one thyristor to be turned on at any one time in either of the two rectifiers.

While control of the output current frequency is determined by the frequency of the i_{Gp} and i_{Gn} signals, control of the output current amplitude is obtained by variation of the phase angle of the P signals relative to the line-to-neutral source voltages. If signal P_1 is considered, which permits the application of voltage v_{AN} to the load circuit via thyristor Q_1, its position may be defined by the phase angle at which it commences. During

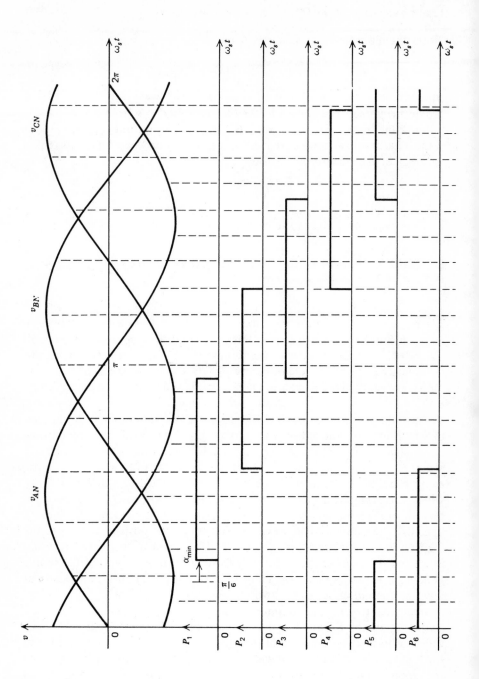

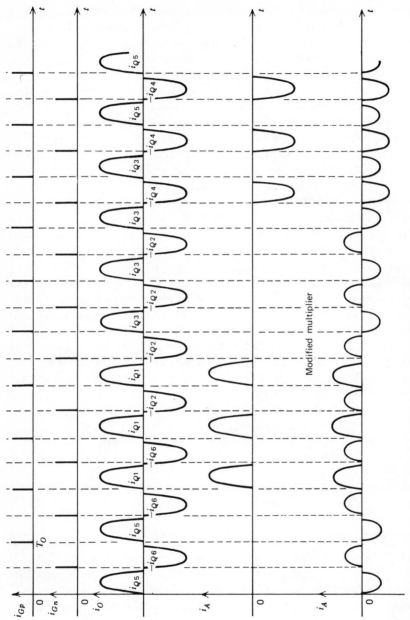

Fig. 8.13 Waveforms of the variables in the circuits of Figs. 8.3 and 8.4.

499

the succeeding one third of the cycle of v_{AN}, thyristor Q_1 may be turned on several times, each time being load-commutated at the end of a positive current pulse. The minimum delay angle for thyristor Q_1 throughout this one third of a cycle is defined by the commencement of the P_1 signal. This angle will be designated α_{min} and is shown in Fig. 8.13.

Further consideration of Fig. 8.3 will show that the voltage applied to the load circuit varies during successive current pulses via the same thyristor. This means that the magnitude of the resulting current pulses also should vary in a manner depending not only on the source voltage at the instant of turn on, but also on the magnitude of the voltage v_C across the capacitor C that has been left there at the end of the preceding current pulse. The regular output-current waveform of Fig. 8.3 is therefore an approximation. For any given set of operating conditions the exact waveform of i_O could be determined, but only at the expense of a great deal of computation. It is therefore desirable to arrive at an approximation that will permit reasonably simple design of a converter and rating of its components.

An approximation that yields good results near the maximum output condition, that is, when α_{min} is zero, may be obtained by considering that the positive and negative source voltages alternately applied to the load circuit remain constant at a value given by the average output voltage of a three-phase half-wave rectifier for which $\alpha = \alpha_{min}$. By the definition of delay angle given in Section 5.2.1, α_{min} is zero at $\omega_s t = \pi/6$. Since V is the line-to-neutral voltage of the source the average value of the output voltage of the rectifier is then given by

$$V_\alpha = 3\left[\frac{1}{2\pi} \int_{\alpha_{min}}^{(2\pi/3) + \alpha_{min}} \sqrt{2}\, V \sin\left(\omega_s t + \frac{\pi}{6}\right) d(\omega_s t) \right]$$

$$= \frac{3\sqrt{3}}{\sqrt{2}\,\pi} V \cos\alpha_{min} \quad \text{V} \tag{8.77}$$

If this approximation is accepted, the analysis of the converter circuit becomes closely similar to that of the series inverter of Section 7.7.2. It may be carried out employing the two equivalent circuits, one for each half wave of the output current, shown in Fig. 8.14a and b.

After several cycles of i_O have taken place, a steady alternation of capacitor voltage v_C will have been established. At the end of one negative half cycle, let

$$v_C = V_{C1} < 0 \quad \text{V} \tag{8.78}$$

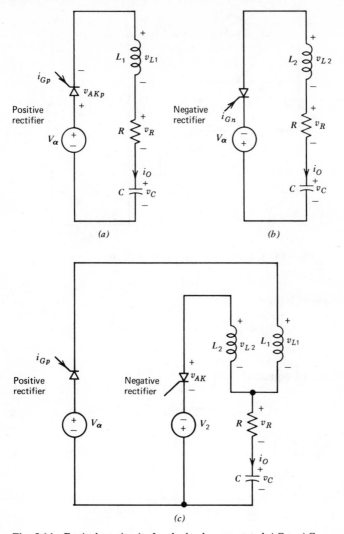

Fig. 8.14 Equivalent circuits for the load-commutated AC-to-AC converter.

As in the case of the series inverter, the cycle of operation of this converter may be divided into four intervals.

Interval I During this interval, the positive half wave of i_O takes place, and thyristor Q_1, Q_3, or Q_5 is conducting. From Fig. 8.15a,

$$Ri_O + L\frac{di_O}{dt} + \frac{1}{C}\int_0^t i_O \, dt + V_{C1} = V_\alpha \quad \text{V} \tag{8.79}$$

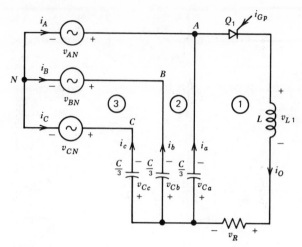

Fig. 8.15 Circuit of Fig. 8.4. Thyristor Q_1 conducting.

and by differentiation

$$\frac{d^2 i_O}{dt^2} + \frac{R}{L}\frac{di_O}{dt} + \frac{i_O}{LC} = 0 \quad \text{A/s}^2 \tag{8.80}$$

The solution of equation 8.80 is

$$i_O = \frac{(V_\alpha - V_{C1})}{\omega_r L}\epsilon^{-\zeta t}\sin\omega_r t \quad \text{A} \tag{8.81}$$

where

$$\zeta = \frac{R}{2L} \quad s^{-1} \tag{8.82}$$

$$\omega_r = [\omega_o^2 - \zeta^2] \quad \text{rad/s} \tag{8.83}$$

$$\omega_o = \frac{1}{\sqrt{LC}} \quad \text{rad/s} \tag{8.84}$$

The remaining circuit variables during this interval may now be determined. They are

$$v_O = v_R = Ri_O = \frac{R(V_\alpha - V_{C1})}{\omega_r L}\epsilon^{-\zeta t}\sin\omega_r t \quad \text{V} \tag{8.85}$$

$$v_{L1} = L\frac{di_O}{dt} = \frac{\omega_o}{\omega_r}(V_\alpha - V_{C1})\epsilon^{-\zeta t}\cos(\omega_r t + \psi) \quad \text{V} \tag{8.86}$$

where

$$\psi = \tan^{-1} \frac{\zeta}{\omega_r} \quad \text{rad} \qquad (8.87)$$

$$v_C = V_\alpha - v_{L1} - v_R$$

$$= V_\alpha - \frac{\omega_o}{\omega_r}(V_\alpha - V_{C1})\epsilon^{-\zeta t}\cos(\omega_r t - \psi) \quad \text{V} \qquad (8.88)$$

Interval I ends when i_O falls to zero at

$$t = \frac{\pi}{\omega_r} = t_1 \quad \text{s} \qquad (8.89)$$

Thus

$$v_O = v_R = 0 \quad \text{V}: \qquad t = t_1^- \quad \text{s} \qquad (8.90)$$

$$v_{L1} = (V_\alpha - V_{C1})\epsilon^{-\zeta\pi/\omega_r} \quad \text{V}: \qquad t = t_1^- \quad \text{s} \qquad (8.91)$$

$$v_C = V_\alpha + (V_\alpha - V_{C1})\epsilon^{-\zeta\pi/\omega_r} = V_{C2} \quad \text{V}: \qquad t = t_1^- \quad \text{s} \qquad (8.92)$$

Interval II At $t = t_1^+$ s, when i_O and di_O/dt are both zero, all currents in the circuit are zero, and from equation 8.92

$$v_C = V_{C2} \quad \text{V}: \qquad t = t_1^+ \quad \text{s} \qquad (8.93)$$

The duration of interval II is the time available for turn-off of the thyristors, which may be expressed as

$$t_q = \frac{T_O}{2} - \frac{\pi}{\omega_r} \geqslant t_{\text{off}} \quad \text{s} \qquad (8.94)$$

where T_O is the periodic time of the output current.

Interval III During this interval, the negative half wave of i_O take place, and thyristor Q_2, Q_4, or Q_6 is conducting. From Fig. 8.14b,

$$Ri_O + L\frac{di_O}{dt'} + \frac{1}{C}\int_0^{t'} i_O\, dt' + V_{C2} = -V_\alpha \quad \text{V} \qquad (8.95)$$

where

$$t' = t - \frac{T_O}{2} \quad \text{s} \qquad (8.96)$$

By differentiation of equation 8.95,

$$\frac{d^2 i_O}{(dt')^2} + \frac{R}{L}\frac{di_O}{dt'} + \frac{i_O}{LC} = 0 \quad \text{A/s}^2 \tag{8.97}$$

The initial conditions in the circuit of Fig. 8.14b are

$$i_O = 0 \quad \text{A}: \qquad t' = 0^+ \quad \text{s} \tag{8.98}$$

$$v_C = V_{C2} \quad \text{V}: \qquad t' = 0^+ \quad \text{s} \tag{8.99}$$

and the solution of equation 8.97 is

$$i_O = -\frac{(V_\alpha + V_{C2})}{\omega_r L}\epsilon^{-\zeta t'}\sin\omega_r t' \quad \text{A} \tag{8.100}$$

Since the current waveform has alternating symmetry, it follows from equations 8.81 and 8.100 that

$$V_{C1} = -V_{C2} \quad \text{V} \tag{8.101}$$

so that from equations 8.92 and 8.101,

$$V_{C2} = \frac{V_\alpha(1 + \epsilon^{-\zeta\pi/\omega_r})}{1 - \epsilon^{-\zeta\pi/\omega_r}} = -V_{C1} \quad \text{V} \tag{8.102}$$

8.3.1 Component Ratings Determination of component ratings may be modeled on the procedure employed for the series inverter in Section 7.7.3. The only points to be borne in mind are that it is necessary to substitute V_α for V and ω_O for ω. The value of V_{C1} also differs in the two cases. Thus from equation 7.203, the average input power provided by the three-phase source is

$$P_s = \frac{2}{T_O}\int_0^{T_O/2} V_\alpha i_{Q1}\, dt \quad \text{W} \tag{8.103}$$

so that from equation 7.204,

$$P_s = \frac{\omega_O}{\pi}\frac{V_\alpha(V_\alpha - V_{C1})}{\omega_O^2 L}[1 + \epsilon^{-\zeta\pi/\omega_r}] \quad \text{W} \tag{8.104}$$

From equation 7.206 the rms value of the load current is

$$I_R = \sqrt{\frac{P_s}{R}} \quad \text{A} \tag{8.105}$$

and this is also the rms current rating of the capacitor C. The rms value of the inductor current is then

$$I_L = \frac{I_R}{\sqrt{2}} \quad \text{A} \tag{8.106}$$

The inductors must not saturate at the peak value of i_O which, from equation 7.210 is

$$i_{Q\text{peak}} = \frac{V_\alpha - V_{C1}}{\omega_o L} \epsilon^{-(\zeta/\omega_r)(\pi/2 - \psi)} \quad \text{A} \tag{8.107}$$

this giving also the peak value of the thyristor current. The rms value of the thyristor current is

$$I_{QR} = \frac{I_R}{\sqrt{6}} \quad \text{A} \tag{8.108}$$

Owing to the poor form factor, the average value of the thyristor current is very low and is therefore not of significance in choosing a suitable component. From equation 7.211, the peak capacitor voltage is

$$v_{C\text{peak}} = V_{C2} \quad \text{V} \tag{8.109}$$

The peak thyristor voltage may be determined from Fig. 8.14c, in which it is assumed that the thyristor of the "positive rectifier" is conducting, and in which therefore v_{L2} is zero. For the circuit mesh formed by the two branches containing the sources it is seen that

$$v_{AK} = 2V_\alpha - v_{L1} \quad \text{V} \tag{8.110}$$

and the peak value of this voltage will occur when v_{L1} has its peak negative value. If the expression in equation 8.86 for v_{L1} is differentiated and equated to zero, then it is found that the minimum value occurs at instant $t = t_b$ when

$$\omega_r t_b = \pi - 2\psi \quad \text{rad} \tag{8.111}$$

so that from equation 8.110

$$v_{AK\,peak} = 2V_\alpha + (V_\alpha - V_{C1})\epsilon^{-(\zeta/\omega_r)(\pi - 2\psi)} \quad \text{V} \tag{8.112}$$

The secondary rating of the three-phase transformer represented by the three-phase source of Fig. 8.3 is

$$S_2 = 3\frac{V}{\sqrt{3}} I_R = \sqrt{3}\, VI_R \quad \text{VA} \tag{8.113}$$

The three sources and the load circuit of Fig. 8.3 constitute a three-phase four-wire system, in which the current in the neutral wire, that is, the load current i_O, is a zero-sequence current that will be shared equally between the three sources. This zero-sequence component will not appear in the primary currents of the three-phase transformer, so that if the transformer turns ratio is n, then the rms value of the primary current will be

$$I_1 = \frac{1}{n}\left[\left(\frac{I_R}{\sqrt{3}}\right)^2 - \left(\frac{I_R}{3}\right)^2\right]^{1/2} = \frac{\sqrt{2}\,I_R}{3n} \quad \text{A} \tag{8.114}$$

The primary transformer rating is

$$S_1 = 3nVI_1 = \sqrt{2}\, VI_R \quad \text{VA} \tag{8.115}$$

which is less than S_2. The secondary transformer rating may be reduced, however, by the circuit modifications discussed in Section 8.3.2.

The power factor at the input terminals of the transformer is

$$\text{PF} = \frac{P_O}{S_1} = \frac{RI_R^2}{\sqrt{2}\, VI_R} = 0.707\frac{RI_R}{V} \tag{8.116}$$

Example 8.3 In the cycloinverter circuit of Fig. 8.3, the line-to-neutral source voltage is 300 V rms, $R = 1.2\,\Omega$, $L = 100\,\mu\text{H}$, and $C = 110\,\mu\text{F}$. The output frequency is 1000 Hz. For maximum output operation, determine the following:

(a) The time available for turn-off t_q.
(b) The peak thyristor voltage $v_{AK\,peak}$.
(c) The power output P_O.
(d) The rms thyristor current I_{QR}.

(e)   The derating factor D_R.

(f)   The power factor at the secondary terminals of the three-phase transformer.

(g)   The power factor at the primary terminals of the three-phase transformer.

Solution

(a)   The natural frequency of the load circuit is

$$\omega_o = \frac{1}{\sqrt{LC}} = \frac{10^4}{\sqrt{1.1}} = 9535 \quad \text{rad/s}$$

The damping factor

$$\zeta = \frac{R}{2L} = \frac{1.2}{2 \times 10^{-4}} = 6000 \quad \text{s}^{-1}$$

so that the ringing frequency is

$$\omega_r = [\omega_o{}^2 - \zeta^2]^{1/2} = [9535^2 - 6000^2]^{1/2} = 7411 \quad \text{rad/s}$$

The periodic time of the output current is 10^{-3} s, so that from equation 8.94,

$$t_q = \frac{T_O}{2} - \frac{\pi}{\omega_r} = \frac{10^{-3}}{2} - \frac{\pi}{7411} = 76.1 \times 10^{-6} \quad \text{s}$$

(b) $v_{AK\,\text{peak}}$ may be obtained from equation 8.112 in which, from equation 8.77, for $\alpha_{\min} = 0$,

$$V_\alpha = \frac{3\sqrt{3} \times 300}{\sqrt{2}\,\pi} = 351 \quad \text{V}$$

From equation 8.102

$$-V_{C1} = \frac{V_\alpha(1 + \epsilon^{-\zeta\pi/\omega_r})}{(1 - \epsilon^{-\zeta\pi/\omega_r})}$$

$$\frac{\zeta\pi}{\omega_r} = \frac{6000\pi}{7411} = 2.543$$

$$-V_{C1} = \frac{351(1 + \epsilon^{-2.543})}{1 - \epsilon^{-2.543}} = 411 \quad \text{V}$$

From equation 8.87

$$\psi = \tan^{-1}\frac{\zeta}{\omega_r} = \tan^{-1}\frac{6000}{7411} = 0.6806 \quad \text{rad}$$

Thus from equation 8.112,

$$v_{AK\text{peak}} = 2 \times 351 + (351 + 411)\epsilon^{(-6000/7411)(\pi - 2\times 0.6806)}$$

$$= 882 \quad \text{V}$$

(c)   The power output may be obtained from equation 8.104, in which the output angular frequency is

$$\omega_O = 2\pi \times 10^3$$

so that from equation 8.104,

$$P_s = \frac{2\pi \times 10^3}{\pi} \times \frac{351(351 + 411)}{9535^2 \times 10^{-4}}[1 + \epsilon^{-2.543}]$$

$$= 63.46 \times 10^3 \quad \text{W}$$

(d)   The rms output current is

$$I_R = \sqrt{\frac{P_s}{R}} = \left[\frac{63460}{1.2}\right]^{1/2} = 230 \quad \text{A}$$

so that the rms thyristor current is

$$I_{QR} = \frac{I_R}{\sqrt{6}} = \frac{230}{\sqrt{6}} = 93.9 \quad \text{A}$$

(e)   The derating factor is, for six thyristors,

$$D_R = \frac{6 \times v_{AK\text{peak}}I_{QR}}{P_s} = \frac{6 \times 882 \times 93.9}{63460} = 7.83$$

(f)   The power factor at the transformer secondary terminals is

$$PF_2 = \frac{P_s}{S_2} = \frac{P_s}{\sqrt{3}\,VI_R} = \frac{63460}{\sqrt{3}\times 300 \times 230} = 0.531$$

(g) The power factor at the transformer primary terminals is

$$PF_1 = \frac{P_s}{S_1} = \frac{P_s}{\sqrt{2}\, VI_R} = \frac{63460}{\sqrt{2} \times 300 \times 230} = 0.650$$

Example 8.4 In the system of Example 8.3 the ringing circuit parameters are changed to $L = 300\,\mu H$, $C = 57\,\mu F$.

Recalculate the quantities determined in Example 8.3 and compare them with the results of Example 8.3.

(a)

$$\omega_o = \frac{10^4}{\sqrt{3 \times 0.57}} = 7647 \quad \text{rad/s}$$

$$\zeta = \frac{1.2}{2 \times 3 \times 10^{-4}} = 2000 \quad \text{s}^{-1}$$

$$\omega_r = [7647^2 - 2000^2]^{1/2} = 7380 \quad \text{rad/s}$$

$$t_q = \frac{10^{-3}}{2} - \frac{\pi}{7380} = 74.4 \times 10^{-6} \quad \text{s}$$

Thus t_q is scarcely affected by the change in parameters, so that the conduction interval for each thyristor is virtually unaltered.

(b)

$$V_\alpha = 351 \quad \text{V}$$

$$\frac{\zeta \pi}{\omega_r} = \frac{2000\pi}{7380} = 0.8514$$

$$-V_{C1} = \frac{351(1 + \epsilon^{-0.8514})}{1 - \epsilon^{-0.8514}} = 874 \quad \text{V}$$

$$\psi = \tan^{-1}\frac{2000}{7380} = 0.2646 \quad \text{rad}$$

$$v_{AK\,peak} = 2 \times 351 + (351 + 874)\epsilon^{(-2000/7380)(\pi - 2 \times 0.2646)}$$

$$= 1305 \quad \text{V}$$

(c)

$$P_s = \frac{2\pi \times 10^3}{\pi} \times \frac{351(351+874)}{7647^2 \times 3 \times 10^{-4}}[1+\epsilon^{-0.8514}]$$

$$= 69.94 \times 10^3 \quad W$$

(d)

$$I_R = \left[\frac{69.94 \times 10^3}{1.2}\right]^{1/2} = 241 \quad A$$

$$I_{QR} = \frac{241}{\sqrt{6}} = 98.6 \quad A$$

(e)

$$D_R = \frac{6 \times 1305 \times 98.6}{69.94 \times 10^3} = 11.03$$

(f)

$$PF_2 = \frac{69.94 \times 10^3}{\sqrt{3} \times 300 \times 241} = 0.558$$

(g)

$$PF_1 = \frac{69.94 \times 10^3}{\sqrt{2} \times 300 \times 241} = 0.682$$

Comparison of the two sets of results shows that a slightly increased maximum power output at a slightly higher power factor has been achieved by the change in circuit parameters. This has been done, however, at the cost of a much higher peak thyristor voltage, resulting in a higher derating factor, and much larger reactive components. The parameters of Example 8.3 are therefore to be preferred.

8.3.2 Modified Load-Commutated Multiplier The circuit of a modified frequency multiplier is shown in Fig. 8.4. It may be analyzed on the basis of the approximation concerning source voltage expressed in equation 8.77 and employed with the unmodified multiplier. In Fig. 8.15 is shown the active part of the circuit of Fig. 8.4 when thyristor Q_1 is conducting. In this circuit

$$v_{AN} + v_{BN} + v_{CN} = 0 \quad V \tag{8.117}$$

$$i_a + i_b + i_c = i_O \quad A \tag{8.118}$$

For mesh 1,

$$Ri_O + L\frac{di_O}{dt} + \frac{3}{C}\int i_a \, dt + V_{Ca1} = 0 \quad \text{V} \tag{8.119}$$

where V_{Ca1} is the voltage at the instant $t=0$ across the terminals of the capacitor $C/3$ connected to phase A of the source. Differentiation of equation 8.119 yields

$$\frac{d^2i_O}{dt^2} + \frac{R}{L}\frac{di_O}{dt} + \frac{3}{LC}i_a = 0 \quad \text{A/s}^2 \tag{8.120}$$

For mesh 2,

$$v_{AN} - v_{BN} = \frac{3}{C}\int i_b \, dt + V_{Cb1} - \frac{3}{C}\int i_a \, dt - V_{Ca1} \quad \text{V} \tag{8.121}$$

From which

$$\frac{d}{dt}(v_{AN} - v_{BN}) = \frac{3}{C}(i_b - i_a) \quad \text{V/s} \tag{8.122}$$

Similarly, for mesh 3,

$$\frac{d}{dt}(v_{BN} - v_{CN}) = \frac{3}{C}(i_c - i_b) \quad \text{V/s} \tag{8.123}$$

From equations 8.122 and 8.123,

$$\frac{d}{dt}(2v_{AN} - v_{BN} - v_{CN}) = \frac{3}{C}(i_b + i_c - 2i_a) \quad \text{V/s} \tag{8.124}$$

and substitution from equations 8.117 and 8.118 in equation 8.124 yields

$$i_a = \frac{i_O}{3} - \frac{C}{3}\frac{dv_{AN}}{dt} \quad \text{A} \tag{8.125}$$

This expression for i_a may be substituted in equation 8.120, yielding

$$\frac{d^2i_O}{dt^2} + \frac{R}{L}\frac{di_O}{dt} + \frac{i_O}{LC} = \frac{1}{L}\frac{dv_{AN}}{dt} \tag{8.126}$$

But since it is assumed that

$$v_{AN} = V_\alpha \quad \text{V} \tag{8.127}$$

then

$$\frac{dv_{AN}}{dt} = 0 \quad \text{V/s} \tag{8.128}$$

and the solution of equation 8.126 becomes the same as that of equation 8.80. Thus so far as the output current is concerned, the performance of the converter circuit of Fig. 8.4 differs in no way from that of the circuit of Fig. 8.3. Thyristor and inductor ratings for the two circuits are therefore identical. In addition, since the load current i_O no longer flows as a zero-sequence component in the secondary windings of the three-phase transformer, the transformer secondary rating becomes equal to the primary rating expressed in equation 8.115.

An idea of the waveform of the transformer secondary current may be obtained by means of further approximations. From equation 8.122,

$$\frac{C}{3}\frac{d}{dt}(v_{AN} - v_{BN}) = i_b - i_a \quad \text{A} \tag{8.129}$$

and since $C/3$ is a small capacitance, and the rate of change of line-to-line source voltage also is small in this context, then the expression on the left-hand side of equation 8.129 is very small, and

$$i_a \cong i_b \quad \text{A} \tag{8.130}$$

From equation 8.123 it may similarly be concluded that

$$i_b \cong i_c \quad \text{A} \tag{8.131}$$

so that from equations 8.118, 8.130, and 8.131, approximately

$$i_a = i_b = i_c = \frac{i_O}{3} \quad \text{A} \tag{8.132}$$

and

$$i_A = i_O - i_a = \frac{2i_O}{3} \quad \text{A} \tag{8.133}$$

It follows that when thyristor Q_4 is conducting

$$i_A = -\frac{2i_O}{3} \quad \text{A} \tag{8.134}$$

and when a thyristor in phase B or phase C is conducting,

$$i_A = \pm\frac{i_O}{3} \quad \text{A} \tag{8.135}$$

The resulting waveform of i_A is shown in Fig. 8.13 under the heading "Modified Multiplier." A comparison of the waveforms of i_A for the unmodified and modified multiplier shows a substantial improvement in form factor for the latter, and this results in a reduction in the secondary transformer rating. The primary transformer rating and input power factor are identical with those of the unmodified multiplier. Even for the modified multiplier the transformer primary current waveform will be such that an input filter circuit will be essential.

PROBLEMS

8.1 The single-phase to single-phase cycloconverter of Fig. 8.5a is supplying a resistive load, and the system of control is such that the delay angle α has the same value for all four thyristors. Contrary to inequality 8.2, the ratio $\omega_s/\omega_O = 3$. The transformer ratio $n = 1$. If $\alpha = 0$:

(a) Sketch approximately to scale the waveforms of:

(i) The gating signals.
(ii) Output voltage v_O.
(iii) Source current i_1.

(b) State whether it is necessary to take any precautions against short-circuit of the source under these conditions and justify your answer.

8.2 Repeat Problem 8.1 with $\alpha = \pi/2$.

8.3 Repeat Problem 8.1 with $\omega_s/\omega_O = 2$.

8.4 For the basic cycloconverter circuit of Fig. 8.1a, v_O and i_O are assumed to be sinusoidal. Sketch one cycle of the waveforms of these variables and show for the different parts of the cycle which controlled rectifier is delivering current and whether it is doing so as a rectifier or as an inverter for each of the following cases:

(a) $\cos\phi_O = 1$.
(b) $\cos\phi_O = 0.866$ lagging.
(c) $\cos\phi_O = -0.5$ lagging.

In each case, state whether the average power flow is from the converter to the load circuit or conversely.

8.5 The single-phase to single-phase cycloconverter of Fig. 8.5a is supplying a sinusoidal output current at unity power factor, and sinusoidal modulation of the delay angles α is employed. Ratio $\omega_s/\omega_O = 9$. The transformer ratio $n = 1$.

For maximum output current sketch approximately to scale the waveforms of:

(a) Output voltage v_O.
(b) Gating currents i_{G1} to i_{G4}.
(c) Source current i_1.

8.6 Repeat Problem 8.5 for an output current that is one half of the maximum value.

8.7 Repeat Problem 8.5 if the load circuit has an impedance angle ϕ_O of 45° lagging at the output frequency, and indicate clearly on the diagram which rectifier is delivering current and whether it is operating as a rectifier or an inverter.

8.8 The single-phase to single-phase cycloconverter of Fig. 8.5a is to supply an RL series load circuit, and sinusoidal modulation of the delay angles α is to be employed. The required output frequency is 1 Hz, and the inductive reactance of the load circuit at this frequency is $\omega_O L = 0.8 \, \Omega$. The load circuit resistance is $1.2 \, \Omega$. A 220-V 60-Hz power supply is to be employed. If the ideal output voltage wave is defined as

$$v_O^* = 110\sqrt{2} \, \sin 2\pi t \quad \text{V}$$

   (a) Determine the required transformer ratio to give the maximum power factor at the primary terminals and calculate that power factor.
   (b) Determine

     (i) The peak thyristor voltage.
     (ii) The thyristor rms current rating.

8.9 The three-phase to single-phase cycloconverter of Fig. 8.11a is supplying a resistive load, and the system of control is such that the delay angle α has the same value for all thyristors. Contrary to inequality 8.35, the ratio $\omega_s / \omega_O = 2$. If $\alpha = 0$:

   (a) Sketch the waveforms of:

     (i) The gating signals.
     (ii) Output voltage v_O.

   (b) State whether it is necessary to take any precautions against short circuit of the source under these conditions and justify your answer.

8.10 Repeat Problem 8.9 with $\alpha = \pi/2$.

8.11 The three-phase to single-phase cycloconverter of Fig. 8.11a is supplying an RL series load circuit, and sinusoidal modulation of the delay angles α is employed. Ratio $\omega_s / \omega_O = 2$. The impedance angle of the load circuit at output frequency is $\phi_O = 60°$. For maximum output current sketch the waveforms of:

   (a) Output voltage v_O.
   (b) Gating currents i_{G1} to i_{G6} and i'_{G1} to i'_{G6}.

8.12 The three-phase to single-phase cycloconverter of Fig. 8.11a is to supply an RL series load circuit, and sinusoidal modulation of the delay angles α is to be employed. The required output frequency is 2 Hz and the inductive reactance of the load circuit at this frequency is $\omega_O L = 0.8\,\Omega$. The load circuit resistance is $1.2\,\Omega$. A three-phase 60-Hz 220-V line-to-line power supply is to be employed directly without a transformer. If the ideal output voltage wave is defined as

$$v_O^* = 110\sqrt{2}\ \sin 4\pi t \quad \text{V}$$

determine:

(a) The peak thyristor voltage $v_{AK\,\text{peak}}$.
(b) The rms thyristor current I_{QR}.
(c) The power delivered to the load circuit P_O.
(d) The rms source current.
(e) The input power factor at the source terminals.

8.13   The three-phase to single-phase frequency tripler shown in Fig. 8.2a is employed to supply a resistive load of $1.5\,\Omega$. The three-phase source represents the secondary windings of a wye-to-wye 60-Hz isolating transformer for which the turns ratio $n = 1$. The source is 3-wire, 110-V line-to-line. If $\alpha = 150°$:

(a)   Sketch approximately to scale the waveforms of v_{AN}, i_O, i_A and the current of the corresponding transformer primary winding i_1.
(b)   Calculate

(i)   RMS thyristor current I_{QR}.
(ii)   Peak thyristor voltage $v_{AK\,\text{peak}}$.
(iii) Time available for turn-off t_q.
(iv)   Primary input power factor PF_1.
(v)   Derating factor D_R.

8.14   Show how a balanced, three-phase 60-Hz voltage source can be converted into a balanced nine-phase 60-Hz source by means of a single transformer with several secondary windings.

8.15   Show how a balanced, three-phase 60-Hz voltage source can be converted by means of a line-commutated multiplier, to a single-phase 540-Hz source to supply a series RL load circuit. Also derive an expression for the power factor at the terminals of the three-phase source.

8.16   Show how the circuit in Fig. 8.2a could be employed as a cycloconverter to convert the three-phase input to a single-phase output at a frequency lower than the source frequency. If sinusoidal modulation of the delay angles were employed with this converter, derive an expression for the maximum rms value of the ideal output-voltage waveform V_{Om}^*.

8.17   In the cycloinverter circuit of Fig. 8.3, the rms line-to-neutral source voltage is 300 V, $R = 1.2\,\Omega$, $L = 30\,\mu\text{H}$, and the output frequency is 1000 Hz. Turn-off time

t_{off} for the thyristors is 75 μs. For maximum output operation determine:

   (a)   The value of the capacitance C.
   (b)   The peak thyristor voltage $v_{AK\text{peak}}$.
   (c)   Power output P_O.
   (d)   Derating factor D_R.
   (e)   The power factor at the transformer secondary terminals.
   (f)   The power factor at the transformer primary terminals.

(Compare these results with those of Examples 8.3 and 8.4).

8.18   For the cycloinverter of Fig. 8.4, the three-phase source represents the secondary windings of a three-phase transformer, of which the rms line-to-neutral voltage is 266 V. t_{off} for the thyristors is 30 μs, the load circuit resistance is 1 Ω, and the output frequency is 3000 Hz.

   (a)   Calculate the values of L and C required if the peak thyristor voltage is to be 780 V.
   (b)   For the values of L and C determined in a, calculate:

   (i)   Output power P_O.
   (ii)   Power factor PF_2 at the transformer secondary terminals.
   (iii)   Power factor PF_1 at the transformer primary terminals.

8.19   If a balanced, four-phase 60-Hz source is available, show how a line-commutated frequency doubler can be built to supply a single-phase load circuit.
   If the load circuit is resistive and t_{off} for the thyristors is 100 μs, sketch approximately to scale the thyristor gating current waveforms and the waveform of the output current i_O at maximum power output.

BIBLIOGRAPHY

The following books are recommended for further reading and reference purposes.

1. Gentry, F. E., F. W. Gutzwiller, N. Holonyak, and E. E. Von Zastrow, *Semiconductor Controlled Rectifiers*, Prentice-Hall, 1964.
2. Bedford, B. D., and R. G. Hoft, *Principles of Inverter Circuits*, John Wiley & Sons, 1964.
3. Schaefer, J. *Rectifier Circuits, Theory and Design*, John Wiley & Sons, 1965.
4. Takeuchi, T. J., *Theory of SCR Circuits and Application to Motor Control*, Tokyo Electrical Engineering College Press, 1968.
5. Heumann, K. and A. C. Stumpe, "Thyristoren," *AEG*, 1969.
6. Kusko, A., *Solid State DC Motor Drives*, The M.I.T. Press, 1969.
7. Pelly, B. R., *Thyristor Phase Controlled Converters and Cycloconverters*, John Wiley & Sons, 1971.
8. Davis, R. M., *Power Diode and Thyristor Circuits*, Cambridge University Press, 1971.
9. McMurray, W., *The Theory and Design of Cycloconverters*, The M.I.T. Press, 1972.
10. Harnden, J. D., and F. B. Golden, *Power Semiconductor Applications*, Vols. I and II, IEEE Press, 1972.
11. Ramshaw, R. S., *Power Electronics*, Chapman and Hall, 1973.
12. Mazda, F. F., *Thyristor Control*, John Wiley & Sons, 1973.
13. Murphy, J. M. D., *Thyristor Control of AC Motors*, Pergamon Press, 1973.
14. Rice, L. R., *Westinghouse Silicon Controlled Rectifier Designers Handbook*, Second Edition, Westinghouse Electric Corporation, 1970.
15. Grafham, D. R., and J. C. Hey *General Electric SCR Manual*, Fifth Edition, General Electric Company, 1972.
16. Hoft, R. G., *International Rectifier SCR Applications Handbook*, 1st Printing, International Rectifier Corporation, September, 1974.

INDEX

NUMERICAL ANSWERS TO PROBLEMS*

CHAPTER 2.   CIRCUITS WITH SWITCHES AND DIODES

2.1 $i = 100\epsilon^{-10^4 t} : v_C = 100(1 - \epsilon^{-10^4 t})$.

2.2 $i = 100(1 - \epsilon^{-1000 t}) : i_L = i_D = 63.2\epsilon^{-100 t} : i \neq 0, v_O = 100; i = 0, v_O = 0$.

2.3   (a) 10^7 : (b) 10^7 : (c) 17.5×10^6 : (d) 10^7 : (e) 5×10^6 : (f) ∞ : (g) ∞.

2.4 $i_1 = 0 : i_2 = \dfrac{10^6 t}{9} - \dfrac{100}{3} : v_1 = -\dfrac{100}{3} : v_2 = -100$.

2.5 $i = \dfrac{V}{\omega L} \sin \omega t : v_L = V \cos \omega t : v_C = V(1 - \cos \omega t) :$ where $\omega = (LC)^{-\frac{1}{2}}$.

2.7 $I_O = 4.18 : I_R = 6.16 : V_O = 41.8 : V_R = 81.2$.

2.8 $v(t) = 2\sqrt{2}\ V[\dfrac{1}{\pi} + \dfrac{2}{3\pi} \cos 2\omega t - \dfrac{2}{15\pi} \cos 4\ \omega t. . .]$.

2.9   (a) $I_O = 9.90 : V_O = 99.0$: (c) $I_{DAVE} = 7.16 : I_{lAVE} = 0 : I_{DRMS} = 7.16: I_{IRMS} = 10.1$: (d) 0.918.

2.10   (b) $I_{lAVE} = 33.3 : I_{IRMS} = 57.7 : V_O = 78.4 : V_R = 79.6$: (c) 0.676.

2.12 $I_O = 4.33$

2.13 $V_O = 45.0 : I_O = 25.0$

CHAPTER 3.   POWER SEMICONDUCTOR SWITCHES

3.1   (a) $i = 10(1 - \sqrt{2} \cos 120\pi t)$: (b) $i_{peak} = 24.1 : t = 1/120$.

3.2   (a) $I_O = 9.33, 1.56 : I_R = 17.5, 6.22$: (b) $I_O = 5.74, 0.70 : I_R = 9.04. 1.57$.

3.3   (a) 105 : (b) 165 : (c) $I_O = 0.71 : I_R = 1.70$: (d) 79.5 : (e) 0.468.

3.4   (a) to (d) 0.

3.5   (a) 114 : (b) 166 : (c), (d), (e) indistinguishable from 3.3.

3.6   (a) $I_O = 17.2 : I_R = 24.6$: (b) $P = 687 : PF = 0.279$.

3.7 22.5

3.8 $15°C$.

3.11   (b) 95.3×10^{-6}.

3.15 5.5×10^{-6}.

CHAPTER 4.   AC VOLTAGE CONTROLLERS

4.1 $2/3$

*Note: Those numerical answers obtained by the use of graphs printed in the text are necessarily approximate.

524

4.2 (a) $\alpha \geqslant \pi/2$: 325 : (b) I_{QR} = 70.7 : I_Q = 45.0 : (c) 8.33×10^{-3} : (d) 114 : (e) 45.3

4.3 (a) $\pi/2 < \alpha\,\pi$: 325 : (b) 70.7 : (c) 45.0 : (d) $\pi/2 < \alpha < \pi$.

4.4 (a) $\pi/2 < \alpha < \pi$: 325 : (b) I_{QR} = 50 : I_Q = 31.8 : (c) $\pi/2 < \alpha < \pi$.

4.5 ∞ : 53.3×10^3 : 37.7×10^3.

4.8 (a) V_{AR} = 135 : I_{QR} = 20.4 : I_Q = 13.0 : (b) $0 < \alpha < 5\pi/6$: (c) 23.

4.9 (a) V_{AR} = 135 : I_{QR} = 8.66 : I_Q = 5.51 : (b) $65 < \alpha < 150$: (c) $\to 0$.

4.10 I_Q = 75.8 : I_{QR} = 119 : v_{AK} = 269.

4.13 (a) v_{AK} = $\pm$ 156 : I_{QR} = 15.0 : I_Q = 9.54 : (b) $65 < \alpha < 180$: (c) 110.

4.14 I_Q = 43.7 : I_{QR} = 68.8 : v_{AK} = 311.

4.17 (a) 130 : (b) 269 : (c) $31.8 < \alpha < 150$.

4.18 (a) 75.2 : (b) 311 : (c) $31.8 < \alpha < 180$.

CHAPTER 5.   CONTROLLED RECTIFIERS

5.2 (a) 13.0 : (b) 15.0 : (c) I_Q = 6.50 : I_{QR} = 10.6 : (d) 0.850.

5.3 (a) 19.7 : (b) 20.1 : (c) I_Q = 9.83 : I_{QR} = 14.2 : (d) 0.669.

5.4 (a) α = 45.0 : PF = 0.636 : (b) α = 135 : PF = -0.636.

5.5 4.14

5.6 179.6

5.7 (b) V_O = 84.5 : I_O = 36.4 : (c) 0.888.

5.8 (b) 0.768.

5.9 (a) 155 : (b) $0 < \alpha < 180$: (c) I_{QR} = 122 : I_Q = 77.7 : (d) +244, -488 : (e) 111.

5.10 As for 5.9 except (d) +122, -244.

5.13 (a) $I_O = I_R$ = 1.62 : (b) 0.936.

5.17 (a) 4.81 : (b) 5.0 : (c) I_Q = 1.60 : I_{QR} = 2.89 : (d) 0.0607.

5.18 (a) 65.6 : (b) 65.7 : (c) I_Q = 21.9 : I_{QR} = 37.9 : (d) 0.00196.

5.19 (a) 49 : (b) 0.627 : (c) 950.

5.20 α = 119 : PF = -0.467 : Effy = 0.99.

CHAPTER 6.   DC – TO – DC CONVERTERS

6.1 (a) $V_O = (t_{ON}/T)V$: $I_O = V_O/R$: (b) $I_{max} = V/R$: (c) zero : (d) $(t_{ON}/T)^{\frac{1}{2}}V$:
     (e) $I_Q = I_O$: $I_{QR} = V_R/R$.

6.2 (a) $V_O = (t_{ON}/T)V$: $I_O = V_O/R$: (b) $I_{max} = I_O$: (c) $I_D = [(T - t_{ON})/T]I_O$: $I_{DR} =$
     $[T - t_{ON})/T]^{\frac{1}{2}}I_R$: (d) $(t_{ON}/T)^{\frac{1}{2}}V$: (e) $I_Q = (t_{ON}/T)I_O$: $I_{QR} = (t_{ON}/T)^{\frac{1}{2}}I_R$.

6.4 (a) V_O = 375 : I_O = 117 : (b) 180 : (c) V_{IR} = 250 : I_{IR} = 38.6.

6.6 (a) V_O = 414 : I_O = 143 : (b) 260 : (c) V_{IR} = 208 : I_{IR} = 95.7.

6.7 (a) 1.08×10^{-3}.

6.8 There is no steady–state operation.

6.10 (a) 17.41×10^{-6} : (b) 54.8×10^{-6} : (c) I_{CPEAK} = 394 : $v_{AKAPEAK}$ = 450.

6.11 $L_1 = 57.6 \times 10^{-6}$: $C = 25.6 \times 10^{-6}$.

6.12 (a) t_{off} is too small and the circuit will not commutate.
6.14 (a) 26.9×10^{-6} : (b) 119×10^{-6} : (d) 38.4×10^{-6}.
6.15 $L_1 = 49.1 \times 10^{-6}$: $C = 6.91 \times 10^{-6}$.
6.16 $I_{OI} = 183$: $P_R = 282$.
6.17 (a) $I_O = 117$: $V_O = 375$: (b) $I_{max} = 180$: $I_{min} = 45.2$: (c) $I_{save} = 72.9$.
6.18 (a) $I_O = 117$: $V_O = 375$: (b) $I_{max} = 258$: $I_{min} = -92.1$: (c) $I_{save} = 72.9$.
6.19 (a) $I_O = 0$: $V_O = 101$.

System is not operating as a chopper.

CHAPTER 7.   INVERTERS

7.1   (b) $I_Q = 84.4$: $I_{QR} = 133$: $I_D = I_{DR} = 0$, (c) Zero. Required.
7.2   (b) $I_Q = 3.17$: $I_{QR} = 59.5$: $I_D = 10.6$: $I_{DR} = 29.3$, (c) Not required.
7.3   (b) $I_Q = 31.7$: $I_{QR} = 59.5$: $I_D = 10.6$: $I_{DR} = 29.3$, (c) 115. Required.
7.4 $0.505 < R < \infty$
7.7   (a) $57.7\mu H$: $30.2\mu F$, (b) (i) $61.8\mu s$ (ii) $10.8\mu s$
7.8   536.
7.9   (a) $t_q = 28.5 \times 10^{-6}$: $t_C = 96.6 \times 10^{-6}$.
7.11 (c) Zero.
7.14 $0.505 < R < \infty$.
7.16 (a) $0.637 : 0.127 : 0.091$.
7.19 185.
7.20 (a) 0.0169, (b) 0.0169.
7.21 (a) $60°$: 0.694, (b) $9°$: 0.435.
7.26 (c) 66.8.
7.28 (a) $3.87 \times 10^{-6} H$: $11.3 \times 10^{-6} F$, (b) 10.3×10^3 : 4.88.

CHAPTER 8.   AC – TO – AC CONVERTERS

8.8   (a) $1.27 : 0.529$, (b) (i) 489, (ii) 53.9.
8.12 (a) ±311, (b) 31.1, (c) 6980, (d) 62.3, (e) 0.294.
8.13 (b) (i) 5.08, (ii) ± 156, (iii) 8.33×10^{-3}, (iv) 0.208, (v) 11.8.
8.17 (a) 73.4×10^{-6}, (b) 807, (c) 36.2×10^3, (d) 9.47, (e) 0.403, (f) 0.489.
8.18 (a) 26.5×10^{-6} H : 42.6×10^{-6} F, (b) (i) 57.5×10^3, (ii) 0.638, (iii) 0.638.